G000074961

*Proceedings of the Royal Institution of Great Britain*

# *Proceedings of the Royal Institution of Great Britain*

Volume 70

*Edited by*
P. DAY

THE ROYAL INSTITUTION

OXFORD
UNIVERSITY PRESS

Great Clarendon Street, Oxford OX2 6DP
Oxford University Press is a department of the University of Oxford
and furthers the University's aim of excellence in research, scholarship,
and education by publishing worldwide in

Oxford New York

Athens Auckland Bangkok Bogotá Buenos Aires Calcutta
Cape Town Chennai Dar es Salaam Delhi Florence Hong Kong Istanbul
Karachi Kuala Lumpur Madrid Melbourne Mexico City Mumbai
Nairobi Paris São Paulo Singapore Taipei Tokyo Toronto Warsaw

and associated companies in Berlin Ibadan

Oxford is a registered trade mark of Oxford University Press

Published in the United States
by Oxford University Press Inc., New York



First published 1999



British Library Cataloguing in Publication Data

Data available

Library of Congress Cataloging in Publication Data
(Data applied for)

ISBN 0 19 850539 6

Typeset by EXPO Holdings, Malaysia

Printed in Great Britain
on acid-free paper by
Bookcraft Ltd., Midsomer Norton, Avon

# PREFACE

Annually the Royal Institution brings to its members texts of Faraday Evening Discourses given over the previous year, so that those who are unable to be present can savour in written form what they missed, and those who were in the Lecture Theatre on the evening in question can recapture in permanent form some very memorable occasions. The present volume is a landmark in two respects. Although written accounts of Discourses go back far into the nineteenth century, the present series of Proceedings has now reached its seventieth volume. Furthermore, it appears in the year of the Royal Institution's Bicentenary (which, of course, is being marked in a wide variety of other ways too). I hope that the contents of this volume will be found to contain just as great a spread of topics as ever: science undoubtedly continues to fulfil Vanevar Bush's famous epithet in his report, to the US Government after the Second World War, as 'the endless frontier'.

For the present Editor this is the last Proceedings that I shall be associated with (except, perhaps, as a future contributor), having stood down as Director in the Autumn of 1998. The seven volumes I have been responsible for have encompassed an astonishing kaleidoscope of subjects and my earnest hope is that they have proved as rewarding to Members as they have to their Editor. In conclusion, I would like to thank the staff of Oxford University Press for their gentle guidance and great skill in seeing the last six volumes into print, and a succession of secretaries (Heather Pickett, Sarah Cripps, Evie Jamieson, and Dawn Hillman) for keeping the Editor and contributors in order.

*London* P. D.
February 1999

# CONTENTS

# PLATES

*The colour plate section falls between pages 260 and 261*

**Plate 1** Six scenes from an early fourteenth century manuscript showing stages in the surgical treatment of a fractured skull. Reproduced by courtesy of the British Library in association with The Wellcome Institute for the History of Medicine ('Medieval Medical Miniatures' by Peter Murray Jones, 1984, Plate IX).

**Plate 2** The radiographic image of Frau Röntgen's hand produced by W.C. Röntgen in November 1895 and featured on the front cover of *Nature* in the centenary year 1995. Reproduced by courtesy of *Nature* (5 Jan 1995).

**Plate 3** A 41p postage stamp issued in 1994 illustrating a CT scan of the thorax. Reproduced by courtesy of the Post Office.

**Plate 4** In clockwise order: (a) virtual endoscopy of the bronchial tree obtained from a three-dimensional high resolution block of CT attenuation values; (b) transverse; (c) sagittal; and (d) coronal sections from the same data.

**Plate 5** Doppler ultrasound demonstration of blood flow in the femoral artery and vein with directional colour coding.

**Plate 6** Transrectal Doppler ultrasound image of a vascular tumour in the prostate gland.

**Plate 7** Three-dimensional MR 'cut away' image of the head and brain. Reproduced by courtesy of *Diagnostic Imaging* (Feb 1995).

**Plate 8** MR signal changes in the occipital cortex with contrast enhancement following optic stimulation, first demonstrated by Belliveau in 1991. Reproduced by courtesy from *Science*, 1991, **254**, 621.

**Plate 9** PET scan of the left hemisphere of a human subject performing a series of intellectual tasks related to words. Blood flow shifts to different locations depending on the task. Reproduced by courtesy of Marcus E. Reichle, Washington University School of Medicine.

**Plate 10** Three-dimensional colour rendered view of a patient with a large acoustic neuroma. The tumour from MR imaging is green, the blood vessels from MR angiography are red and the bone from CT is grey. Reproduced by courtesy from *Radiology*, 1994, **191**, 447.

**Plate 11** Images from the Visual Human Dataset pioneered by the University of Colorado and now available on the Internet.

# CONTRIBUTORS

**Anthony T. Barker**,
Consultant Clinical Scientist,
Dept of Medical Physics and Clinical Engineering,
Royal Hallamshire Hospital,
Sheffield

**Simon Conway Morris** FRS
Professor of Evolutionary Palaeobiology,
University of Cambridge

**Peter Day** FRS
Fullerian Professor of Chemistry,
The Royal Institution of Great Britain, London

**A. Hamnett**
Pro-Vice Chancellor,
University of Newcastle

**Stephen T. Holgate**
MRC Clinical Professor of Immunopharmacology,
School of Medicine, University of Southampton

**David M. Howard**
Professor of Electronics and Head of the Department of Electronics,
University of York

**Ian Isherwood** CBE
Emeritus Professor of Diagnostic Radiology, University of Manchester

**Lord Lewis of Newnham** FRS
Warden, Robinson College,
Cambridge

**Robert A. J. Matthews**
Grueschart, France

**Peter Melchett**
Executive Director,
Greenpeace UK,
London

**Salvador Moncada** FRS
Wolfson Institute for Biomedical Research
University College, London

**Paul Nurse** FRS
Director General,
Imperial Cancer Research Fund

**Nancy J. Rothwell**
Professor of Physiology,
School of Biological Science
University of Manchester

**Russell Stannard**
Former Professor and Head of Physics,
Open University

**Crispin Tickell**
GCMG, Cirencester,
Gloucestershire

**C. E. Webb** FRS
Professor,
Department of Physics,
University of Oxford

# *The philosopher's tree: Faraday today at the Royal Institution*

PETER DAY

If the Royal Institution could be said to have a patron saint, then that person would have to be Saint Michael: not the familiar symbol of one of our long-standing and valued Corporate Members (Marks & Spencer) but, of course, Michael Faraday. He it was who, quite apart from all his remarkable discoveries in so many disparate fields of physical science, created the Royal Institution that we still recognize, through the kind of activities we pursue, and the way we go about them. In a phrase, he set the agenda, of which we are all the inheritors. In the *Four Quartets*, T.S. Eliot put the relation between past and present in words of quite startling simplicity, as poets will: 'Time past and time present are all perhaps present in time future, and time future contained in time past.'

But the agenda that Faraday established, and pursued so single-mindedly and effectively throughout his life did not consist only of a programme still faithfully followed by one institution over a 150-year time span. It encompasses a whole approach to the world around us: inanimate, animate, and even social. It contains three elements, and I want to touch on all of them briefly as I relinquish the post which he held for so long and with such unique distinction. The starting point in his approach to the world was vigorous, enthusiastic, imaginative experimentation, asking simple direct questions of nature to discover how the world works: what, in other words, are the 'rules of the game'. The second step was to put the knowledge acquired in front of those who may be most receptive to it, and that means, especially, but not exclusively, young people. The final step is to ensure that society as a whole has these values embedded in it, especially when decisions have to be made on how to proceed with issues where some acquaintance with nature's rules is decisively important (and that, as we know, can mean nearly all issues). We might call that a higher form of education. Faraday gave us striking examples of all three of these elements, and I want to share with you some examples, juxtaposing past and present, and in particular by hearing Faraday's own voice, unfortunately not directly, because sound recording had not been invented in his time, but by what he wrote.

I hope to convince you, too, that among all his other manifold virtues, Faraday had a fine way with words, and it is that which provides me with the title at the head of this article. You may have wondered what a philosopher's tree is. Philosopher was the word commonly used till the middle of the last century to denote what we now call a scientist, but to understand the significance of the word 'tree', consider the following letter written by Michael Faraday at the age of 20, describing how he wished to write.

> *It is my wish, if possible, to become acquainted with a method by which I may write ... in a more natural and easy progression. I would, if possible, imitate a tree in its progression from roots to a trunk, to branches, twigs and leaves, where every alteration is made with so much ease and yet effect that, though the manner is constantly varied, the effect is precise and determined.*

The extracts that follow will enable you to judge how well he succeeded.

As I have indicated, Faraday's programme takes its starting point from carefully, persistently observing the world as it is, probing it, prodding it, and drawing only conclusions that are supported by those observations—that is, by experiment. Faraday described his approach when writing to an old friend, quite late in his life, and his words also serve to remind us of his remarkable beginnings:

> *I entered the shop of a bookseller and bookbinder at the age of 13, in the year 1804, remained there 8 years, and during the chief part of the time bound books. Now it was in these books, in the hours after work, that I found the beginnings of my philosophy. There were two that especially helped me; the Encyclopaedia Britannica, from which I gained my first notions of Electricity and Mrs. Marcet's conversations on chemistry, which gave me my foundation in that science. I believe I had read about phlogiston etc in the Encyclopaedia, but her book came as the full light in my mind. Do not suppose that I was a very deep thinker or was marked as a precocious person. I was a very lively, imaginative person, and could believe in the Arabian nights as easily as the Encyclopaedia. But facts were important to me & saved me. I could trust a fact, but always*

*cross examined an assertion. So when I questioned Mrs. Marcet's book by such little experiments as I could find means to perform, & found it true to the facts as I could understand them, I felt that I had got hold of an anchor in chemical knowledge & clung* fast *to it.*

But we should not forget that in the young Michael Faraday's life, looking at the world around him was no chore but on the contrary, was great fun. Imagine, if you will, a rainy evening in London. Two young friends had been spending a weekend afternoon together, and it was time to go home. The following day the 19-year-old Michael wrote to his friend;

*Dear Abbott,*

*Were you to see me, instead of hearing from me, I conceive that one question would be how did you get home on Sunday evening? I suppose this question because I wish to let you know how much I congratulate myself upon the very pleasant walk (or rather succession of walks, runs and hops) I had home that evening, and the truly Philosophical reflections they gave rise to.*

*I set off from you at a run and did not stop until I found myself in the midst of a puddle and quandary of thoughts respecting the heat generated in animal bodies by exercise. The puddle, however, gave a turn to the affair and I proceeded from thence deeply immersed in thoughts respecting the resistance of fluids to bodies precipitated into them.*

*My mind was deeply engaged on this subject, and was proceeding to place itself as fast as possible in the midst of confusion, when it was suddenly called to take care of the body by a very cordial, affectionate & also effectual salute from a spout. This of course gave a new turn to my ideas and from thence to Blackfriars Bridge it was busily bothered amongst Projectiles and Parabolas. At the Bridge the wind came in my face and directed my attention as well as earnestly as it could go to the inclination of the Pavement. Inclined Planes were then all the go and a further illustration of this point took place on the other side of the Bridge, where I happened to proceed in a very smooth, soft, and equable manner for the space of three or four feet. This movement, which is vulgarly called slipping, introduced the subject of friction, and the best method of lessening it, and in this frame of mind I went on with little or no interruption for some time except occasional and actual experiments connected with the subject in hand, or rather in head.*

*The Velocity and Momentum of falling bodies next struck not only my mind but my head, my ears, my hands, my back and various other parts of my body, and tho I had at hand no apparatus by which I could ascertain those points exactly, I*

> *knew that it must be considerable by the quickness with which it penetrated my coat and other parts of my dress. This happened in Holborn and from thence I went home Sky-gazing and earnestly looking out for every Cirrus, Cumulus, Stratus, Cirro-Cumuli, Cirro-Strata and Nimbus that came from above the Horizon.*

The jaunty air of this letter is almost worthy of Gene Kelly in *Singing in the Rain*. For the young Michael Faraday, experiments could be done anywhere, even at home: in another letter to Benjamin Abbott he describes in triumphant terms his success in making a galvanic cell:

> *I have lately made a few simple galvanic experiments merely to illustrate to my self the first principle of science. I was going to Knights to obtain some Nickle, & bethought me that they had Malleable Zinc. I enquired & bought some. Have you seen any yet? The first portion I obtained was in the thinnest pieces possible; observe it in a flattened state. I obtained it for the purpose of forming discs, with which & copper to make a little battery. The first I completed contained the immense number of seven pairs of Plates! and of the immense size of halfpennies each! I, Sir, covered them with seven half-pence and I interposed between seven or rather six pieces of paper, soaked in a solution of Muriate of Soda!—but to laugh no longer Dear A _____, rather wonder at the effects this trivial power produced. It was sufficient to produce the decomposition of the Sulphate of Magnesia; an effect which extremely surprised me, for I did not, (I could not) have any idea that the agent was component to the purpose.*

Such an experiment could easily be done at home to this day by any clever teenager who gets the pieces from a local hardware shop. Nowadays we use electrochemistry at the Royal Institution for rather different purposes, for growing crystals, but still in quite a similar way as can be seen from Fig. 1. The compounds crystallizing from such cells in the basement of 21 Albemarle Street in 1998 are superconductors, but made from molecular components. The molecules in question are made in a laboratory on the third floor that, while not exactly like the one (Fig. 2a) where Faraday made his, but maybe not so very different (Fig. 2b).

Indeed, the new field of molecular-based magnets and superconductors that is fascinating the members of my own research group at the present time brings to mind another conjunction between past and present in our laboratory. Faraday made much of what he called 'electromagnetic rotations', based on the fact that the current carried by a wire induces a magnetic field in its neighbourhood, as discovered by

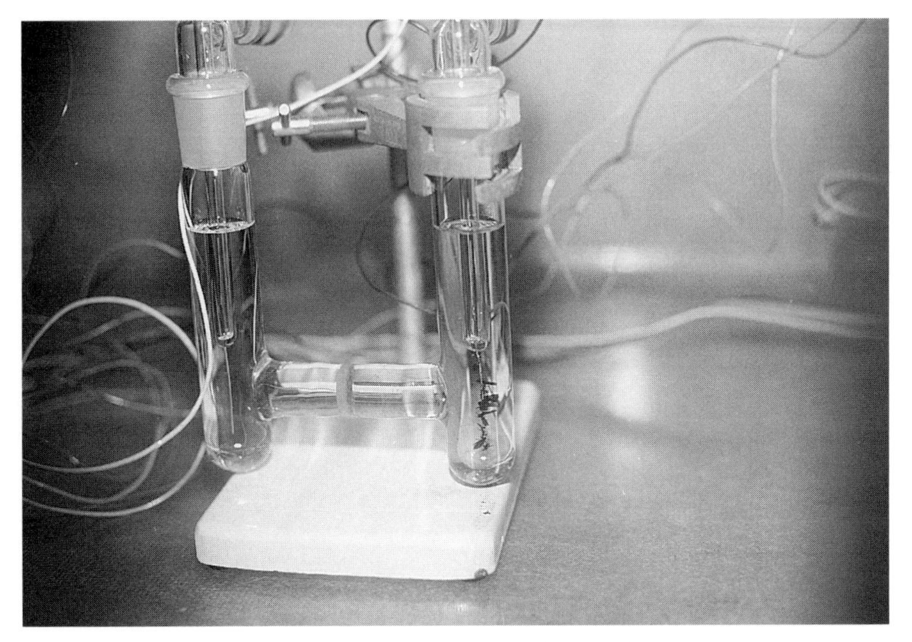

**Fig. 1** Electrochemical cells in use at the Royal Institution for growing crystals of molecular charge transfer salts.

Oersted. The fact is that the direction of the current determines the polarity of the field, so the arrangement is *chiral*, i.e. it is a kind of helix. Chirality in nature is a fascinating subject, and indeed a series of Royal Institution Christmas Lecture was devoted to it only a few years ago. However, we have been drawn to look at chirality in crystal structures quite recently, in relation to the molecular superconductors made in our electrochemical cells.

The compounds in question are made up from alternate layers of organic cations, which carry the superconducting current and inorganic anions that contain unpaired electrons, and hence give rise to paramagnetism. (In passing, it is with noticing that the words 'cation' and 'anion' themselves were originally coined by Michael Faraday.) In the present context the important feature of these unusual compounds is that the anions are chiral, actually mimicking three-bladed propellers. What is even more remarkable is that two different materials, that differ only in the arrangement of the two kinds of chiral anion have dramatically different properties, one being a superconductor, the other a semiconductor.

It is not very widely known that semiconductivity, too was discovered by Faraday, as the following extract from his laboratory notebook makes

(a)

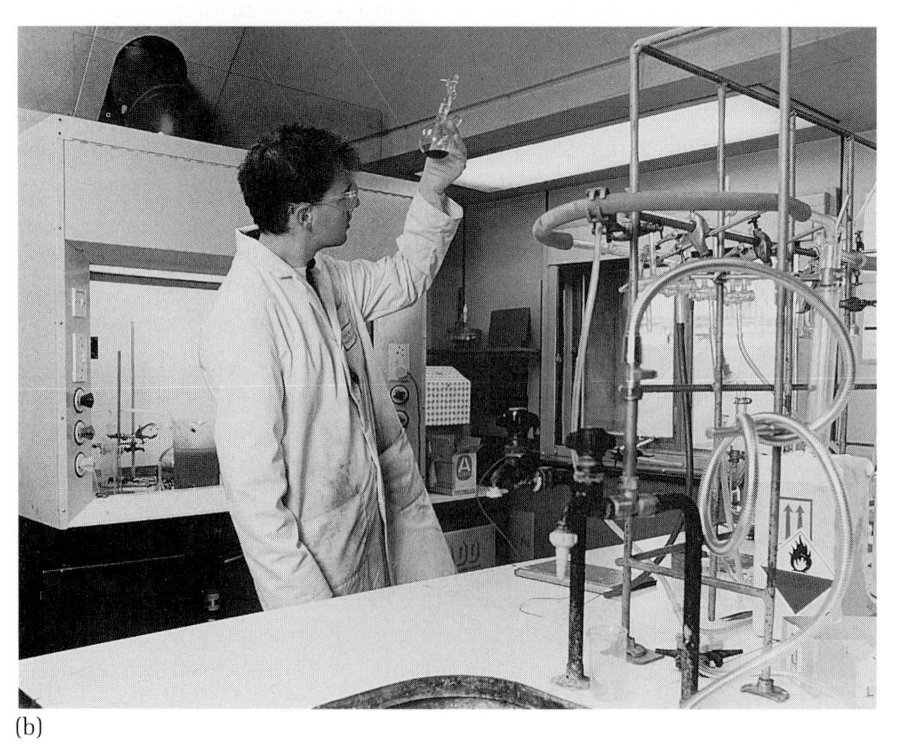

(b)

**Fig. 2** Chemistry laboratories in the Royal Institution (a) in Faraday's time; (b) today.

clear (sulphuret of silver is what we would nowadays call silver sulphide, $Ag_2S$):

> *21ST FEBY, 1833*
> Sulphuret of Silver—very extraordinary. *At first on piece of glass flask in air, but afterwards in tube, fuzed into its place in fire. When all was cold conducted a little (by galvanometer) and if quite cold at first conducting power did not increase. But if battery current strong or if sulphuret continued to increase in conducting power* ... The heat rose as the conducting power increased *(a curious fact), no other source of heat than the current being present. Yet I do not think it became high enough to* fuze the sulphuret ... *The whole passed whilst in the solid state. The hot sulphuret seems to conduct as a metal would, and could get sparks with wires at the end and a fine spark with charcoal.*

Figure 3 shows the sketch, from the notebook, of the simple piece of apparatus for making the measurements.

Perhaps the most perceptive (and, indeed, seminal) of all Faraday's experiments were in the field of magnetism, that enigmatic force, generated spontaneously by only a few substances, but also, as Faraday demonstrated, by electric currents.

Figure 4 is his first published picture of the lines of magnetic flux around a bar magnet detected (as school children often do nowadays) with iron filings. Many of Faraday's experiments on this subject were carried out in the 'magnetic laboratory' now reinstated as a museum in the basement of the Royal Institution. In the same small room will be found the 'great electro magnet' that he used. Nowadays we use something a bit more elaborate to investigate the magnetic properties of the compounds that we make, as may be seen from Fig. 5, which shows the sensitive magnetometer based on a superconducting magnet, and a detector known as a SQUID (not seafood, but a 'superconducting

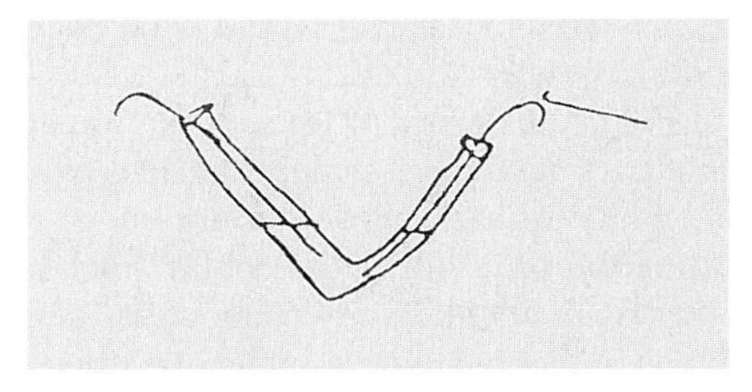

**Fig. 3** Faraday's apparatus for measuring the electrical resistance of silver sulphide.

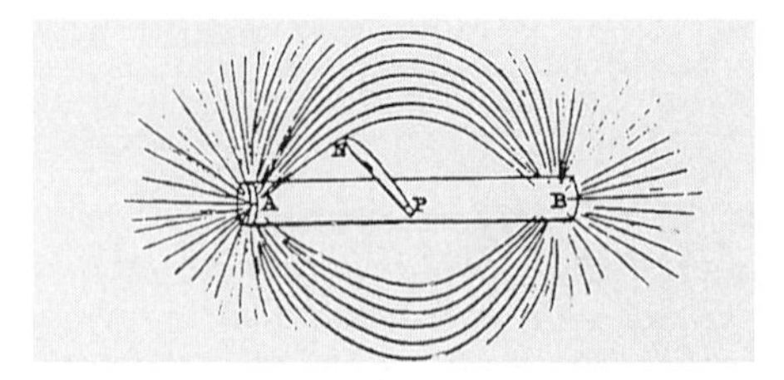


**Fig. 4** The first published picture (1839) of the lines of magnetic flux around a magnet.

**Fig. 5** Magnetometer using a superconducting magnet.

quantum interference device'). Why should we still be studying magnetism? Surely we know all about magnets, and what does a chemist have to do with magnets anyway?

The magnet in Fig. 4 was a bar of iron, but nowadays chemists can make magnets that are transparent, such as the one shown in Fig. 6, that was discovered in my research group some 20 years ago, or even soluble, by building the solid out of molecules. Such efforts have also led to magnets that do not contain any metal atoms at all, only the elements C, H, N, such as the one whose molecular structure is shown in Fig. 7, although up till now they have only been found to be magnetic at very low temperatures.

**Fig. 6** An optically transparent ferromagnetic material.

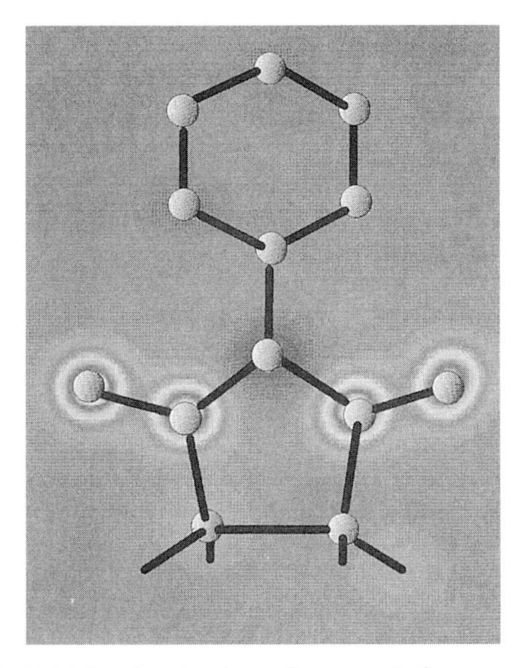

**Fig. 7** Molecular structure of an organic ferromagnet.

One reason for being interested in transparent magnets goes back to one of Faraday's most seminal experiments (as always, a very simple one), which he described in his noteboook.

> *A piece of heavy glass ... being a silico borate of lead, and polished on the two shortest edges, was experimented with. It gave no effects when the same* magnetic poles *or the* contrary *poles were on opposite sides (as respects the course of the polarized ray)—nor when the same poles were on the same side, either with the constant or intermitting current—BUT, when contrary magnetic poles were on the same side, there was an* effect produced on the polarized ray, *and thus magnetic force and light were proved to have relation to each other. This fact will most likely proved exceedingly fertile and of great value in the investigation of both conditions of natural force.*

It is worth drawing attention to the word BUT, written in capital letters and underlined several times, in contrast to the rest of the text, written in his usual immaculate copperplate handwriting. Clearly, he realized at once how significant the result was, as we can also observe from his final comment, one of the most remarkable pieces of understatement in the history of science.

Only a very small minority of solids are spontaneously magnetic, but another great discovery of Faraday was that all substances have the property of what he called diamagnetism, i.e. that the lines of magnetic flux diverge instead of converging, within the material. His experiments designed to illustrate the universality of the property lead to some amusing asides in his notebooks:

> *It is curious to see such a list as this of bodies presenting on a sudden this remarkable property, and it is strange to find a piece of wood, or beef, or apple, obedient to or repelled by a magnet. If a man could be suspended, with sufficient delicacy, and placed in the magnetic field, he would point equatorially.*

Many years after Faraday carried out his experiments on diamagnetism it turned out that this property is one of those characterizing the superconducting state, about which we spoke earlier. Indeed, a superconductor is a perfect diamagnet, which brings about the possibility of levitating objects by posing an array of magnets above a superconducting plate. At least up till now, superconductivity remains inherently a low temperature property, but with the advent of materials showing the property above the boiling point of liquid nitrogen, some quite spectacular demonstrations become possible (Fig. 8). Indeed, several years ago, when giving a Friday Evening Discourse on Superconductivity I was able (quite appropriately) to levitate a bust of Michael Faraday! Magnets made from superconducting wire are now very big business as high magnetic fields now form the basis of many advanced measurement and diagnostic techniques. The one that has made the greatest impact on the public is body imaging, called magnetic resonance imaging, which is

**Fig. 8** Magnet levitating above a superconducting ceramic in liquid nitrogen.

now found in many hospitals. The quality of the images obtained can be gauged from Fig. 9, which holds a special interest for me, as the vertebrae in question are mine!

Returning to the theme of Faraday's life and work, we may ask what way of life it was that enabled him to accomplish all these great discoveries. The answer is, a very constrained one (constrained, that is, in a geographical and social sense, not of course an intellectual one). The

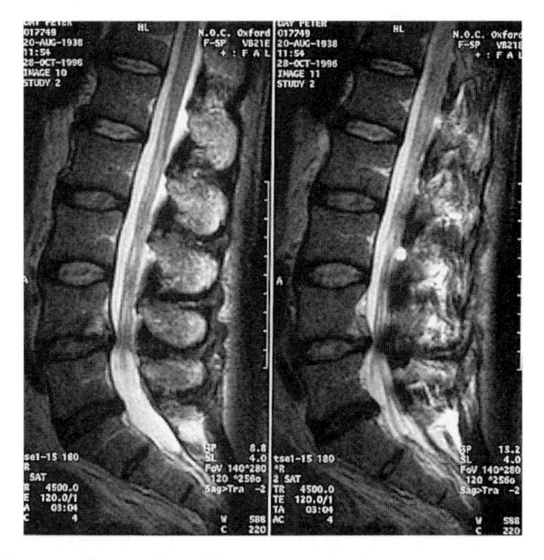


**Fig. 9** The author's backbone seen by magnetic resonance imaging.

strength of his attachment to the Royal Institution is nicely brought out in one of his letters to the American physicist Joseph Henry:

> *Your accounts of your transits over the world, and changes in the position of your family, almost startle & shame me, for I feel as if I could have shewn none of the energy and perseverance which carries you through all these things. I have been here so attached to the Royal Institution that I feel as if I were a limpet on a rock, and that any chance which might knock me from my position would leave me but little hopes of attaching myself anywhere again.*

Having have seen his laboratory it is pertinent to look at his office. Fortunately for us a friend of the Faradays, Harriet Moore was a water-colour painter and her work still hangs upstairs in the Director's Flat, and Fig. 10 juxtaposes the 1850s with 1998. Eagle eyes may notice that some of the furniture remains the same, although the quantity of books and papers has grown.

A key to Faraday's success in accomplishing so much in this modest domestic environment was his preoccupation with time, and how to use it to best effect. This trait developed earlier in life, as we see from the following letter to his teenage friend Benjamin Abbott:

> *Dear Abbott,*
>
> *What is the longest, and the shortest thing in the world: the swiftest and the most slow: the most divisible and the most extended: the least valued and the most regretted: without much nothing can be done: which devours all that is small: and gives life and spirits to everything that is great?*
>
> *It is that, Good Sir, the want of which has till now delayed my answer to your welcome letter. It is what the Creator has thought of such value as never to bestow on us mortals two of the minutest portions of it all at once. It is that which with me is at the instant very pleasingly employed. It is Time.*

Later in life he was equally trenchant about time wasting on official business, especially committees (we may well sympathize!):

> *With respect to committees, as you would perceive I am very jealous of their formation. I mean* working committees. *I think business is always better done by few than by many. I think also the working few ought not to be embarrassed by the idle many and, further, I think the idle many ought not to be honoured by association with the working few. I do not think that my patience has ever come nearer to an end than when compelled to hear (in the examination of witnesses etc, etc. in committee) the long rambling malapropros enquiries of members who still have nothing in consequence to propose that shall advance the business. But in all this too, I will promise to behave as well as I can.*

(a)

(b)

**Fig. 10** The Director's study at the Royal Institution, (a) in Faraday's time; (b) in 1998.

By a natural progression, work in the laboratory also led to the lecture theatre, not so much the formal professional lecture as the popular exposition. For Faraday, as for his successors at the Royal Institution, telling people about science is as important as doing it in the first place. And the means that he used were the same as the tools of his research, direct simple experiments and clear, approachable explanations. Nowhere is that more evident than in the Christmas Lectures, now as much as then. 'Then' is symbolized by the famous picture of the 1862 lectures, given before the Prince Consort and the two royal princes (Fig. 11a). 'Now' is symbolized by the ubiquitous presence of television (Fig. 11b) although the two occasions are strikingly similar.

One of the most eloquent of his hymns in praise of science comes, not from his most famous series of Christmas Lectures, The Chemical History of a Candle, but the less well known 'Various Forces of Nature', which concentrates on physics. In the following, one can almost hear the voice.

> *I shall here claim, as I always have done on these occasions, the right of addressing myself to the younger members of the audience. And for this purpose, therefore, unfitted as it may seem for an elderly infirm man to do so, I will return to second childhood and become, as it were, young again amongst the young.*
>
> *Let us now consider for a little while, how wonderfully we stand upon this world. Here it is we are born, bred, and live, and yet we view these things with an almost entire absence of wonder to ourselves respecting the way in which all this happens. So small, indeed, is our wonder, that we are never taken by surprise; and I do think that, to a young person of ten, fifteen, or twenty years of age, perhaps the first sight of a cataract or a mountain would occasion him more surprise that he had ever felt concerning the means of his own existence, —how he came here; how he lives; by what means he stands upright; and through what means he moves about from place to place. Hence, we come into this world, we live, and depart from it, without our thoughts being called specifically to consider how all this takes place. And were it not for the exertions of some few inquiring minds, who have looked into these things and ascertained the very beautiful laws and conditions by which we* do *live and stand upon the earth, we should hardly be aware that there was anything wonderful in it.*

Two extremely simple demonstrations about cohesive forces will serve to illustrate his approach. The first uses a towel as a syphon:

> *When you wash your hands you take a towel to wipe off the water; and it is by that kind of wetting, or that kind of attraction which makes the towel become wet with water, that the*

(a)

(b)

**Fig. 11** Royal Institution Christmas Lectures, (a) in 1862; (b) in 1993.

> *wick is made wet with the tallow. I have known some careless boys and girls (indeed, I have known it happen to careful people as well) who, having washed their hands and wiped them with a towel, have thrown the towel over the side of the basin, and before long it has drawn all the water out of the basin and conveyed it to the floor, because it happened to be thrown over the side in such a away as to serve the purpose of a syphon.*

The second demonstrates how adding salt to water lowers its freezing point:

> *I remember once, when I was a boy, hearing of a trick in the country alehouse; the point was how to melt ice in a quart-pot by the fire, and freeze it to the stool. Well, the way they did it was this: they put some pounded ice in a pewter pot and added some salt to it, and the consequence was, that when the salt was mixed with it, the ice in the pot melted (they did not tell me anything about the salt) and they set the pot by the fire, just to make the result more mysterious. And in a short time the pot and the stool were frozen together, as we shall very shortly find it to be the case here. And all because salt has the power of lessening the attraction between the particles of ice. Here you see the tin dish is frozen to the board. I can even lift this little stool up by it.*

In the more famous 'Candle' series we also have instances of the clearest kind of exposition, not all of whose lessons have been learnt up to the present day. For example, take the eloquent account of respiration by plants:

> *All the plants growing upon the surface of the earth, like that which I have brought here to serve as an illustration, absorb carbon. These leaves are taking up their carbon from the atmosphere to which we have given it in the form of carbonic acid, and they are growing and prospering. Give them a pure air like ours, and they could not live in it; give them carbon with other matters, and they live and rejoice. This piece of wood gets all its carbon, as the trees and plants get theirs, from the atmosphere, which, as we have seen, carries away what is bad for us and at the same time good for them: what is disease to the one being health to the other. So are we made dependent, not merely upon our fellow-creatures, but upon our fellow-existers, all Nature being tied together by the laws that make one part conduce to the good of another.*

As a recent television programme revealed by interviewing newly graduated students from one of the world's most prestigious academic institutions, Massachusetts Institute of Technology, this simple lesson is still, not universally acknowledged, even among the most highly educated. Such misconceptions lead us to wonder very seriously about the state of education of the public at large. In his time, too, Faraday had cause to wonder at the state of education of the public at large about the laws of nature, especially in the 1850s when a craze of spiritualism, exemplified by table turning swept London. He even wrote a letter to *The Times* about it:

> *Permit me to say that I have been greatly startled by the revelation which this purely physical subject has made of the condition of the public mind. No doubt there are many persons who have formed a right judgement or used a cautious reserve. But their number is almost as nothing to the great body who have believed and borne testimony, as I think, in the cause of error. I do not here refer to the distinction of those who agree with me and those who differ. By the great body, I mean such as reject all consideration of the equality of cause and effect, who refer the results to electricity and magnetism, yet know nothing of the laws of these forces,—or to attraction, yet show no phenomena of pure attractive power,—or to the rotation of the earth, as if the earth revolved around the leg of a table,—or to some unrecognised physical force, without inquiring whether the known forces are not sufficient—or who even refer them to diabolical or supernatural agency, rather than suspend their judgement, or acknowledge to themselves that they are not learned enough in these matters to decide on the nature of the action. I think the system of education that could leave the mental condition of the public body in the state in which this subject found it, must have been greatly deficient in some very important principle.*

This concern for the state of public education about what Faraday calls 'cause and effect' is still with us, but it led him to examine what role a knowledge of science can bring—appropriately enough in a Friday Evening Discourse on the relation between basic science and the electric telegraph. (Shades of today's debates about pure and applied science!) His words are more eloquent than I could muster, so I will quote again:

> *If the term 'education' may be understood in so large a sense as to include all that belongs to the improvement of the mind, either by the acquisition of the knowledge of others or by increase of it through its own exertions, we learn by them what is the kind of education science offers to man. It teaches us to be* neglectful *of nothing;—not to despise the small beginnings, for they precede of necessity all great things in the knowledge of science, either pure or applied. It teaches a continual comparison of the* small *and* great, *and that under differences almost approaching the infinite: for the small as often contains the great in principle as the great does the small, and thus the mind becomes comprehensive. It teaches to deduce principles*

> *carefully, to hold them firmly, or to suspend the judgement:—to discover and obey* law, *and by it to be bold in applying to the greatest what we know of the smallest. It teaches us first by tutors and books to learn that which is already known to others, and then by the light and methods which belong to science to learn for ourselves and for others, so making a fruitful return to man in the future for that which we have obtained from the men of the past. The beauty of electricity, or of any other force, is not that the power is mysterious and unexpected, touching every sense at unawares in turn, but that it is under* law, *and that the taught intellect can even now govern it largely. The human mind is placed above, not beneath it, and it is in such a point of view that the mental education afforded by science is rendered supereminent in dignity, in practical application, and utility.*

That people who believe themselves educated should know little about science was a source of disappointment, and disquiet to Faraday, as we see from the evidence he gave to the Royal Commission on Education in 1862:

> *The phrase 'training of the mind' has to me a very indefinite meaning. I would like a profound scholar to indicate to me what he understands by the training of the mind; in a literal sense, including mathematics. What is their effect on the mind? What is the kind of result that is called the training of the mind? Or what does the mind learn by that training? It learns things, I have no doubt. But does it learn that training of the mind what enables a man to give a reason in natural things for an effect which happens from certain causes? Or why in any emergency or event he does (or should do) this, that, or the other? It does not suggest the least thing in these matters. It is the highly educated man that we find coming to us again and again, and asking the most simple question in chemistry or mechanics; and when we speak of such things as the conservation of force, the permanency of matter, and the unchangeability of the laws of nature, they are far from comprehending them, though they have relation to us in every action of our lives. Many of these instructed persons are as far from having the power of judging these things as if their minds had never been trained.*

Or, in more colloquial vein, in answer to one of the Commissioners. You can hear the irony in his voice,

> *I am not an educated man, according to the usual phraseology and therefore can make no comparison between languages and natural knowledge, except as regards the utility of language in conveying thoughts. But that the natural knowledge which has been given to the world in such abundance during the last*

> *50 years should remain untouched, and that no sufficient attempt should be made to convey it to the young mind growing up and obtaining its first views of these things, is to me a matter so strange that I find it difficult to understand. Though I think I see the opposition breaking away, it is yet a very hard one to overcome. That it ought to be overcome, I have not the least doubt in the world.*

This kind of education, acknowledging the relation between cause and effect, is borne in on all of us from our earliest years as children, and becomes the basis of an intuitive feeling for how the world works. What science does is to inform us, after diligent enquiring and sifting of evidence, that such intuitions are not always valid. As intuition is a poor guide, we come back to systematic science, to observing the world, prodding it and summing up what we find in general laws, not immutable but provisional, although valid till altered:

> *The laws of nature, as we understand them, are the foundation of our knowledge in natural things. So much as we know of them has been developed by the successive energies of the highest intellects, exerted through many ages. After a most rigid and scrutinizing examination upon principle and trial, a definite expression has been given to them. They have become, as it were, our belief or trust. From day to day we still examine and test our expressions of them. We have no interest in their retention if erroneous. On the contrary, the greatest discovery a man could make would be to prove that one of these accepted laws was erroneous, and his greatest honour would be the discovery.*
>
> *Let us go out into the field and look at the heavens with their solar, starry, and planetary glories; the sky with its clouds; the waters descending from above or wandering at our feet; the animals, the trees, the plants; and consider the permanency of their actions and conditions under the government of these laws. The most delicate flower, the tenderest insect, continues in its species through countless years, always varying, yet ever the same. These frail things are never-ceasing, never changing, evidence of the law's immutability.*

What a man, what a patron saint, what an inspiration!

## Acknowledgements

In preparing this text I have had many fascinating conversations with, and much help from, Irena McCabe and Frank James, both long-standing members of the Royal Institution staff, whose knowledge of the Institution, and of Michael Faraday in particular, is second to none.

## Further reading

No detailed footnotes are given, but all the extracts quoted may be found in *The correspondence of Michael Faraday*, Vols 1–4, F.A.J.L. James (ed.), Institution of Electrical Engineers, 1992–98; *The chemical history of a candle*, M. Faraday (many editions); *On the various forces of nature*, M. Faraday, F. Warne, 1896. Many more examples from Faraday's writing are presented in *The philosopher's tree*, P. Day (ed.), IOP Press, 1999.

PETER DAY

Born 1938 in Kent and educated at the local village primary school and nearby grammar school at Maidstone. An undergraduate at Wadham College, Oxford, of which he is now an Honorary Fellow. His doctoral research, carried out in Oxford and Geneva, initiated the modern day study of inorganic mixed valency compounds. From 1965 to 1988 he was successively Departmental Demonstrator, University Lecturer and An Hominem Professor of Solid State Chemistry at Oxford, and a Fellow of St John's College (Honorary Fellow 1996). Elected Fellow of the Royal Society in 1986; in 1988 he became Assistant Director and in 1989 Director of the Institut Laue-Langevin, the European high flux neutron scattering centre in Grenoble. In October 1991, he was appointed Director of The Royal Institution and Resident Professor of Chemistry, and Director of the Davy Faraday Research Laboratory, and in September 1994, he became Fullerian Professor of Chemistry. His present research centres on the synthesis and characterization of (mainly molecular) inorganic and metal-organic solids in the search for unusual magnetic and electron transport (including superconducting) properties.

# *The fossils of the Burgess Shale and the Cambrian 'explosion': their implications for evolution*

SIMON CONWAY MORRIS

**Abstract**

About 540 million years ago the fossil record of animals, both in terms of their skeletons and the traces they made in the seabed, indicates an extraordinary diversification referred to as the Cambrian 'explosion'. The magnitude and significance of this evolutionary event have been further emphasized by the study of the soft-bodied fossils. These are best known from the Burgess Shale (Middle Cambrian), but other deposits from North Greenland (Sirius Passet) and South China (Chengjiang) are also yielding remarkable new finds. These Burgess Shale-type faunas reinforce earlier perceptions that the Cambrian 'explosion' is one of the most significant events in the history of life, but in doing so they reopen questions concerning both the origins of this event and its implications. It now seems likely that even though animals diversified so spectacularly in the Cambrian, there was an extended prior history. Evidence for this in the fossil record includes the Ediacaran faunas, but their time of abundance at c. 580 million years is incongruent with data from molecular biology that points to a much older origination of animals. A consequence of the Cambrian 'explosion' was the populating of the oceans by a wide variety of animals, representing an apparently enormous number of bodyplans. An orthodox view has emerged that the origin of novel bodyplans is a problem of macroevolution that is not amenable to explanations derived from the study of microevolution. In fact, the fossil record suggests otherwise, and the analysis of such fossils as the halkieriids provides hypotheses to explain the origin of phyla as apparently disparate as the annelids and brachiopods. Another item of received wisdom is that the Cambrian 'explosion' was a time of unrestrained experimentation in biological design, with the corollary that the end-points of this evolutionary process—including humans—are unpredictable products of a welter of contingently driven processes operating through the history of life. Such a chaotic view is not supported by the fossil record, which points to major constraints being imposed on biological form. While the ultimate emergence of humans from this evolutionary process may indeed be unpredictable, the rise of intelligent species may be assigned a much higher probability. What happened on Earth, should happen elsewhere in the Galaxy. Or should it? This chapter closes with an unfashionable review of the improbability of finding intelligent life beyond the solar system.

## Introduction

Let me begin by making a simple set of assumptions. First that my ancestry (and yours) can be traced back to at least 3.8 billion years (Byr) ago, that is shortly after the origin of life. Next let us suppose that for the first 1.8 Byr we were bacteria, with let us say an average generation time of a day. Then let us suppose that for the slightly shorter period of 1.4 Byr we were single-celled eukaryotes. Let us allow a typical generation time of 3 months. Finally, and now we are on the home stretch, we have 600 million years (Myr) of animal evolution to go through. Perhaps an average generation time has risen to 2 years. Thus approximately $10^{12}$ generations separate us from the origin of life, but of these nearly all were as bacteria.

Viewing life from this perhaps unfamiliar perspective raises a number of questions, few of which are easy to answer. One of these is why, despite the immense number of generations, especially as bacteria, did it take evolution so long to produce organismal complexity at the macroscopic rather than biochemical level? Another question, which will be the focus of our enquiry here goes as follows: To what extent are the evolutionary pathways leading to the end results of the present-day world—ignoring for the moment what might lie in the future—an inevitable outcome from a starting point of bacteria about 3.8 Byr ago? From our human stance it is understandable how evolution may indeed appear as a grand unfolding, even a drama. One does not have to be a follower of Pierre Teilhard de Chardin and his near-Jungian concept of the noosphere to acknowledge that even in the absence of humans the world today is certainly more complex in morphology, ecology, and behaviour than it was in the Archaean (Fig. 1). Nevertheless, such a view has received its

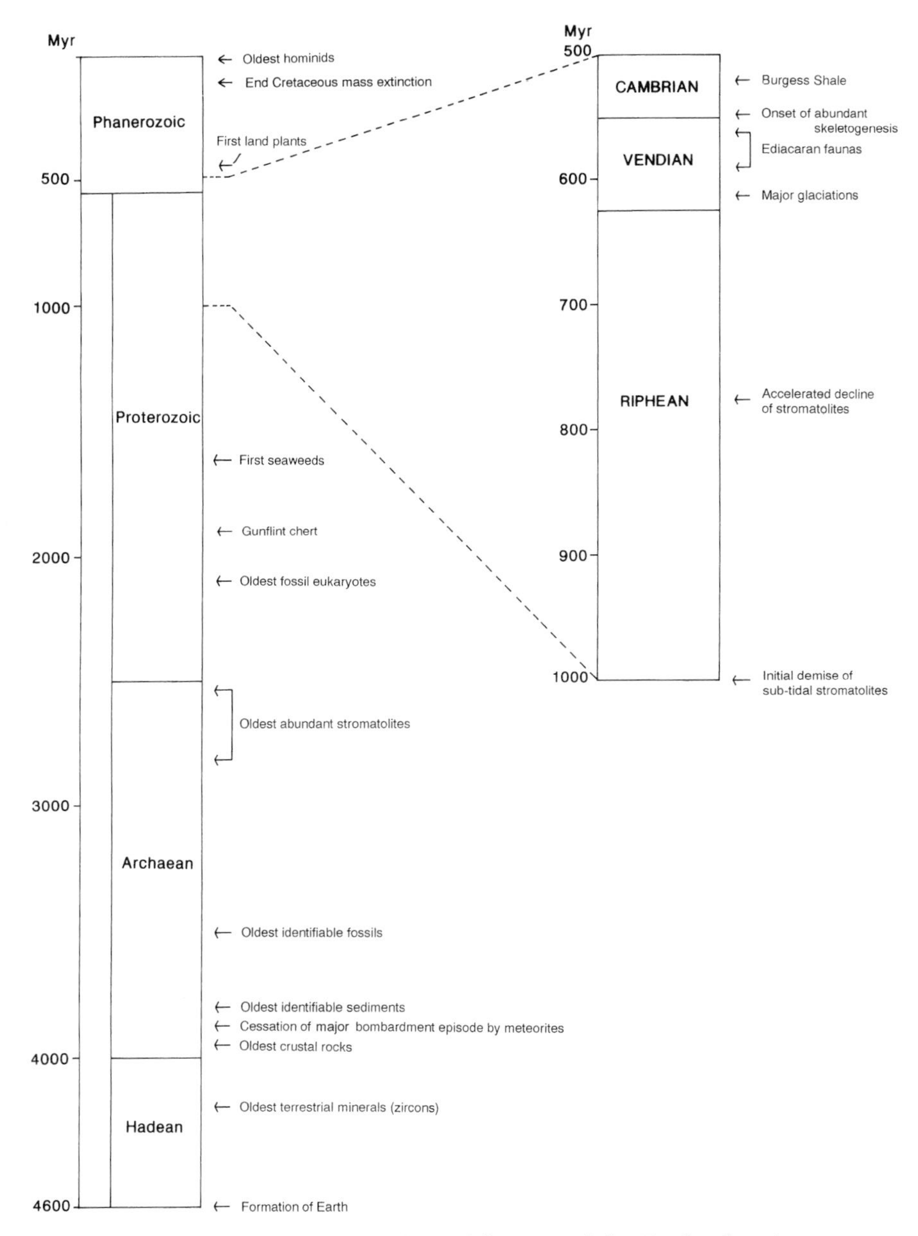


**Fig. 1** Outline of the geological history of the Earth, showing the principal divisions of time (left column) and an enlargement of the interval that encompasses the Cambrian 'explosion' (right column).

fair share of ridicule. Recent attempts to portray evolution as an effectively quirky process, largely dependent on contingent events that may in themselves be rather trivial,[1] has a subtly corrosive effect. This is because such a view removes any possibility of recognizing common themes and principles, which even if they do not deserve the name of laws can at least be seen as regularities. This is not a trivial point. One of the boasts of the physicist is that if she is standing on an Earth-like planet on another galaxy, then the apple she tosses to her male colleague will follow a trajectory whose course may be determined by precise equations. The biologist, however, could not even begin to address the probability of there being sexes, let alone hands and apples on other worlds.

## Setting the scene: the path to the Cambrian

In this chapter I will use a rather unusual fossil assemblage, known as the Burgess Shale fauna, as a vehicle for a journey in the hinterland of evolution (see also Ref. 2). The immediate importance of this fauna is that it is exceptionally informative about an evolutionary event known as the Cambrian 'explosion'.[3] This episode has attracted a great deal of attention, and rightly so because it is one of the most important in the history of life. Although the Cambrian 'explosion' does not mark the actual origination of animals, which may have been much earlier (see below), as the term suggests it is an astonishing diversification that not only saw the emergence of many major bodyplans but also set the seal on new systems of ecology. Why the Burgess Shale is so important is because in contrast to the normal circumstances of fossilization where only hard skeletal remains occur (excluding trace fossils, the makers of which generally are not known), in the Burgess Shale a far more complete cross-section of Cambrian life is available because the great majority of species preserved in this fauna are soft-bodied. At first sight this advantage might appear to be outweighed by the fact that the Burgess Shale is Middle Cambrian in age and thus about 35 Myr younger than the initiation of the Cambrian 'explosion' at c. 545 Myr.[4] This fauna, however, occupies relatively deep water and there is some evidence that rates of taxonomic turnover are substantially lower in comparison with the faunas living in shallow waters.[5] In other words in terms of evolution the Burgess Shale is probably a relatively conservative assemblage. Not only does it provide an echo of the Cambrian 'explosion', but it is a clear echo.

The nature of the Burgess Shale, its stratigraphical context and palaeogeographic setting, and its importance to Cambrian palaeontology, have been reviewed at length elsewhere.[2,6] Much of the fauna has now been redescribed, although the reasons for the exceptional preservation are still incompletely understood.[7] The Burgess Shale was discovered almost 90 years ago, by the great American geologist Charles Walcott. At almost the same time a French geologist, Henri Mansuy, discovered some similar fossils near Yiliang in Yunnan province, South China. These he described briefly in 1912.[8] The real importance of these Chinese fossils only became clear in the 1980s when excavations at nearby Chengjiang, especially the locality known as Maotianshan, yielded an abundance of wonderfully preserved fossils. This Chengjiang biota[9] has some striking similarities to that of the Burgess Shale.[5] The former assemblage, however, is Lower Cambrian in age and probably about 10–15 Myr older than the Burgess Shale. Furthermore, Chengjiang was located on a continent (known as the South China craton) that was separated by a wide ocean from the Laurentian (more or less equivalent to North America) craton where the Burgess Shale occurs. Then, as now, the super-island of Greenland formed part of the Laurentian craton. It is from the area known as Peary Land that yet another Burgess Shale-like fauna comes. This is the Sirius Passet assemblage, which is about the same age as Chengjiang, and possibly even slightly older. Rather curiously, the fossil fauna of Sirius Passet is not particularly similar to either the Burgess Shale or Chengjiang assemblages. In contrast to the Burgess Shale, research into these other two soft-bodied assemblages is proceeding very actively. Recent work has both forced us to revise some earlier ideas[10,11] and opened up new areas of study.[12,13] This research and the earlier studies on the Burgess Shale help us to place these faunas in the wider context of Cambrian life, and to emphasize yet further the dramatic nature of the Cambrian 'explosion'.

## *The Precambrian longueur*

The term 'explosion' presupposes that at least in principle there could have been a preceding period of quiescence, if not dormancy, in the history of life. Is this correct? The answer appears to be 'Yes', although it is predicated on the assumption that the fossil record for the Precambrian is complete enough to allow reliable inferences. The outlines of what at present we know are indicated on Fig. 1. This shows that the first fossils may be as old as 3.5 Byr, although some of the identifications are controversial.[14,15] Thereafter, much of the fossil record is of bacteria, especially cyanobacteria. The relative richness of their fossil record appears to be largely due to two factors.[16] First, cyanobacterial cells are enveloped in a layer of mucilage that fortuitously has an enhanced potential for fossilization. Secondly, cyanobacteria are tolerant of a very wide range of environments, including hypersaline intertidal areas where few other organisms can flourish. In the sediments of such an environment diagenetic chert ($SiO_2$, the flint from Chalk is perhaps a more familiar example) often forms very quickly from porewaters rich in dissolved silica. Sometimes chert growth is so rapid that cyanobacterial communities are engulfed and preserved.[17] Because of this bias in favour of cyanobacteria, we must be careful not to draw too sweeping conclusions about the pace of change in the Precambrian. Evolution was not completely static. For example, recently rather convincing evidence of specialized cells, known as akinetes, in c. 1500-Myr-old cyanobacteria was published by S. Golubic and co-workers.[18] Nevertheless, it remains true that in comparison with later events the rate of microbial evolution appears to have been cripplingly slow. Such a sweeping statement, based as it is on morphology, takes no account of possibly dramatic changes in microbial biochemistry. On this area the fossil record must remain almost, but not quite completely, silent. For example, there are two episodes in the late Archaean (c. 3.1 and 2.6 Byr ago) where carbonates in many parts of the world have conspicuously light values of $\delta^{13}C$ (that is they are relatively enriched in the isotope $^{12}C$ as against $^{13}C$).[19] The best explanation for this geochemical anomaly in carbon isotopes appears to be the hypothesis that there was a peculiar abundance of a group of archaebacteria known as the methanogens, which otherwise have no fossil record. As their name suggests these bacteria produce methane as an end-product of their metabolism, a process that involves extreme fractionation of $^{12}C$ against $^{13}C$ and the subsequent involvement of another group of bacteria, known as the methylotrophs. What is very obscure is why methanogens, which are common in many types of sediment, rose to such abundance in the distant past.

So with these provisos let us accept that if evolution in the Precambrian was not stagnant, it showed marked torpor. The view of Precambrian life being literally aeons of unchanging organic sameness may have lulled us into an uncritical scientific lassitude. After all, the current emphasis in evolutionary theory is on the opportunism, the jury rigging of organismal design, and the possibilities of runaway processes, especially in sexual selection. This endless potential for adaptive advantage creates an apparent paradox, because if periods of accelerated evolution are common why did the Precambrian appear to show such an evolutionary longueur? One possibility is that we simply underestimate the sheer difficulty of moving out of the microbial rut. Maybe the complexities of assembly of the eukaryote cell by both endosymbiosis and *de novo* invention of other organelles[20] really require hundreds of millions of years to achieve. Similarly, the sticking together of cells to form multicellular organisms may be a tediously slow process, requiring a precisely controlled chemistry for cell adhesion, the development of mechanisms for cell communication, and genetically inspired differentiation. The fact that bacteria are inventive when it comes to adapting to new substrates or developing resistances to antibiotics may not be particularly relevant to any argument that tries to address the onset of macroscopic complexity.

Another popular proposal to explain evolutionary stagnation is linked to the low levels of atmospheric oxygen inferred for much of the Precambrian. For example, it could be significant that what appear to be the remains of the earliest eukaryotes at about 2 Byr[21] coincides with evidence for a significant increase in atmospheric oxygen.[22] As we will see below, however, there is evidence that at least in the late Precambrian oxygen levels may have fluctuated quite markedly, but their effects on evolution are not well understood. Yet another possibility is that once a particular microbial system is established it may prove very resilient, and display enormous resistance to being dislodged.

## *The first animals*

What of the evolution of animals? When did they first appear, and what did they look like? These apparently simple questions are surprisingly difficult to answer. About the only item that is beyond doubt is that the animals (or to give them their technical name Metazoa) had appeared by the beginning of the Cambrian, that is about 550 Myr ago (Fig. 1). Many palaeontologists, however, believe that the record of animals can be traced further back, into the late Precambrian in the guise of the Ediacaran assemblages.[23] As we will see below this idea is controversial. Beyond that, back towards 1000 Myr which is agreed by almost everyone to be the approximate limit for the earliest animals, everything is debatable and an arena for conflicting positions. The fossil record is uniformly unhelpful, and the plethora of continuing claims for very ancient animals[24,25] should rouse scepticism because earlier reports have consistently failed to withstand scrutiny. Why then, in the absence of a credible fossil record, does the debate still show signs of life? The reasons are twofold. The first line of evidence comes from structures known as stromatolites. These are exceptionally common in Precambrian sediments. Stromatolites are built by microbial mats, and their characteristic laminations (Fig. 2) are a direct product of the way they form. In brief, the top surface of the mat is rich in photosynthetic microbes,

**Fig. 2** A fossil stromatolite, consisting of a series of parallel columns. Note the laminated structure, representing progressive upward migration of microbial mats. Age ca. 800 Myr; locality uncertain.

notably the cyanobacteria. Periodically, the mat is buried beneath a layer of sediment, perhaps deposited by tidal scour. Cut off from sunlight the mats would die, were it not for their remarkable property of being able to move upwards and so re-establish the sea-floor community. The repeated episodes of burial and then upward movement produce the stromatolitic laminations. Although stromatolites must have covered huge areas of the Precambrian seafloor, it has long been appreciated that the overall variety of stromatolites began to decline about 1000 Myr ago.[26] A number of workers have proposed that this decline marks the appearance of animals who promptly began to graze the mats and burrow into them. Such activities, so the argument goes, disrupted the microbial ecosystems and hence the types of stromatolite they could build. A 1000-Myr-old origination for animals is also consistent with the evidence from molecular biology. This approach is based on the so-called 'molecular clock'.[27–29] In outline the 'clock' assumes that if the rate of substitution of one building block by another in the chain of either DNA or a protein (nucleotides and amino acids, respectively) can be shown to occur at a fairly regular rate, then extrapolation to account for the greatest degree of substitution observed in the molecule will indicate the point of origination in geological time. In the context of the origin of animals not everyone agrees that such exercises in extrapolation are valid,[30] but if the principle of the molecular 'clock' is accepted then on the data available an appearance of animals at least 700 Myr ago is reasonable.

So if we accept these lines of evidence—the decline of stromatolites and molecular 'clocks'—Where are the animals themselves? If they were present, they must have been very small, because otherwise they would have left obvious traces as they moved through the sediments. These trace fossils have not been found. We must assume then that the early metazoans were a millimetre or less size. This means that they would have approached in size and ecology the protozoans, that is single-celled eukaryotes characterized by *Amoeba*. Indeed it is sometimes forgotten how closely certain protozoans can approach the Metazoa. Some ciliates, for example, are remarkably complex and with such structures as holdfasts and tentacle-like extensions they resemble tiny animals. In fact, mistakes of identification can occur. For example, recently some supposed rotifers (a phylum of metazoans also known as the wheel animacules, that are especially characteristic of lake faunas) on closer examination transpired to be ciliated protozoans.[31] Miniaturized animals, little bigger than some protozoans, are well known today in what is known as the meiofauna.[32] Typically they inhabit the interstitial spaces between sand grains. For those familiar with the hydroids, holothurians, sipunculans, priapulans, polychaetes, or crustaceans, it seems quite remarkable that such scaling down can be achieved while still retaining a recognizable bodyplan, albeit one adapted to exigencies of life between the sand grains. Indeed, it has been argued that perhaps these meiofaunal species are primitive[33] so that the Cambrian 'explosion' is simply a scaling up, equivalent to Alice's famous change in size as she nibbled the mushroom. The evidence, while not dismissing the possibility of the meiofauna housing some primitive relicts, nevertheless points to much of it being secondarily reduced in dimensions from much larger animals.

Another venerable suggestion, recently revived, is that the earliest metazoans were similar to living marine larvae, minute in size and spending their time either swimming through the water column or crawling across the seabed. Only later, as a newly formulated argument proposes,[34] did the larvae set aside special cells that in due course were employed for the differentiation of the adult. Unfortunately, with some notable exceptions,[35,36] the chances of finding fossil examples of larvae are small. But they might not be non-existent. What are interpreted as arthropod embryos have been described from the Middle Cambrian of China,[37] and further discoveries in this area are currently being documented. It is worth noting, however, that even if the fossil record of eggs and larvae is much richer than appears to be the case, by no means everyone is convinced that modern larvae are in any way a guide to the nature of primitive animals.[38]

It is not my intention to belittle any of the hypotheses that look to either the meiofauna or larval types, but in the context of the search for the first animals I am inclined to mutter 'So what?'. If early animals spent literally hundreds of millions of years as microscopic creatures, then many of the features that are believed to have played a key part in animal evolution, e.g. body cavities, biomineralization, nervous systems, and predation, would have been either impossible or scaled down to literally embryological dimensions. Such metazoans, if they ever existed, for all intents and purposes would have been just another group of protozoans, albeit one with a most interesting future.

## The Ediacaran world

So where are the specimens that allow the palaeontologist to murmur 'At last a convincing fossil animal'? The focus of interest is presently centred on the fossils that existed near the end of the Precambrian in an interval known as the Vendian (Fig. 1). These remains are generally referred to as the Ediacaran fauna. The central paradox in understanding these fossils involves their preservation. In brief, the problem is as follows. For the most part Ediacaran fossils resemble groups of animals such as certain cnidarians, notably the jellyfish and seapens, and various sorts of worm[23,39] (Figs 3 and 4). In very few cases, however, is the comparison precise, and more often there are worrying discrepancies. These fossils show no sign of having possessed hard skeletal parts, similar for example to those that formed the calcareous skeleton of a trilobite, yet despite their effectively soft-bodied preservation the Ediacaran fossils not only have a very wide distribution, but at many localities they may be abundant. To complete the conundrum the fossils usually occur in sandstones and siltstones, which any palaeontologist will tell you is the last place to expect to find soft part preservation. In a bold hypothesis that aimed to resolve this paradox a German palaeontologist, Adolph Seilacher, proposed that the similarities to animals were quite misleading. Seilacher[40,41] argued that the Ediacaran organisms had a unique bodyplan. This construction not only happened to confer a

**Fig. 3** An Ediacaran fossil *Cyclomedusa*. The disc-like shape and radiating structures have invited comparison with the jellyfish, but this type of fossil is now more widely interpreted as the basal holdfast of a seapen-like organism.

high potential for fossilization, but he proposed that it was a type of eukaryote that although multicellular represented a quite separate evolutionary lineage to the animals. In recognition of their distinctiveness he proposed they be called the Vendobionta. It was a clever solution to a difficult problem, but it is probably wrong. Here is why.

The key comes from three fossils collected from the Burgess Shale by Charles Walcott,[42] perhaps in his 1917 season. When I came across the specimens in the Smithsonian Institution it was clear that Walcott had considered them, because in the same drawer were a set of rather faded photographs. As involvement with administration and the nurturing of American science took more and more of his time, Walcott never found the time to publish an account of the fossils. In 1993 I named the specimens as *Thaumaptilon walcotti*, that is Walcott's wonderful feather.[43] Figure 5 shows why the generic name was chosen. The largest specimen

**Fig. 4** An Ediacaran fossil *Dickinsonia*. Although the specimen is incomplete, it is clearly bilaterally symmetrical. Note the well-developed 'segments'. Many palaeontologists interpret *Dickinsonia* as some sort of worm, but this is not universally accepted.

is almost 20 cm in length, and is especially informative. There is a basal holdfast, succeeded by an axial region from which arise numerous branches. On each of these there are abundant pustule-like structures.

*Thaumaptilon walcotti* has a striking similarity to a group of living marine animals, known as the seapens. These are colonial animals, and are in fact close relatively to the corals and sea anemones. The pustules on the branches of *Thaumaptilon* are assumed to be the remains of the individual animals. In living seapens there is also a complex system of internal canals that allows intercommunication among the individuals of the colony, and also is used to maintain an internal fluid pressure

**Fig. 5** The Burgess Shale seapen (Anthozoa: Cnidaria) *Thaumaptilon walcotti*, a fossil species with clear similarities to a number of Ediacaran 'fronds' such as *Charniodiscus*.

that helps to keep the colony upright. Similar canals can be seen *Thaumaptilon*.[43] But this Burgess Shale animal has one interesting difference to the most similar of its modern day counterparts. In these latter animals the branches arise from the axis, but are separated from each other to permit the capture of suspended food particles by the flow of seawater between the branches. In *Thaumaptilon*, however, all the branches appear to be attached to a common base. What is significant in this context is that frond-like fossils are a common component in many Ediacaran assemblages, and several have a remarkable resemblance to the younger *Thaumaptilon*. So if we accept that this Burgess Shale fossil is effectively an Ediacaran survivor, then Seilacher's[40,41] idea of the Vendobionta begins to look decidedly questionable. This is because there is no reason to suppose that the basic structure and composition of these frond-like organisms, which we can now assume to be pennatulaceans, changed from Ediacaran to Cambrian times. Yet the frond-like fossils in Ediacaran sediments are preserved in exactly the same way as the co-occurring types of fossil (Figs 3 and 4). Accordingly, attributing a uniquely resistant composition to the Ediacaran fossils (as part of their peculiar bodyplan) in order to explain their preservation as fossils seems to be rather implausible.

So if the original composition of at least the Ediacaran seapens was little different from those of the Cambrian and today, this still leaves unanswered the question: Why, despite their soft-bodied composition, are Ediacaran fossils so common, and why do they occur in lithologies, such as sandstones, that otherwise are quite atypical for occurrences of softpart fossil preservation? There is no simple answer to this question. Conceivably, the explanation lies with the proposed absence of predators and scavengers, and perhaps the restricted degree of burrowing in the sediments. And must we abandon entirely Seilacher's ingenious concept of the Vendobionta? It remains the case that in the Ediacaran biotas a few types of fossil appear to be so peculiar that it is probably worth reserving the idea of the vendobionts, at least for the time being. Moreover, even if *Thaumaptilon* shows a possible way forward for understanding one group of Ediacaran fossils, fitting many of the others (Fig. 4) into a general phylogenetic scheme that will permit easy connections with the Cambrian radiation is still proving very difficult. We should not underestimate the problematic nature of the Ediacaran world. New eyes might give us new ideas.

## The Cambrian 'explosion'

For many years it was thought that the Ediacaran faunas disappeared some time before the onset of the Cambrian. While this remains true of some parts of the world, recent evidence from Namibia suggests that at least here it was otherwise.[44] Via a set of arguments that are circuitous but reasonable, the evidence based on radiometric age determinations suggests that at least in Namibia the Ediacaran faunas reached right up to the Vendian–Cambrian boundary (Fig. 1). Now let us step past this seemingly innocuous barrier. As we enter the Cambrian we are on the threshold of joining one of the greatest evolutionary roller coasters of all time: the Cambrian 'explosion'. At first things seem to move rather slowly. In the lowest beds of the Cambrian, such as the Manykay horizon in northern Siberia, there are only a few skeletal fossils. Most of them look pretty scrappy, but in fact they appear to belong to several quite distinct groups. In any event the pace soon picks up and within a few millions of years we have a truly bewildering array of skeletal fossils.[45] Not only that, but in various parts of the world in sediments of the same age, there is a dramatic increase in the diversity of trace fossils.[46] This is an important observation because most trace fossils are made by animals either with very delicate skeletons or ones that are entirely soft-bodied. The deposits that actually yield the soft-bodied fossils themselves are found slightly higher in the stratigraphic column, most famously from the Chengjiang biota in South China, which is approximately the same age as the Sirius Passet fauna from North Greenland. There is good reason to think that the extraordinary range of fossils which these localities in China and Greenland yield did not appear immediately before the actual sediments at these localities were deposited, but evolved earlier in geological time, closer to the Vendian–Cambrian boundary. The implication of everything we have learnt so far about the Cambrian 'explosion' is that evolution must have proceeded at a truly remarkable rate. Some of the key steps in this evolutionary diversification may have been achieved in the Ediacaran faunas,[47] but a glance at the sheer variety of Lower Cambrian animals, especially from Chengjiang and Sirius Passet, leaves one reeling. Where on earth did all these animals come from? How was this event triggered? Did evolution proceed via pathways that have no equivalent today? Indeed, in some sense did Nature exhaust itself so that the rest of animal history is merely a tired playing out of predictable themes?[1] Let us examine the evidence.

## *Triggering the Cambrian 'explosion'*

With an event as dramatic as the Cambrian 'explosion', it is scarcely surprising that a multitude of triggering mechanisms has been proposed. In one recent review[48] some 20 different candidates are listed, albeit under a series of main headings. Such a prolixity of possible causes, however, hints that we may be some way from asking the right sort of question. And what do I mean by a 'trigger' in this sort of context? Evolution is historical and contingent. It builds on previous events and at first sight it seems reasonable to suppose that specific events—such as the rise of animals—are only likely or even possible when certain prior conditions are fulfilled. Unfortunately, what those conditions are is not at all clear. In this sense we are threatened by a regress, not quite infinite given the origin of life occurred at some distant but specific point in time, but still one that may be so labyrinthine as to make the search for a specific trigger redundant. This would leave us to air opinions rather than to sustain hypotheses. In the end, of course, this bears on a point already briefly raised and one to which I will return at the end of this review: To what extent are evolutionary trajectories, that from our perspective halfway through the life of the Solar System (or so we presume) we see as end-points, the inevitable consequences of given starting points? To put it baldly: Did the first living cell prefigure human consciousness?, or if you want to take a truly cosmic view did the initial conditions of the Big Bang that seemingly made the stellar synthesis of carbon and heavier elements inevitable also encapsulate within it all future history? This is not to deny the role of contingency, but there still exist natural laws. The extent to which these latter impinge on organic evolution is extraordinarily uncertain, but I would suggest that abandoning the notion of purpose may be premature. If, as one distinguished cosmologist has said, the Universe appears to be 'a put-up job', then can we be sure that the evolution of life is exempt from similar principles?

But I should return to the problem of evolutionary triggers, and what may be the best way to approach them. Maybe we should simply try to consider local factors that might precipitate an event at one time rather than another. As is the way in many areas of science, some of the proposed mechanisms alternate between bouts of intense interest and popularity, before sinking into obscurity, only to re-emerge subsequently as once again serious contenders. In the case of the Cambrian 'explosion' this is exemplified by the proposed role of atmospheric oxygen. The landmark for recent discussions was the influential paper by Berkner and Marshall.[49] They connected the earliest history of the Earth, where oxygen values were effectively nil, to present day where levels of atmospheric oxygen are 21%, in order to explore some of the biological consequences of this increase. In the Precambrian the intense flux of ultraviolet radiation, a consequence of a negligible ozone screen, precluded life not only on land, but in the shallower waters. Oxygen levels, however, continued to climb owing to photosynthesis, so that it is conjectured that the rise of animal life was tied to the passing of a critical threshold, perhaps 1% of present atmospheric levels. While the idea of 'oxygen oases' receives little support today, the consequences of the latter idea have been explored in a number of interesting directions. Towe[50] pointed out that the protein collagen, which is largely characteristic of the metazoans (until recently it was thought to be unique to metazoans, but is now known to occur in the Fungi[51]), and is vital for their structural integrity (a good example in humans are our tendons) can only be synthesized in the presence of free oxygen. No oxygen, no collagen, and certainly no large metazoans. Several authors[52] have also stressed that some Ediacaran fossils appear well adapted for gas diffusion, which could be consistent with low oxygen levels. Thus some Ediacaran species are conspicuous for their high surface area to volume ratio, or are inferred to be similar to Recent cnidarians where the bulk of the body is composed of metabolically inert mesoglea and the living tissue forms little more than a veneer on either side. Such a feature could also be consistent with depressed levels of oxygen. Another approach is an ingenious exploration of the relationship between oxygen and metazoan diversity, especially with respect to skeletons.[53] In a number of submarine basins, e.g. the Black Sea, the bottom waters are anoxic, poisoned by hydrogen sulphide and home only to anaerobic bacteria. The shallower waters, of course, are well oxygenated. Thus, one can imagine a transect along the seabed from basin floor to the shore, which in progressively shallower waters will record falling levels of hydrogen sulphide and increasing oxygen. In such marine basins as soon as there is enough oxygen the first animals appear, and these are soft-bodied worms. Only when oxygen levels climb appreciably do skeletonized animals appear. By substituting geological time for distance along this transect so the pattern of occurrences crudely mimics events across the Precambrian–Cambrian boundary, with the soft-bodied Ediacaran fauna being succeeded by the skeletal assemblages of the Cambrian.

Matters, however, are not quite so simple. Estimating oxygen levels in the geological past is not straightforward, but some evidence indicates that during the late Precambrian values swung quite markedly,[54] and in Ediacaran times they may have been even higher than the present atmospheric level of 21%.[55] In addition, the distribution of animals on the margins of anoxic waters is now known to be more complex than once thought,[53] such that species with skeletons can occur in waters much depleted in oxygen.[56] Moreover, this hypothesis of monotonically increasing levels of atmospheric oxygen may not have much bearing on the dramatic increase in trace fossils diversity[46] that runs in parallel to that of the skeletal record. Nevertheless, some data do suggest that very close to the Precambrian–Cambrian boundary there was a sudden increase in oxygen. Hence, one might argue that if the scene was already set, then a rather abrupt rise in oxygen could provide the necessary trigger for the Cambrian 'explosion'.

More recently much interest has been expressed in the possible role of genetics for triggering the Cambrian 'explosion'.[57] At its simplest the hypothesis would propose either the evolution of a gene for a specific compound, such as collagen, or a significant change in developmental mechanisms. This may sound reasonable, but our present knowledge allows little more than conjecture. The most urgent task, perhaps, is to discover what genetic differences, especially in terms of developmental cues, there are between protozoans and metazoans, or fungi and metazoans if the evidence for a link between the latter two groups is accepted.[58] Only then will it be clear what the significant differences are, because overall I suspect we will be more struck by the similarities. It has also been conjectured that the Cambrian 'explosion' was largely a result of a series of genetic revolutions that led to the evolution of the major bodyplans. But this is beginning to look decidedly less likely. The reason is that, rather unexpectedly, seemingly very different animals—that one would expect to have radically different genetic instructions—actually have an extraordinarily similar genetic architecture.[59] This has been dramatically shown in the case of the *Hox* genes, where to the first approximation there is a one-to-one correspondence between the genes that are crucial in the development of the fly (Phylum: Arthropoda) to those in the mouse (Phylum: Chordata). Some of these genes have also been detected in more primitive animals, including flatworms, cnidarians, and perhaps even sponges. In fly and mouse the *Hox* genes have various functions, but they are especially important in the differentiation of various parts of the body, and their expression can be mapped out in the developing embryo with a high degree of precision. Unfortunately, rather little is known yet about the exact function of these genes in the more primitive metazoans, but it is likely that some have an important part in the definition of body axes. It should be apparent that if at least the common ancestor of arthropods and vertebrates had this complement of *Hox* genes to specify its principal divisions of the body, then the obvious multitude of differences in the descendant forms must depend on genetic differences, but ones that are unlikely to have arisen by profound changes in the genetic architecture.

Thus, if the basic genetic architecture is already in place by the earliest Cambrian and quite possibly the Ediacaran, then some other motor for the 'explosion' may have to be found. It is here that we look to ecology, and specifically the interactions between organisms. The link between the appearance of hard parts in the Cambrian and predators goes back many years and has been rearticulated by G.J. Vermeij.[60] It can be pointed out, of course, that hard parts serve many more functions than protection. In addition, animals may protect themselves in other ways, e.g. camouflage and toxins, that do not involve hard parts. There are also plenty of examples where hard parts may be reduced or lost, as in the various pelagic gastropods,[61] so their possession need not be regarded as an unmitigated advantage. Nevertheless, for the Cambrian animals the primary role of their hard parts seems to be one of protection. What evidence supports this assertion? Until quite recently it was thought that these early animal communities were free from predators. This still appears to be almost entirely the case for the Ediacaran faunas. Work on the Burgess Shale, however, unequivocally shows predation to have been highly significant. The most graphic illustration of this are the gut contents, such as hyoliths in the priapulid *Ottoia*[62] and crushed skeletal debris in the arthropod *Sidneyia*.[63] Ancillary evidence is also available. This includes the arrangements of mouth parts and jaws, and what appear to be defensive arrays of spines in such animals as the lobopod *Hallucigenia* and stem-group polychaete *Wiwaxia*, as well as animals from other deposits such as *Ecnomocaris*.[64] Looking beyond examples of exceptional preservation, the roster of evidence for predation includes bite-marks,[65] boreholes,[66] and digging traces intercepting worm burrows.[67]

Other evidence for ecological interaction is also beginning to emerge. From Lower Cambrian sediments in the North-West Territories of Canada exquisitely preserved pieces of arthropod limbs have been extracted.[68] Many of these appendages had setose extensions that looked well-suited for harvesting phytoplankton, of which the Cambrian fossil record largely consists of a group known as the acritarchs, the affinities of which are not very clear. Workers on acritarchs have noted that during the Cambrian their variety and degree of ornamentation e.g. spines, both increase significantly. Butterfield[68] suggested that the changes in acritarch morphology were driven by the rise of such arthropodan filter-feeders. Why this should be is quite obscure. However, a considerable amount is known about the mechanics of arthropod filter-feeding,[69] and it should be possible to devise experiments to see what effect ornamentation or other features of the plankton have on their capture rates.

There was also a wide variety of other animal suspension feeders in the Cambrian that caught generally smaller particles in the seawater. These included the abundant brachiopods, but perhaps even more important were the sponges. In many Burgess Shale-type faunas sponges are a conspicuous component,[9,70,71] while in reefal habits the calcareous archaeocyaths flourished.[72] Examination of deeper-water sediments from the Cambrian often reveals an enormous abundance of spicules, suggesting that some parts of the sea-floor must have been carpeted with sponges. While sponges were present in Ediacaran faunas, certainly their abundance in the Cambrian suggests that huge volumes of seawater were being filtered for food.

There is a further twist in this story of Cambrian filterers, which apart from its possible relevance to understanding the Cambrian 'explosion' is also a useful reminder that insights often come from other disciplines. Granted that the effect of filterers on Cambrian ecology must have been considerable, could they have even been crucial for its initiation? This is the claim made by Logan and co-workers,[73] which stemmed from some ingenious work on the isotopic ratios of carbon ($^{12}C : {}^{13}C$, or $\delta^{13}C$ in the notation against an agreed standard) in late Precambrian and Cambrian sediments. In brief, they observed that there were marked differences in the values of $\delta^{13}C$ in certain types of organic matter (specifically hydrocarbons such as *n*-heptadecane) according to which side of the Boundary the samples came from. The explanation they offered was that the isotopic ratio for Precambrian carbon was consistent with it being subject to extensive bacterial attack before being buried in the seafloor. In contrast, that from the Cambrian retained an isotopic signature that suggested that it was little altered from the algal material which was harvested by consumers and so incorporated into the first step of the marine food web. To explain this surprising discrepancy Logan and co-workers[73] argued that in the Precambrian the organic matter produced remained finely particulate, as a sort of marine 'snow'. Being so minute rates of sinking through the water column were very slow. Accordingly, the bacteria had plenty of time to recycle the carbon and so impose their own isotopic signature. In the Cambrian, however, the appearance of arthropod grazers,[68] broadly similar to the living krill, resulted in the organic matter being packaged in faecal pellets. These fell through the water column much more quickly, coming to accumulate as part of the sediment on the seafloor. In doing so they avoided extensive bacterial degradation, and so retained the algal carbon isotope signature. Logan and co-workers[73] pointed out that if the water column was filled with marine 'snow', populated by huge bacterial populations, then because of their metabolic activity there would be intense demand on any available oxygen in the seawater. This in turn would lead towards a strong tendency towards anoxia, especially in deeper waters away from the mixing zone of the surface. This would result in chemical cycles within the ocean very different from those that characterize a well-oxygenated ocean. These workers went on to suggest that the Cambrian 'explosion' was initiated by the appearance of the grazers, and so the rapid transfer of organic material to the seafloor by faecal pellets. A testable consequence of this proposed model is that faecal pellets should be recoverable from late Precambrian sediments. Nevertheless, this hypothesis leaves me wondering what prompted the evolution of these grazers in the first place.

Many other areas of the ecology during the Cambrian 'explosion' require specific study. What, for example, was the effect of the greatly increased infaunal activity as is evident from the dramatic radiation of trace fossils?[46] How, in particular, were those animals that eat organic-rich sediment—the deposit feeders—affected? Similarly, what may we infer from a proliferation of grazers, scraping the surface films of bacteria and algae? Presumably, they too provoked ecological changes, of which the development of calcareous skeletons in many algae might be the most obvious consequence (although the interested reader should consult the work by Knoll and co-workers[74] for a cautionary note concerning algal calcification and the role of oceanic carbonate saturation).

Although there is no shortage of ecological experimentation in Recent communities—deliberate exclusion, controlled introductions and so forth—it remains the case that it is still impossible to model an event such as the Cambrian 'explosion', except in the most intuitive terms. In the Burgess Shale fauna I attempted to apply theories of resource partitioning and niche division.[75] The majority of ecological categories I recognized fell into the so-called log–normal distribution, the ecological significance of which is still debatable. It would, nevertheless, be interesting to compare other well-documented faunas, notably those of Chengjiang[9] and Sirius Passet,[13] to see if niche allocation and trophic structure show any systematic changes. No general theory of ecology exists, but in any event in this context it might be a mistake to think in terms of cause and effect. The ecology of Cambrian seas is more likely to have been driven by feedbacks. What was the resilience of such systems? One can imagine almost endless alternatives. Did the appearance of a new group of predators lead to a cascade of consequences through the food web, or could an apparently innocuous change in the floating community of phytoplankton restructure entire communities? But Cambrian ecology also needs to be placed in the context of the multitude of animal designs that were occupying or more realistically defining niches. How did they evolve, and does the speed and range of this diversification call into question current evolutionary theory based on the precepts of neodarwinism?

## *The origin of novelty*

Let me begin with a quotation from the paper by Bergström:[76] 'The morphic or genetic gaps between animal phyla were as wide in the Cambrian as they are today. Various problematic fossils therefore are unlikely to be intermediate between phyla; most likely they belong to distinct phyla'. Encoded in this statement are a series of assumptions that may be difficult to justify. At first sight the opening sentence seems eminently justifiable. Take a bivalve mollusc from your nearest sandy beach. Now visit a museum of palaeontology and study the bivalves of successively older ages. They will be found in all sorts of shapes and sizes, but even as far back as the Lower Cambrian,[77] the bivalves have kept a fundamental identity. On the other hand if we examine a number of other animals from the Cambrian they do indeed look extremely peculiar, or at least at first sight. These are referred to as the problematica. They have attracted considerable attention as it was realized that the Burgess Shale in particular housed a significant number of species whose phylogenetic affinities were, to put it mildly, controversial. Some looked so strange that it seemed reasonable to suggest that they might indeed represent distinct bodyplans. These bodyplans in turn might be interpreted as extinct phyla, ones seemingly confined to the Cambrian. The implication of Bergström's statement, however, is that not only was the Cambrian enriched in phyla relative to the present day, but if the so-called problematica were not intermediates between phyla, then so far as the fossil record is concerned we would never be able to observe the origin of phyla. This may be literally true in terms of documenting the evolution of a new phylum by tracing the precise lineage of intermediate species, but Bergström's statement might also be taken to imply that the evolutionary mechanisms that led to the appearance of the phyla were different from those operating at lower taxonomic levels, notably those that give rise to species themselves. In terms of evolutionary theory the claim would be that macroevolution, the origin of phyla, is separate from the origin of species which emerge by the processes of microevolution.

This problem of studying the origin of phyla has been brought into even sharper focus by recent work on the molecular biology of representatives of many extant phyla. Studies have tended to concentrate on a molecule known as ribosomal ribonucleic acid (rRNA or the rDNA). This has been shown to be a powerful tool for phylogenetic analysis,[78,79] although in principle either other molecules can be employed[80] or a search made for particular 'signatures' of genetic arrangement.[81] While there are a number of difficulties in using these molecules the underlying assumptions seem reasonable: in closely related species the sequence of nucleotides or amino acids along the molecular chain that goes to make the entire molecule will be very similar. A well-known example is the haemoglobin of humans and chimpanzees, in which the sequence of amino acids is identical. In more distantly related taxa the extent of substitution of one nucleotide or amino acid by another at various points along the chain will be proportionally greater. Some of these molecular data confirm hypotheses that have been accepted by nearly all zoologists. For example, all agree that despite their obvious anatomical differences chordates, e.g. humans, and echinoderms, e.g. sea urchin, are closely related. Both are usually placed in a larger grouping known as the deuterostomes (where they are joined by another group, the hemichordates). In other cases the molecular data revived an hypothesis previously only held by a minority. One example was the phylogenetic position of the annelids, best known as the earthworm, but familiar also as the leech and the marine bristle worm (polychaetes). This phylum was generally allied with the arthropods, whereas an unfashionable view, now strongly supported by molecular biology, argues that annelids are much closer to the molluscs.[82,83] In yet other cases the conclusions that must be drawn from molecular data run headlong into established thought. Take, for example, the marine brachiopods, well known as fossils but still abundant in some areas today. They are possessors of a bivalved shell and characteristic feeding organ known as the lophophore. Almost without exception zoologists had placed this group close to the deuterostomes. Molecular biology, on the other hand, strongly indicates a position adjacent to the annelids and molluscs.[84–86]

The one drawback of molecular biology, however, is that invaluable as the data are, it is inevitably based on living species. Thus, such data tell us nothing about what the common ancestor, say of an annelid and a mollusc, looked like. Neither can they explain how the transitions to each phylum were achieved, nor in what ecological and functional contexts the evolution of these groups actually happened. In other words molecular biology is a uniquely important data source for recording the results of history, but it is uninformative about the actual narrative.

## *The role of fossils*

It is now emerging that the fossil record is far from silent on these issues, and in some cases appears to be satisfactorily congruent with the molecular data. But by no means all problems are solved; there is certainly much to be learned, especially in such key areas as the origins of flatworms[87,88] and also the deuterostomes.[89] But if the case example I explain below of the fossil record throwing light on the origin of phyla gains acceptance, then future workers might be better to accept it as one of general principle rather than let it remain as an exception. The alternative, in my opinion, is to continue to follow will o'wisp ideas of mysterious macroevolutionary forces the precise formulation of which continually recedes as the investigator advances, into the mire.

My chosen example concerns a Lower Cambrian group known as the halkieriids. These fossils were first recognized almost 30 years ago in Cambrian rocks from the Baltic island of Bornholm.[90] As more examples became available, especially from areas such as Siberia and Australia, it was realized that these fossils (Fig. 6) were actually individual skeletal elements (sclerites) that originally formed some sort of external armour. Unfortunately, upon death the skeleton rapidly disarticulated, so in the absence of complete fossils the arrangement of the sclerites in life was necessarily conjectural. One clue, however, came from a Burgess Shale animal known as *Wiwaxia*.[91] Here the sclerites are preserved *in situ* (Fig. 7), in an arrangement referred to as its scleritome. The relationships of *Wiwaxia* were somewhat controversial. Its discoverer, Charles Walcott, had interpreted it as an annelid,[92] specifically an example of the major marine group known as the polychaetes. On the other hand, I had been

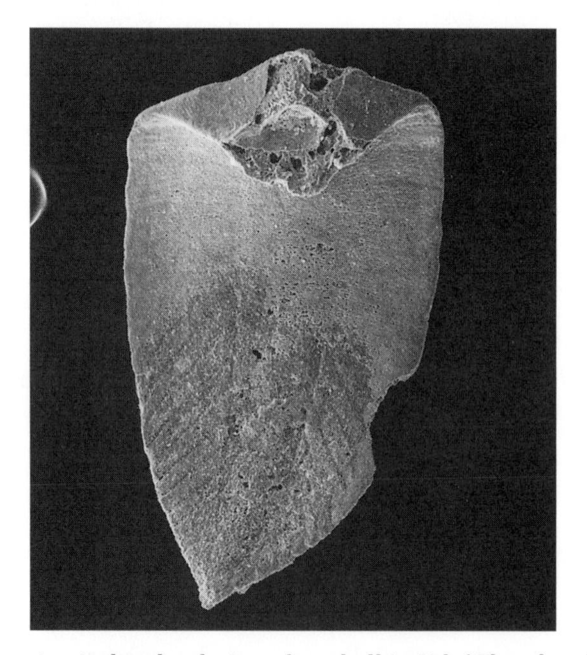

**Fig. 6** An isolated sclerite of an halkieriid (*Thambetolepis delicata*) from the Lower Cambrian of South Australia. The hard parts of the sclerite were originally preserved as calcium carbonate, but have since been replaced by secondary deposits of calcium phosphate. The base of the sclerite, at the top of the figure, shows the foramen through which a stalk projected, originally attaching the sclerite to the body. Faintly visible on the main part of the sclerite are traces of the internal system of canals.

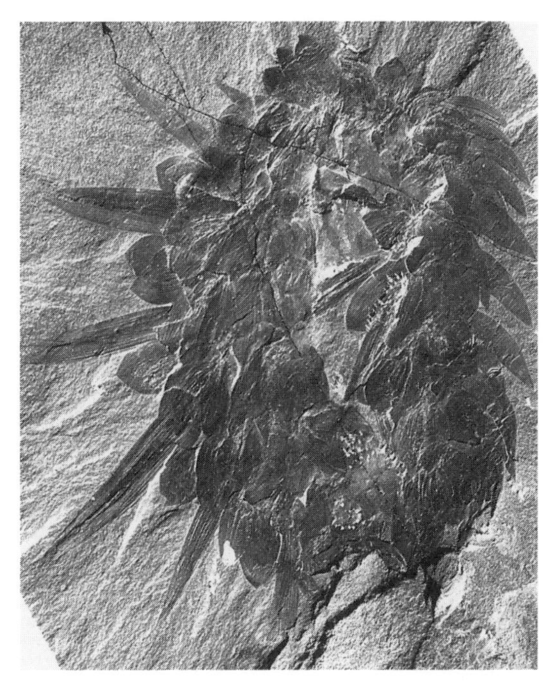

**Fig. 7** A specimen of *Wiwaxia corrugata* from the Middle Cambrian Burgess Shale of British Columbia, Canada. The individual sclerites that coated much of the body are visible, and it should be apparent that they occur as a number of different shapes. On the right-hand side the recurved siculate sclerites are visible, while projecting from the left-hand side are the elongate spines whose primary purpose appears to have been defensive.

more impressed by the similarities between *Wiwaxia* and the molluscs.[91] At that time the new molecular data[82] supporting the link between the annelids and molluscs had not been published, but had it been available then the apparent divergence of opinion between Walcott and myself on the affinities of *Wiwaxia* would have been seen to be more apparent than real.

The question of the appearance of the halkieriid scleritome was solved with the discovery in 1989 of articulated specimens (Fig. 8) from the Sirius Passet fauna of North Greenland.[13] To the first approximation, these finds supported the conjectured arrangement of the sclerites according to their comparable disposition in *Wiwaxia* (Fig. 7). A totally unexpected feature of the halkieriids, however, was a prominent shell at either end of the body (Fig. 8). The undersurface of the animal was evidently a soft foot, presumably used for locomotion in the same manner

**Fig. 8** An articulated specimen of a halkieriid (*Halkieria evangelista*) from the Lower Cambrian Buen Formation of Peary Land, North Greenland. The individual sclerites of the scleritome form a densely imbricated armour over the upper surface of the body. The posterior shell is clearly visible, whereas the anterior shell is less developed, but can be seen at the front end of the specimen where the body is bent to the right-hand side.

as such molluscs as the snails and a more primitive group known as chitons The dorsal armour of calcareous sclerites in halkieriids also recalled the likely arrangement of the incipient skeleton in the first molluscs. This arrangement persists in a group generally accepted as the most primitive molluscs, the aplacophorans, as well as occurring in the marginal girdle of the chitons which forms a skirt-like arrangement around the edge of the animal. Halkieriids, however, cannot simply be forced into a molluscan strait-jacket; for one thing the manner in which they secreted their sclerites was decidedly different from any known mollusc.[13]

What at first appeared to be a further complication arose from work by Butterfield.[93] Using a technique of acid digestion he extracted sclerites of *Wiwaxia* from the Burgess Shale. Thus isolated they revealed a microstructure extraordinarily similar to that found in the chitinous bristles (chaetae) of annelids. Butterfield concluded that as the sclerites were no more than chaetae, *Wiwaxia* must be a marine annelid, that is a polychaete, just as Walcott[92] had originally said. The arrangement and anatomy of the sclerites in the Sirius Passet halkieriid, however, certainly supported a relationship with *Wiwaxia*, and this suggested that not only was the story more complicated, but also more interesting.[13] In brief, it now seems more likely that *Wiwaxia* is on the way to becoming a polychaete, the group generally regarded as the most primitive within the annelids, rather than being a 'true' or 'genuine' annelid. Now, this may seem to be a semantic point: Surely we can expand our definition of annelids to encompass *Wiwaxia*? Indeed we can, but by doing so we are in danger of losing sight of its greater importance which is to show how annelids could have arisen in the first place.

To explain this it is necessary first to review the basic anatomy of the polychaete annelids. 'True' annelids have a body divided into a series of more or less identical segments, each of which bears two bundles of chitinous chaetae on either side. In the primitive forms the upper bundles of chaetae (notochaetae) form a protective felt, while the lower bundles (neurochaetae) are involved with the cycle of walking movements. In *Wiwaxia* (Fig. 7) the precursors of both notochaetae and neurochaetae can be identified, but they still form a more-or-less continuous protective covering with the other sclerites. In particular, it seems unlikely that the precursors of the neurochaetae could have been involved with locomotion. In this respect *Wiwaxia* appears to have been much more mollusc-like, crawling on a muscular foot rather than employing the stepping motion of polychaetes. Finally, *Wiwaxia* has a jaw-like structure, different from that of annelids but similar to the molluscs. It should now be clear how it is more sensible to view *Wiwaxia* as, so to speak, 'half a polychaete', just as the famous *Archaeopteryx* from the Jurassic of Germany is really just 'half a bird'.

The halkieriids precede *Wiwaxia* in geological time, and thus show the evolution of this group at a yet earlier stage. But their phylogenetic usefulness is not exhausted, because halkieriids also appear to throw light on the origin of another phylum, the brachiopods.[13] Not only that but this hypothesis is interestingly congruent with the molecular data mentioned above.[84,85] A long-recognized curiosity of brachiopods is that around the margin of each valve of the shell there are chitinous bristles. Their microstructure is identical to that of annelid chaetae.[94] For as long as the phylogenetic position of brachiopods was placed near to the deuterostomes, and therefore remote from the annelids, then this very close resemblance of setae and chaetae had to be dismissed as accidental, one that evolutionary biologists would refer to as of convergence. In the halkieriids the shells, especially the posterior one (Fig. 8), are surprisingly similar to the valves of brachiopods. Perhaps now we can understand the origin of brachiopods? Imagine a juvenile halkieriid, perhaps only a few millimetres long. At this stage of life, shortly after hatching, the two shells will not be separated as they are in the adult (Fig. 8), but would have been close to each other, effectively back-to-back. In normal circumstances as the animal grew in size each of these two shells would continue to grow, but would also move further and further apart, as new sets of sclerites were interpolated to ensure a continuous protective coat for the animal. In the juvenile stage, however, it is possible that sometimes the animal folded itself so that one shell was swung beneath the other, perhaps as a defensive reaction. In this way the halkieriid would become bivalved. Recall that around each shell of the juvenile halkieriid there would have been a zone of tiny sclerites. If, as we saw above, these sclerites could evolve via *Wiwaxia* into the chaetae of annelids, then presumably they could just as easily have given rise to the chitinous setae of brachiopods. Perhaps, therefore, this is how the phylum Brachiopoda originated? Incidentally, if such a folding together of the valves sounds an implausible mechanism, then it is worth reminding ourselves of the early development of a primitive living brachiopod known as *Neocrania*.[95] In this brachiopod, at first the larva crawls across the seafloor, but then upon settling (all adult brachiopods are sessile) it rotates the posterior section of the body beneath the anterior section and begins to secrete the two valves, one on top of the other.

This, of course, can only be an outline of what may have happened with the origin of the annelids and brachiopods. A common error is to suppose that the actual fossils in the museum drawer *are* the ancestral forms. This is very unlikely and in the case of either the Sirius Passet halkieriids and brachiopods, or the Burgess Shale *Wiwaxia* and annelids, it is impossible because it is known that the latter group (brachiopods and annelids) in each case appeared considerably earlier in the stratigraphic record than the former groups. Rather, the importance of such fossils as *Wiwaxia* and the halkieriids is to reveal anatomical configurations—now lost to the present-day world—that indicate how apparently very dissimilar phyla such as the annelids or brachiopods could have evolved from animals that at first sight look both entirely unrelated and problematic.

## *The way forward*

The concept of fossil problematica is only useful, therefore, as a statement of ignorance about their wider relationships. Attempts to portray evolution in the Cambrian as a chaotic welter of bodyplans,[1] of which only a handful emerged from this phylogenetic mêlée, really misses the point.[2] To be sure, many of these groups were geologically short-lived, but that is the case throughout the fossil record. It is indeed true that many fossil problematica—and they are not just confined to the Cambrian—remain very difficult to understand, but this is most likely due to insufficient data. The combination of an incomplete fossil record and the relative rapidity of evolution mean that in only a few cases will we be ever fortunate enough to trace with any exactitude the precise course of events that led to a particular bodyplan. Even the story[13] of the origins of the annelids and brachiopods from wiwaxiids and halkieriids is still only based on a handful of species: new discoveries from the Cambrian may well compel radical alterations to these hypotheses. Nor should we be surprised if this happens. At the moment the earliest Burgess Shale-type faunas[5] with abundant specimens post-date the onset of the Cambrian 'explosion' by some millions of years: in terms of evolution much must have already happened by the time the Chengjiang and Sirius Passet faunas were fossilized. As new Burgess Shale-type faunas continue to be documented, so the discovery of species related to the so-called problematica almost invariably reduces the phylogenetic mystery rather than compounding it.

What makes the Cambrian 'explosion' so exceptionally interesting is the apparent rate of diversification. Figures of only a few millions of years for the development of most of the phyla are now being routinely mentioned, although we should remind ourselves that the earlier part of this history of animal radiation may possibly stretch back into Ediacaran times thereby extending the process significantly.[30,47] Nevertheless, whatever the rate of diversification the information emerging on the phylogenetic closeness of certain phyla (from molecular biology), the origin of phyla as seen in the fossil record (palaeontology), and the basic identity of genetic mechanisms for developmental specification of bodyplans, e.g. *Hox* genes, must cast some doubt on appeals to abstruse mechanisms of genomic organization that allow supposedly catastrophic rearrangements of bodyplans that have been thought to characterize the Cambrian 'explosion'. If, to return to earlier arguments, the principal motor for Cambrian diversification was ecological, then much of the evolution of bodyplans would have been concerned with adaptations to, and alterations for, such features as feeding, locomotion or attachment, protection, and reproduction.

What will become increasingly clear in the next few years, I believe, is that the principles and mechanisms of evolution operating during the Cambrian 'explosion' will transpire to be no different from those observed elsewhere in the fossil record. Although unsurprising from what we already know, there is still extraordinary resistance—in some quarters—to the notion that major groups, including phyla, do not appear *de novo* ready formed by macroevolutionary mechanisms to spring into ecological action. Rather, the characters that are chosen to define a phylum are acquired sequentially, so that a series of species will define what in the terminology of cladistics is termed a stem-group. There has been particular success with this concept in the study of vertebrate evolution. Here, fossil documentation exists for the definition of stem-groups in amphibians,[96] Reptiles,[97] Mammals,[98], and birds.[99] In each case the fossils show some, but crucial to understand, not all of the anatomical characters that are taken to define a 'true' example of each vertebrate of these classes. In popular parlance they are 'missing links'. For those obsessed with the neat pigeonholing of animal groups these fossils and stem-groups are the classifiers' nightmare. But for students of evolution they are one of the mainstays of the argument that only fossils can answer one aspect of how to understand the origin of a major group. In the context of the gradual emergence of a particular bodyplan, which certainly need not be restricted to the level of the phylum, Kemp[98] has proposed the useful idea of correlated progression. This has already been applied to some aspects of the Cambrian 'explosion'.[100] Such an approach, of course, neatly undermines the essentialist and static concept of the phylum. Often regarded as a sacrosanct taxonomic concept, defined almost in terms of Platonic Idealism, in fact the phylum is simply the identification of a monophyletic unit—quite possibly with millions of species—whose wider relationships *at present* are uncertain. As a combination of approaches, mostly relying on Cambrian palaeontology and molecular biology,[101] uncover the relationships between phyla, so the notion of the phylum will be reduced to informal usefulness, if not ultimate redundancy.

One other point deserves discussion. This question asks: at what stage of the history of animals can the Cambrian 'explosion' itself be regarded as inevitable? Is it from the first animal, or only at some later point in their evolution? The answer may depend on a more detailed understanding of the developmental genetics in the most primitive metazoans, notably the sponges, cnidarians (corals, seapens, and so forth), and perhaps another group known as the ctenophores (sea combs). As noted above some key developmental genes have been identified in the primitive corals, even though their precise functions are still unclear. Take first the sponges, which are both almost universally regarded as the most primitive of metazoans and are known from the Ediacaran faunas.[102] Given their bodyplan could we predict with any confidence the subsequent diversifications of the Cambrian 'explosion'? I am not sure we could do so. This is because despite their macroscopic size and a degree of body organization that can be relatively sophisticated within the constraints of drawing in seawater for feeding and expelling the waste through the large opening known as the osculum, sponges have neither defined tissues nor a nervous system. Unless we accept such features as the ability for cells to adhere together (in fact such multicellularity has arisen independently many times) and possibly the presence of at least one homeobox gene[103] (although some doubt now appears to accompany these reports; G. Balavoine, personal communication) as necessary to initiate the Cambrian 'explosion'—and these possibilities can by no means be rejected—then it may be that in itself the evolution of the sponges was not enough. After all, if the last 400 Myr of the Precambrian had been 'frozen' in a world of protozoan evolution and ecology, then perhaps it is not fantastic to imagine the next 400 Myr where the only animals were sponges.

With the appearance of the cnidarians, the actual origins of which remain mysterious, further events leading to more complex animals, however, were probably inevitable. For example, even though the nervous system in cnidarians is arranged in a primitive net there are concentrations of nerve in sorts of ganglia, and indeed some jellyfish have well-defined eyes. Moreover, much more advanced groups of animals, such as the echinoderms and some hemichordates, also employ nerve nets. Cnidarians are, of course, capable of muscular activity and in the ctenophores, which are probably as closely related to the cnidarians as anything else, the mesoglea (the gelatinous material that separates outer ectoderm and inner endoderm) contains muscle fibres. This arrangement may prefigure how the bulk of the musculature in the mesoderm of more advanced animals may have arisen.

We should be careful, however, not to concentrate solely on metazoan evolution and the unfolding complexity of body organization from the sponges onwards. While it certainly may transpire that the Cambrian 'explosion' is for all intents and purposes a radiation of the more advanced animals (flatworms and above), taking over from the more primitive cnidarian-dominated ecosystems that flourished in Ediacaran times, it is still possible that the 'fuse' leading to the Cambrian 'explosion' needs to be traced further back than the animals themselves, to earlier eukaryotes. Even if the fungi are the closest relatives of the animals,[58] their common ancestor would have been some sort of protozoan. To date attempts to detect *Hox* genes in the protozoans apparently have met with no success. Nevertheless, a substantial part of the genetic basis of the animals must have been encoded in the protozoans (and fungi) as is clear from the recognition of various shared genetic homologues.[104] In addition, there are specific structures that may deserve closer scrutiny. Consider, for example, those protozoans with eye spots. It would be fascinating to know whether the genes that code for this primitive visual system are similar—or perhaps even identical—to the now famous Pax-6, which is responsible as a master control gene for many, if not all, animal eyes.[105]

## *Post-Cambrian problematica*

Without belittling the excitement of unravelling the Cambrian 'explosion', it is still difficult to argue that it shows evolutionarily unique features that are found nowhere else in the history of life. One recurrent claim, however, is that no new phyla evolved after the Cambrian. Even with the proviso that the phylum as an evolutionary concept actually has only a limited usage, it is worth pointing out that strange animals are by no means confined to the Cambrian. From the Carboniferous, about 200 Myr after the initiation of the Cambrian 'explosion' are animals whose strangeness certainly complements anything from the Burgess Shale. But their exoticism is really of our making. Such animals as *Typhloesus*, with its extraordinary tail-fin and apparently blind gut,[106] or *Tullimonstrum* with a prehensile grasping organ and eyes on elongate stalks,[107] may look the epitome of the bizarre. Each of these animals, however, is known from only a single species and a single stratigraphic horizon. As with the Cambrian problematica, as and when related species are found then some of their oddness will probably slip away as their wider relationships become more clear. Lest this seem no more than a pious hope, then the case of the hitherto enigmatic *Ainiktozoon*, from the Silurian of Scotland, provides a good counter-example. Long regarded as problematic, with possibly vague similarities with the tunicates, it has now been convincingly reinterpreted as an arthropod.[108] Similar resolution surely awaits *Typhloesus* and *Tullimonstrum*, and in any event it seems less likely that the groups represented by these fossils appeared as far back as the Cambrian as part of the initial diversification of animals and then evaded fossilization until the Carboniferous. Exactly when they arose, of course, cannot be determined, but an origination nearer the Carboniferous seems just as possible.

Not are supposedly bizarre animals restricted to the fossil record. One of the most famous of living examples is *Xyloplax*, dredged up from the ocean depths, from sunken driftwood. It is clearly an echinoderm, but it has a very peculiar morphology.[109] Until recently it had been classified as a highly aberrant relative of a group of starfish known as the caymanostellids. Doubt has now been thrown on this interpretation, and some authors have reaffirmed the original view that only by placing *Xyloplax* in its own group will we do justice to its morphological distinctiveness.[110]

## The constraints on evolution

The emphasis on the numbers of animal phyla, whether they all appeared in the Cambrian and whether any subsequently became extinct, has overshadowed what I believe is an equally, if not more, interesting problem for biology. This revolves around the identification of the constraints and limits on the evolution of animal life (or any other Kingdom for that matter). With the plenitude of species that occupy nearly all parts of the continents and oceans in a seemingly endless variety, it may seem paradoxical to talk of constraints rather than the customary celebration of life's exuberance. Such celebrations, however, tend to honour the specific, whereas what we require are organized systems of knowledge that will help us to rationalize what we know so as to prepare the next set of questions. Even to talk about 'constraints and limits' may sound delightfully vague. But there is an approach that shows that even if the species are many, the options open to them are surprisingly few. This is, of course, the study of evolutionary convergence. In at least one respect the importance of convergence is very widely acknowledged, because it is the bugbear of systematists. How can we be sure that a strikingly similar character shared by two species has not evolved independently in response to a common need, rather than been derived from a common ancestor? In many cases it is extraordinarily difficult to decide, not least because almost invariably if a scheme of classification is accepted that takes one set of features as being phylogenetically reliable because of their derivation from a common ancestor, then automatically other sets of characters become convergent. In terms of the main features of metazoan evolution it is now becoming increasingly widely accepted that many features thought to be fundamental for establishing relationships between phyla—such as the type of body cavity (e.g. the coleom) and metameric segmentation—have evolved independently several times. But with respect to any overall synthesis the phenomenon of convergence in evolution has been strangely neglected (but see Ref. 111). At one level what does it matter how many bodyplans evolved in the history of life if the same themes emerge again and again?

In the context of such a broad review I will only give two examples, both pertinent indirectly to the Cambrian 'explosion'. The first involves skeletons. Perhaps not surprisingly, there are only a relatively limited number of ways in which skeletons can be constructed. A pairwise comparison of all skeleton types, as undertaken by Thomas and Reif,[112] reveals 1536 possible combinations, giving 186 pairs. Of these a number (c. 6.5%) are functionally impossible, and thus are never found in Nature. The main feature to emerge from this analysis,[112] however, was that relatively few of the combinations could be regarded as rare (c. 26.9%), whereas the remainder were either frequent or abundant. Moreover, practically no combination that could in theory exist failed to be found in some organism. Clearly, the matrix of skeletal possibilities has been pretty well explored by animals.

Let us now turn to a more specific example. What could be more extraordinary than the recent announcement of not a new species or even family, but an hitherto unknown phylum?[113,114] The animal, *Symbion pandora*, was found living on the appendages of *Nephrops*, the scampi. It is certainly a strange beast with a combination of characters found nowhere else in the Metazoa. Of course, at one level putting *Symbion* in its own phylum, named the Cycliophora,[113] solves nothing because it must be related to something else in the Metazoa. The trouble is as yet we don't know what.

Sooner or later molecular data and/or the discovery of related species with a somewhat different set of anatomical characters will help to pin *Symbion* to the correct branch in the tree of life. But in another sense *Symbion* is already perfectly familiar, because however remarkable its bodyplan may appear to be, even a cursory examination will show that this seemingly strange animal is replete with features that converge on those found in other groups. These similarities may not be of phylogenetic significance, but they do suggest that however novel *Symbion* is in terms of classification, there is little new under the sun or for that matter in the sea.

Let us look at some of the more notable features of *Symbion*. What about attachment to the appendages of lobsters; is that unique? Not at all: remarkably similar morphologies are found in a group of rotifers (the seisonids) and also the protozoan ciliates. There is also a specialized group of barnacles—the octolasmids—that does much the same. *Symbion* feeds by using a ring of cilia, but a similar system is also found in rotifers. The life cycle of *Symbion* is rather extraordinary, although its basic arrangement of asexual and sexual cycles may be a consequence of settlement on a lobster which, because ultimately it will cast its exoskeleton by moulting, will shed any epizoan colonists, including *Symbion*. In any event, other metazoans certainly show alternations of asexual and sexual reproduction. In the sexual cycle *Symbion* has a dwarf male, but so do many other animals, including echiuran worms, barnacles, and angler fish. In the asexual cycle the interior tissues of *Symbion* 'melt down' and then reconstitutes themselves, but this also occurs in some barnacles and also during the pupation of insects. There is nothing else quite like *Symbion*, but on the other hand is there anything in *Symbion* that has not at least some sort of parallel elsewhere?

But in terms of evolution on this planet there is one species without parallel: only we can understand a word of this article. To even begin to trace the rise of human intelligence, so far as at present we understand it, would require at least twice the space available here. Rather, my final question in this essay is a diversion, perhaps even an entertainment. Given our unique position, and note I say nothing of privilege—it is surely more of a responsibility—then what can we discern of our place not in the context of evolutionary antecedents but in the examination of what in principle may transpire to inhabit other worlds, those beyond our Solar System.

## Other worlds

Last sections of papers are occasionally given over to the wilder sorts of speculation: May I have your indulgence? One argument that has arisen from studies of the Burgess Shale proceeds as follows.[1] So great is the range of animal types, encoded in terms of formal taxonomy as a multitude of phyla, in the Cambrian that the 'choices' or 'avenues' open to evolution were very large indeed. Thus, the chances of history repeating itself and following the same pathways to the diversity of present day life—were that possible—are too remote to take seriously. We live in a contingent Universe, and at any step in the history of life a different pathway could have been taken, a different world emerged. This is completely correct, but only in a trivial sense. At any stage of evolution the majority will be weeded out. Visit any museum of paleontology should you entertain residual doubts: practically every species you see there will be extinct. In other words, extinction is the norm. If the history of life was one of untrammelled expansion then the chance deletions of extinction might well provide a vast number of alternative end-points, depending on local contingencies. But evolution is very far from untrammelled. Yes, there may be runaway episodes, like the Cambrian 'explosion' or on a much shorter time-scale human evolution, but life occupies a very finite planet. Even, as seems likely, much of evolution is directed by accommodation between species rather than the constraints of physics (e.g. gravity, viscosity) and chemistry (e.g. availability of trace elements), the limits are self-evident. Evolutionary convergence, as we have seen, is rampant.[2,111]

In one sense, of course, the history of life does repeat itself. Successive waves of animals have embarked on their respective diversifications, and at many levels the end-results are often similar. But history is not static and it changes, even if as obedient moderns naturally we are told to shun the word 'progress'. The biological complexity of today is probably considerably in excess of the Jurassic, let alone the Cambrian. Past faunas and floras are only available, of course, in the rocks. If, however, they could by some sleight of hand or time machine be revived, then I suspect that unless they quickly found a refugia then for most species their time of enjoyment in the world of today would be decidedly brief. Ecosystem and evolution have moved on; in this sense the present is not the key to the past.

But one day, in principle, all these speculations might be put to the test. This is because the best possible experiment of rerunning the Cambrian 'explosion' would be to build a spaceship and visit a 5-Byr-old Earth-like planet to see what had happened. As the spaceship door opens would we be greeted by a friendly postman on his way to deliver invitations to a Friday Evening Discourse, or would there only be a sea of tentacles and slug-like objects, or indeed would the entire planet be a howling, sterile desert? Why should we care? Some draw comfort from the belief that whatever disasters of environmental catastrophe face us here on Earth there is not one planet with one rain forest, but scattered through the Universe millions through which flit billions of species of butterfly (or the next nearest thing). The Earth may be ruined by our stupidity, but up in the sky Eden after Eden basks in the light of alien suns. And not only is there a very widespread assumption that extraterrestrial life is abundant, but so too it is supposed that on at least some of these remote planets there also reside intelligent species, hopefully sympathetic to our desires and hopes. And is not all this eminently reasonable? The building blocks of life are well understood, and in many cases their biological synthesis has been achieved. We should not mar the argument by pointing out that despite intense experimentation the generation of a replicating cell seems a very remote prospect. Courage, more money, more imagination. Let us invoke a multitude of inhabited worlds, and reassuring ourselves with the inevitability of convergent evolution, seek extraterrestrial civilizations.

The conclusion I will reach—outrageous to the SETI (Search for Extraterrestrial Intelligence) industry—is that at least as far as extraterrestrial human-like intelligences are concerned, we are wasting our time.[2] The discussion has three brief sections: (i) Are Earth-like planets as common as we would like to believe? (ii) If life is ubiquitous then why haven't we been informed? (iii) What are the consequences of accepting the unfashionable view that for all intents and purposes we are alone?

The search for planets beyond our Solar System continues, with several apparently convincing examples. To date, however, only very massive planets, however, can be detected because of the gravitational perturbations they engender. The chances of life evolving on such giant planets remains contentious. How common are Earth-like planets? Calculations of planetary distributions around stars are fraught with imponderables, but there might be a wide range of possibilities.[115] Relative to the Sun, the Earth's orbit is rather finely placed. Small deviations either towards or away from the Sun would probably jeopardize life. And not all evidence points to an abundance of planetary systems. The apparent rarity of comets derived from other planetary systems, which would have hyperbolic orbits, rather than from the Solar System's Oort Clouds, suggests that equivalents to our Solar System might be much scarcer than is thought.[116] The Earth itself may be very peculiar in other ways. It has been suggested that our daughter satellite the Moon imparts a high degree of stability to the axial inclination of the Earth by virtue of the gravitational coupling. Other planets, in contrast, may have undergone major and abrupt shifts in axial inclination, perhaps in ways prejudicial to any life present. Satellites as large as the Moon orbiting Earth-like planets also may be very uncommon, yet could exert important controls on the evolution of life.[117]

But let us overlook these points. The commonly-held view remains that extraterrestrial life is pervasive and while civilizations are rare, at any one time the galaxy houses at least a few. If that is so, is it naive to ask why we have no evidence of visitors? A culture that is capable of travelling across distances equivalent to at least tens of light years is surely capable of leaving the appropriate calling card. If they did, so far we have overlooked it. Nevertheless, the SETI programme grinds on, the enthusiasm of the searchers is undimmed, and yet so far the monitors register complete silence. Of course, only a handful of stars have been investigated, and no doubt our attempts to read the messages that are flying to and fro across the galaxy are hopelessly amateurish.

I list these points not because I fervently believe that only the Earth houses life, but for the sake of an argument. Let us accept the hypothesis that contrary to received wisdom life is unique to this planet. Even if there is only a million in one chance that we are indeed alone, then it surely beholds us to act as more responsible custodians and stewards than at present we are doing. Human beings have the potential to be very great, not least in altruism and intellectual capacity. On a small scale the latter is expressed in the delight of exploring vanished worlds that lead back to the Burgess Shale where we can see the seeds of our own history and more importantly re-contemplate our own destiny.

## Acknowledgements

Sandra Last kindly typed several versions of this manuscript. I also thank Dudley Simons and Hilary Alberti for technical assistance. The work discussed here depended on the good will and collaboration of many individuals, but I would like to thank in particular Stefan Bengtson, John Peel, and Harry Whittington. Support by the Natural Environment Research Council, the Royal Society and St John's College, is warmly acknowledged. This is Cambridge Earth Sciences Publication 5503.

## References

1. S.J. Gould, *Wonderful life. The Burgess Shale and the nature of history*, Norton, New York, 1989.
2. S. Conway Morris, *The crucible of creation. The Burgess Shale and the rise of animals*, Oxford University Press, Oxford, 1998.
3. J.H. Lipps and P.W. Signor (ed.), *Origin and early evolution of the Metazoa*, Plenum, New York, 1992.
4. S.A. Bowring, J.P. Grotzinger, C.E. Isachsen, A.H. Knoll, S.M. Pelechaty, and P. Kolosov, *Science*, 1993, **261**, 1293.
5. S. Conway Morris, *Trans. R. Soc. Edinburgh: Earth Sci.*, 1989, **80**, 271.

6. H.B. Whittington, *The Burgess Shale*, Yale University Press, New Haven, 1985.
7. N.J. Butterfield, *Lethaia*, 1995, **28**, 1.
8. H. Mansuy, *Mem. Service Geol. Indochine*, 1912, **1**(2), 1.
9. J.Y. Chen, G.Q. Zhou, M.Y. Zhu, and K.Y. Yeh, *The Chengjiang biota. A unique window of the Cambrian 'Explosion'*, National Museum of Natural Science, Taiwan, c. 1996 [in Chinese].
10. L. Ramsköld and X-G. Hou, *Nature*, 1991, **351**, 225.
11. X-G. Hou, J. Bergström and P. Ahlberg, *GFF*, 1995, **117**, 163.
12. G. Budd, *Nature*, 1993, **364**, 709.
13. S. Conway Morris and J.S. Peel, *Phil. Trans. R. Soc. London*, 1995, **B347**, 305.
14. R. Buick, J.S.R. Dunlop, and D.I. Groves, *Alcheringa*, 1981, **5**, 161.
15. R. Buick, *Palaios*, 1991, **5**, 441.
16. A.H. Knoll, K. Swett and J. Mark, *J. Paleontol.*, 1991, **65**, 531.
17. N.J. Butterfield, A.H. Knoll, and K. Swett, *Fossils Strata*, 1994, **34**, 1.
18. S. Golubic, V.N. Sergeev, and A.H. Knoll, *Lethaia*, 1995, **28**, 285.
19. M. Schidlowski, J.M. Hayes, and I.R. Kaplan, in *Earth's earliest biosphere. Its origin and evolution* (ed. J.W. Schopf), Princeton University Press, Princeton, 1983.
20. B.D. Dyer, and R.A. Obar, *Tracing the history of eucaryotic cells*, Columbia University Press, New York, 1994.
21. T.M. Han and B. Runnegar, *Science*, 1992, **257**, 232.
22. J.A. Karhu, and H.D. Holland, *Geology*, 1996, **24**, 867.
23. M.F. Glaessner, *The dawn of life. A biohistorical study*, Cambridge University Press, Cambridge, 1984.
24. J.A. Breyer, A.B. Busbey, R.E. Hanson, and E.C. Roy, *Geology*, 1995, **23**, 269.
25. S. Sarkar, S. Banerjee, and P.K. Bose, *Neues Jahrbuch Geol. Paläontol. Mh*, 1996, **H7**, 425.
26. M.R. Walter and G.R. Heys, *Precambrian Res.*, 1985, **29**, 149.
27. B. Runnegar, *Lethaia*, 1982, **15**, 199.
28. B. Runnegar, *J. Mol. Evol.*, 1985, **22**, 141.
29. G.A. Wray, J.S. Levinton, and L.H. Shapiro, *Science*, 1996, **274**, 568.
30. S. Conway Morris, *Curr. Biol.*, 1997, **7**, R71.
31. P.N. Turner, *Invert. Biol.*, 1995, **114**, 202.
32. O. Giere, *Meiobenthology. The microscopic fauna of aquatic sediments.* Springer-Verlag, Berlin, 1993.
33. P.J.S. Boaden, *Zool. J. Linn. Soc.*, 1989, **96**, 217.
34. E.H. Davison, K.J. Peterson, and R.A. Cameron, *Science*, 1995, **270**, 1319.
35. K.J. Müller and D. Walossek, *Trans. R. Soc. Edinburgh: Earth Sci.*, 1986, **77**, 157.
36. D. Walossek and K.J. Müller, *Lethaia*, 1989, **22**, 301.
37. X-G. Zhang and B.R. Pratt, *Science*, 1994, **266**, 637.
38. G. Haszprunar, L.V. Salvini-Plawen, and R.M. Rieger, *Acta Zool. (Stockh.)*, 1995, **76**, 141.
39. B.M. Waggoner, *Syst. Biol.*, 1996, **45**, 190.
40. A. Seilacher, *Lethaia*, 1989, **22**, 229.
41. A. Seilacher, *J. Geol. Soc. London*, 1992, **149**, 607.

42. E.L. Yochelson, *Proc. Am. Philos. Soc.*, 1996, **140**, 469.
43. S. Conway Morris, *Palaeontology,* 1993, **36**, 593.
44. J.P. Grotzinger, S.A. Bowring, B.Z. Saylor, and A.J. Kaufman, *Science*, 1995, **270**, 598.
45. S. Bengtson, S. Conway Morris, B.J. Cooper, P.A. Jell, and B.N. Runnegar, *Mem. Assoc. Australas. Paleontol.*, 1990, **9**, 1.
46. S. Jensen, *Fossils Strata*, 1997, **42**, 1.
47. R.A. Fortey, D.E.G. Briggs, and M.A. Wills, *Biol. J. Linn. Soc.*, 1996, **57**, 13.
48. P.W. Signor and J.H. Lipps, in *Origin and early evolution of the Metazoa* (ed. J.H. Lipps and P.W. Signor), Plenum, New York, 1992.
49. L.V. Berkner and L.C. Marshall, *Proc. Natl Acad. Sci. USA*, 1965, **53**, 1215.
50. K.M. Towe, *Proc. Natl Acad. Sci. USA*, 1970, **65**, 781.
51. M. Celerin, J.M. Ray, N.J. Schisler, A.W. Day, W.C. Stetter-Stevenson, and D.E. Laudenbach, *EMBO J.*, 1996, **15**, 4445.
52. P.E. Cloud, *Paleobiology*, 1976, **2**, 351.
53. D.C. Rhoads and J.W. Morse, *Lethaia*, 1971, **4**, 413.
54. A.H. Knoll, J.M. Hayes, A.J. Kaufman, K. Swett, and I.B. Lambert, *Nature*, 1986, **321**, 832.
55. L.A. Derry, A.J. Kaufman, and S.B. Jacobsen, *Geochim. Cosmochim. Acta*, 1992, **56**, 1317.
56. C.E. Savrda and D.J. Bottjer, in *Modern and ancient shelf continental shelf anoxia* (ed. R.V. Tyson and T.H. Pearson), Geol. Soc. Spec. Publ., 1991, **58**, 201.
57. D. Erwin, J. Valentine, and D. Jablonski, *Am. Sci.*, 1997, **85**, 126.
58. S.L. Baldauf and J.P. Palmer, *Proc. Natl Acad. Sci. USA*, 1993, **90**, 11558.
59. S.B. Carroll, *Nature*, 1995, **376**, 479.
60. G.J. Vermeij, *Palaios*, 1990, **4**, 585.
61. C.M. Lalli and R.W. Gilmer, *Pelagic snails. The biology of holoplanktonic gastropod mollusks*, Stanford University Press, Stanford, 1989.
62. S. Conway Morris, *Spec. Pap. Paleontol.*, 1977, **20**, 1.
63. D. Bruton, *Phil. Trans. R. Soc. London*, 1981, **B295**, 619.
64. S. Conway Morris and R.A. Robison, 1988, *Univ. Kansas Paleontol. Contrib. Pap.*, **122**, 1.
65. L. Babcock, *J. Paleontol.*, 1993, **67**, 217.
66. S. Conway Morris and S. Bengtson, *J. Paleontol.*, 1994, **68**, 1.
67. S. Jensen, *Lethaia*, 1990, **23**, 29.
68. N.J. Butterfield, *Nature*, 1994, **369**, 477.
69. M.A.R. Koehl, *Contemp. Maths.*, 1993, **141**, 33.
70. J.K. Rigby, *Paleontol. Canadiana*, 1986, **2**, 1.
71. M. Steiner, D. Mehl, J. Reitner, and B-D. Erdtmann, *Berliner Geowiss Abh, (E)*, 1993, **9**, 293.
72. R. Wood, A. Yu. Zhuravlev, and A. Chimed Tseren, *Sedimentology*, 1993, **40**, 829.
73. G.A. Logan, J.M. Hayes, G.B. Hieshima, and R.E. Summons, *Nature*, 1995, **376**, 53.
74. A.H. Knoll, I.J. Fairchild, and K. Swett, *Palaios*, 1993, **8**, 512.
75. S. Conway Morris, *Paleontology*, 1986, **29**, 423.
76. J. Bergström, *Ichnos*, 1990, **1**, 3.

77. B. Runnegar and C. Bentley, *J. Paleontol.*, 1983, **57**, 73.
78. R.A. Raff, C.R. Marshall, and J.M. Turbeville, *Annu. Rev. Ecol. Syst.*, 1994, **25**, 351.
79. A.M.A. Aguinaldo, J.M. Turbeville, L.S. Linford, M.C. Rivera, J.R. Garey, R.A. Raff, and J.A. Lake, *Nature*, 1997, **387**, 489.
80. M. Kobayashi, H. Wada, and N. Satoh, *Mol. Phyl. Evol.*, 1996, **5**, 414.
81. J.L. Boore and W.M. Brown, *The Nautilus*, 1994, **2** (Suppl.), 61.
82. M.T. Ghiselin, *Oxford Surv. Evol. Biol.*, 1988, **5**, 66.
83. C.B. Kim, S.Y. Moon, S.R. Gelder, and W. Kim, *J. Mol. Evol.*, 1996, **43**, 207.
84. K.M. Halanych, J.D. Bacheller, A.M.A. Aguinaldo, S.M. Liva, D.M. Hillis, and J.A. Lake, *Science*, 1995, **267**, 1641.
85. S. Conway Morris, B.L. Cohen, A.B. Gawthrop, T. Cavalier-Smith, and B. Winnepenninckx, *Science*, 1996, **272**, 282.
86. S. Conway Morris, *Nature*, 1995, **375**, 365.
87. G. Haszprunar, *J. Zool. Syst. Evol. Res.*, 1996, **34**, 41.
88. G. Balavoine, *C.R. Acad. Sci. Paris (Sci. Vie)*, 1997, **320**, 83.
89. H. Gee, *Before the backbone*, Chapman and Hall, London, 1996.
90. C. Poulsen, *Math.-fys Meddr.*, 1967, **36**, 1.
91. S. Conway Morris, *Phil. Trans. R. Soc. London*, 1985, **B307**, 507.
92. C.D. Walcott, *Smithson Misc. Collect.*, 1911, **57**, 109.
93. N.J. Butterfield, *Paleobiology*, 1990, **16**, 272.
94. L. Orrhage, *Z. Morphol. Ökol. Tiere*, 1973, **74**, 253.
95. C. Nielsen, *Acta Zool. (Stockh.)*, 1991, **72**, 7.
96. M.I. Coates, *Trans. R. Soc. Edinburgh: Earth Sci.*, 1996, **87**, 363.
97. T.R. Smithson, *Trans. R. Soc. Edinburgh: Earth Sci.*, 1994, **84**, 377.
98. T.S. Kemp, *Mammal-like reptiles and the origin of mammals*, Academic Press, London, 1982.
99. M.K. Hecht, J.H. Ostrom, G. Viohl, and P. Wellhofer (ed.), *The beginnings of birds. Proceedings of the International Archaeopteryx Conference Eichstätt 1984*, Freunde der Jura-Museums, Eichstätt, 1985.
100. D.H. Erwin, *Biol. J. Linn. Soc.*, 1993, **50**, 255.
101. S. Conway Morris, *Dev. Suppl.*, 1994, 1.
102. J.G. Gehling and J.K. Rigby, *J. Paleontol.*, 1996, **70**, 185.
103. K. Seimiya, H. Ishiguro, K. Miura, Y. Watanabe, and Y. Kurosawa, *Eur. J. Biochem.*, 1994, **11**, 1815.
104. V. Berteaux-Lecellier, M. Picard, C. Thompson-Coffe, D. Zickler, A. Panvier-Adoutte, and J-M. Simonet, *Cell*, 1995, **81**, 1043.
105. S.I. Tomarev, P. Callaerts, L. Kos, R. Zinovieva, G. Halder, W. Gehring, and J. Piatigorsky, *Proc. Natl Acad. Sci. USA*, 1997, **94**, 2421.
106. S. Conway Morris, *Phil. Trans. R. Soc. London*, 1990, **B327**, 595.
107. R.G. Johnson and E.S. Richardson, *Fieldiana, Geol.*, 1969, **12**, 119.
108. W. van den Brugghen, F.R. Schram, and D.M. Martill, *Nature*, 1997, **385**, 589.
109. F.W.F. Rowe, A.N. Baker, and H.E.S. Clark, *Proc. R. Soc. London*, 1988, **B233**, 431.
110. V.B. Pearse and J.S. Pearse, in *Echinoderms through time* (ed. B. David, A. Guille, J-P. Féral, and M. Roux), Balkema, Rotterdam, 1994, 121.
111. J. Moore and P. Willmer, *Biol. Rev.*, 1997, **72**, 1.

112. R.D.K. Thomas and W-E. Reif, in *Constructional morphology and evolution* (ed. N. Schmidt-Kittler and K. Vogel), Springer-Verlag, Berlin, 1991, 283.
113. P. Funch and R.M. Kristensen, *Nature*, 1995, **378**, 711.
114. S. Conway Morris, *Nature*, 1995, **378**, 661.
115. G.W. Wetherill, *Annu. Rev. Earth Planet. Sci.*, 1990, **18**, 205.
116. J.C. Brandt and R.D. Chapman, *Rendezvous in space: the science of comets*, Freeman, New York, 1992.
117. N.F. Comins, *What if the moon didn't exist?*, Harper Perennial, New York, 1995.

## SIMON CONWAY MORRIS

Born 1951, an undergraduate in Bristol, and apart from four years in the Open University, his academic life has been based in Cambridge. His work on fossils has taken him to Greenland, Mongolia, Canada, and Australia. He has won several prizes, including the Walcott Medal of the National Academy of Sciences, and he was elected a Fellow of the Royal Society in 1990. His first book, on the Burgess Shale, was published in March 1997, by Kodansha, and an enlarged English edition was published by Oxford University Press as *The crucible of creation: the Burgess Shale and the rise of animals* in 1998.

# *Brain killers: understanding degenerative diseases of the brain*

NANCY J. ROTHWELL

Biomedical research over the last 20 years can claim significant successes. Major advances have been made in understanding, and developing treatments for some of the most common killer diseases, such as cancer and heart disease. Unfortunately, there has been considerably less success in other forms of disease particularly those associated with the brain; such as the psychiatric diseases (e.g. schizophrenia and depression), and the neurodegenrative conditions often known as the 'brain killers'.

## Neurodegenerative diseases

Neurodegenerative diseases occur when there is either rapid or prolonged damage and death to cells of the brain. These diseases have a high incidence (Table 1). Stroke affects about 130 000 people in the UK each year. Of these, about one-third will die very soon after the stroke, another third will be severely disabled, and the remaining third, although probably making a good recovery, will be at significant risk of a further stroke. In contrast, brain injury is rather less common, but affects almost exclusively young people, particularly children and adolescents. Brain injury is in fact on the increase in a number of countries, including the USA where gunshot wounds form a major proportion of these devastating traumas. Alzheimer's disease is one of the most prevalent of the neurodegenerative diseases, with approximately three-quarters of a million people affected in the UK at any one time. Of course the incidence of this, like many other brain killer diseases is increasing, simply because

**Table 1.** The size of the problem: incidence of some of the major neurological diseases each year in the UK

| | |
|---|---|
| Stroke | 130 000 |
| Brain injury | 13 000 |
| Alzheimer's | 700 000 |
| Parkinson's | 110 000 |
| Multiple sclerosis | 5 000 |

our population is getting older. We, as are most other Western societies, are living longer, and are now seeing more of the diseases of old age.

Unfortunately, in spite of their devastating effects on mortality and morbidity, there are few, if any, successful treatments for many of these conditions. At present there is no widely successful treatment for stroke. While there are some treatments available (e.g. for multiple sclerosis, for Parkinson's and for Alzheimer's disease), these are not wholly successful and many carry side-effects. Therefore, intense research effort is devoted towards trying to understand these conditions and develop new treatments or preventative medicines.

The brain is unusual compared with other parts of the body in several respects. First, it is enclosed in the skull (the cranium), making it rather inaccessible and difficult to study, although recent developments in imaging techniques have allowed huge advances. The brain contains about 1 billion nerve cells. These cells, known as neurones, are the main communicating cells in the brain which allow us to think, remember, feel emotions, and control most physiological functions. Unlike many cells in the body, once neurones die they are never replaced, so damage to the brain is usually irreversible. Furthermore, while in some cases, large areas of the brain may suffer damage without proving fatal to the patient, damage to other, small areas can be devastating. When neurones are injured they may go through a period of gradual repair and recovery. However, once they have progressed beyond this stage, the damage is irreversible and the neurones will die. It is damage to neurones such as this which is the main underlying feature of a stroke.

## Stroke

A stroke occurs when blood flow and oxygen supply to the brain is reduced. A condition known as cerebral ischaemia. Cerebral ischaemia can result from several factors, e.g. when an artery carrying blood to the brain blocks because of a clot or bleeds, resulting in cerebral haemorrhage. Other related conditions in which cerebral ischaemia can also cause neuronal death, include poor general circulation, heart failure, impaired respiration or drowning, and the condition known as birth asphyxia, when newborn babies are deprived of oxygen. Cerebral ischaemia is also an important secondary factor in head injury which can lead to delayed neuronal damage. The most common cause of stroke in humans is blockade of a middle cerebral artery which supplies areas of the brain in the cortex and underlying basal ganglia.

Brain imaging studies can now reveal the gradual development of cell death (infarction) following occlusion of such a major artery. If this blood supply can be restored quickly, much of the tissue is probably still viable even several hours after the original clot. This viability has been elegantly demonstrated with a new class of drugs known as 'clot busters'. One of these, tissue plasminogen activator (tPA) has recently proven successful in clinical trials of patients in the USA who recently suffered a stroke. tPA and other similar compounds diffuse the clot, allowing blood to reflow again into the ischaemic brain region. If the drug is given sufficiently early, there can be almost full recovery of the area deprived of oxygen and blood and of the patient. However, such drugs do carry significant risks, because they can be associated with haemorrhage around the site of the clot. When a clot adheres to the blood vessel, it may damage the wall of the vessel so that when the clot is dissolved the vessel may be prone to bleed. Furthermore, when blood is allowed to reflow again after a blockage, the reflow itself can cause damage by providing high levels of oxygen to compromised tissue. This effect, known as reperfusion injury, can be worse than the original ischaemia. Therefore, while drugs such as tPA will be of therapeutic use, they may be applicable only to a relatively small number of patients. For the remainder we must look towards other potential treatments.

A major breakthrough in understanding the factors which cause brain damage after stroke was the realization that the infarct develops quite slowly, often many hours after the initial stroke. Even when a major artery blocks, very few areas of the brain are totally deprived of blood and oxygen as most receive some blood supply from adjacent blood vessels, known as collaterals. Thus, these areas of the brain may be able to survive on low oxygen for long periods of time, but die gradually, partly as a result of the ischaemia and partly because of toxins released from the damaged brain tissue itself.

Understanding these processes has been limited by the difficulties in carrying out research on human brain. Therefore, most advances have derived from studies on experimental animals where cerebral ischaemia can be induced experimentally in a manner similar to that seen in humans. Thus, for example in rodents, surgical occlusion of the middle cerebral artery results in a pattern of infarction very similar to that seen in patients with blockage of the artery. Studies on experimental animals have revealed that a relatively small area of damage develops within the first hour or so after a stroke, but the vast majority of the damage occurs more gradually over the following hours. It is now believed that this secondary damage is the result of so-called 'brain killers' (neurotoxins) released from the damaged tissue (Fig. 1).

Many of these toxic molecules, produced not only by neurones in the brain but also by other cells in the brain particularly glia (see below), are essential in normal brain function. For example, the neurotransmitter glutamate, is important as a chemical signal between neurones and is involved particularly in learning and memory. Under normal conditions, glutamate is stored in nerve terminals and is released in only very small quantities which are subsequently rapidly taken up again into the nerve terminals. Therefore, the concentrations of glutamate outside cells in the brain are very tightly regulated. However, in disease and injury such as stroke, there may be extensive release of glutamate from the nerve terminals, and the reuptake of glutamate is impaired in conditions of low oxygen (Fig. 2). It is this excessive release of glutamate, in

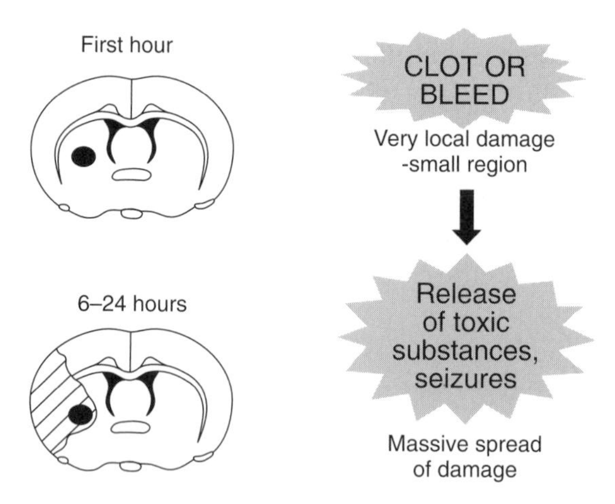


**Fig. 1** Damage caused by a stroke occurs over several hours after the result. The initial damage (shown as a dark region on the brain section) may be quite localized, but then areas over the following hours, damage spreads to include major brain regions (hatched area).

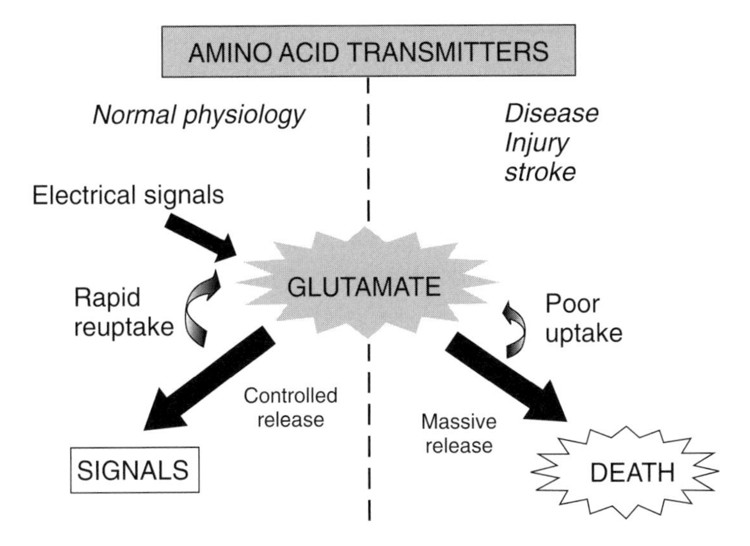


**Fig. 2** Glutamate is important in normal brain when its release and reuptake are carefully controlled (left side of figure). In contrast after injury, there is massive release of glutamate and low reuptake that leads to very high extracellular glutamate concentrations which can cause tissue damage or death.

the brain after a stroke (and indeed after brain injury), which is one of the main contributing factors in the subsequent neuronal death. Considerable effort has been devoted to developing molecules which will block the release of glutamate or block its action on neurones. Many of these have been proven to be highly successful in experimental animals, where they can reduce the damage resulting from a stroke by more than 50%. Unfortunately, as yet none have proved successful in clinical trials. Other factors released in excessive quantities in the brain after damage which contribute to neuronal death, include ions such as calcium, free radicals, and molecules normally produced by the immune system.

## Cytokines

The immune system is the body's primary defence against infection and disease. When activated, cells of the immune system can, identify, kill and engulf invading organisms, and produce an array of molecules which help to fight disease. Particularly important among these molecules is a group of proteins known as cytokines. The cytokines include four families of distinct polypeptides, known as interleukins, tumour necrosis factors, growth factors, and interferons, which can be produced by many cells in the body and have a wide array of actions. Most of these actions tend to restore the body to normal homeostasis after infection, injury, and inflammation. However, some effects of cytokines can cause damage directly, particularly when these molecules are present in large quantities.

The brain has always been considered rather different in its immune response to other tissues. For example, it is possible to transplant foreign cells into the brain without the rejection that is normally seen in other parts of the body. Therefore, the brain has been termed an immune-privileged site, which responds very differently to infection and disease. However, recent research has revealed that although the brain clearly is unique in its immune response, it can, and does, produce many of the molecules normally associated with the immune system, including the cytokines.

Interleukin-1 (IL-1) was one of the first identified cytokines. It is involved in many aspects of immunity, and is produced and can act in the brain. IL-1 is not produced by neurones, but rather by glia, the cells surrounding neurones within the brain. The term, glia literally means 'glue', and these cells, which comprise over half the brain matter, were originally believed to simply hold neurones together and provide an architecture for their support. It is now recognized that glia have many important physiological functions, including nutrition, communication, and repair and, interestingly, specific roles in disease and injury. It is the glia that are the primary source of many cytokines such as IL-1 in the brain. In normal healthy brain tissue, IL-1 levels are very low and barely detectable. However, IL-1 production is markedly increased in both animals and humans in a number of diseases (Table 2). Studies on animals now suggest that IL-1 is a primary mediator of the response to brain damage, and this evidence derives from several series of experiments.

First it was shown that injection of very small quantities of IL-1 into the brains of animals after a stroke, increases the amount of damage resulting from stroke by approximately twofold, suggesting that IL-1 can exacerbate brain damage. Secondly, and perhaps more importantly, it has been shown that blocking the production or action of IL-1 in the brain, dramatically reduces the damage caused by stroke by over 50% (Table 3). Inhibition of IL-1 reduces not only the size (volume) of infarction, but also the oedema (swelling) and the invasion of immune cells from the blood. Blocking IL-1, also increases the number of surviving neurones and improves neurological function. Thus, inhibiting IL-1 may be a useful therapeutic strategy for treating stroke in humans.

**Table 2.** Diseases in which brain levels of interleukin-1 are increased.

| | |
|---|---|
| Stroke | Alzheimer's |
| Brain injury | Parkinson's |
| Epilepsy | Motor neurone disease |
| Brain infections | Multiple sclerosis |
| Brain tumours | Down syndrome |

**Table 3.** Experimental conditions in which inhibiting interleukin-1 action reduced damage.

- Focal, global, permanent, reversible ischaemia
- Traumatic injury
- Excitotoxic damage
- Clinical symptoms of EAE (model of multiple sclerosis)
- Heat stroke damage
- Epileptic seizures

These studies have now been extended to other forms of experimental brain damage. For example, in head injury it is known that many of the same mechanisms contribute to neuronal damage as those seen in stroke. Secondary cerebral ischaemia, increased release of glutamate, calcium, and free radicals are all important in the damaging effects of head injury. As in cerebral ischaemia, there is a marked increase in the production of cytokines such as IL-1 after head injury in animals and humans. Furthermore, increasing IL-1 by administration of the cytokine to animals worsens the damage caused by head injury, while blocking IL-1 reduces the damage. It has also now been shown that excitatory transmitters such as glutamate stimulate the production of IL-1, suggesting that it is produced and acts at a point beyond the action of these transmitters to cause inflammation and neuronal death (Fig. 3).

The exact mechanisms by which IL-1 contributes to neuronal death is not known, but it probably has a number of actions including direct effects on neurones themselves, and indirect actions on glia and on blood vessels within the brain to cause release of toxins, invasion of immune cells, brain swelling, and other responses. What is not yet known is exactly how the neurones die in response to stroke, head injury, or IL-1.

## Murder or suicide?

Cells in the body can die in two different ways, which are, effectively, murder or suicide. Where there is extensive or very rapid damage to a cell

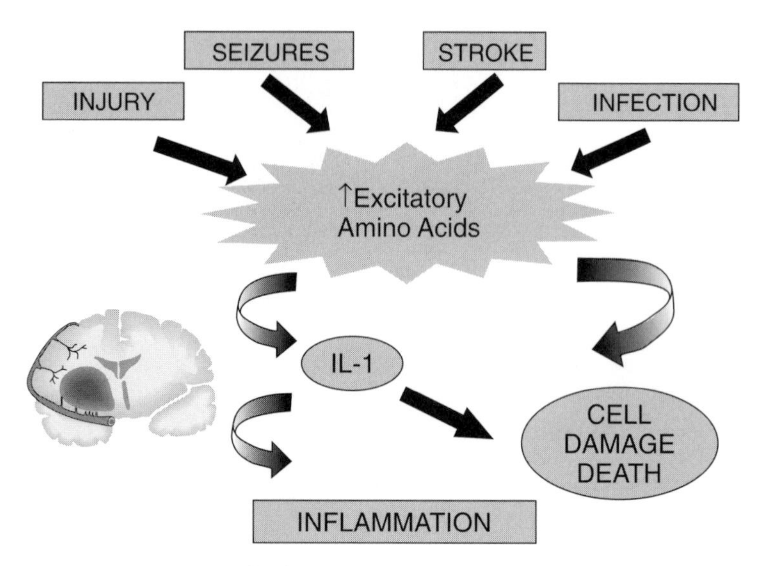


**Fig. 3** How is interleukin-1 (IL-1) involved in damage. Scheme depicting how interleukin-1 may contribute to damage and inflammation in the brain in disease.

it usually dies by a process of necrosis ('murder'). Cells swell and eventually burst releasing their contents and causing local inflammation. In contrast, some stimuli activate a pathway which causes the cell to 'commit suicide'. This process, known as apoptosis (or programmed cell death), involves specific genes and pathways which the cell appears to activate in order to execute cell death. Apoptosis was first identified in cells of the immune system, and is the subject of intense research in cancer. Cancer cells become 'immortal' because they fail to respond to the normal signals which trigger apoptosis. It is now known that neurones and glia can also undergo apoptosis. This process of genetically determined cell death is particularly important during development of the nervous system when too many neurones are produced. Clearly, understanding the factors which cause, and those which might inhibit apoptosis, could provide important clues as to how to treat neuronal cell death. Several of these factors have now been identified, largely on the basis of research on a small worm called *C. elegans*. In *C. elegans* a specific number of neurones die during development, and a set of genes has been identified which either cause or inhibit this cell death. One of these genes, *ced-3* (short for *C. elegans* death 3 gene) causes cells to die by apoptosis. Quite recently, it was found that mammals have a gene which is very similar to *ced-3*. The protein product of this gene in mammals is an enzyme known as ICE. ICE stands for IL-1$\beta$-converting enzyme, the enzyme which is required to

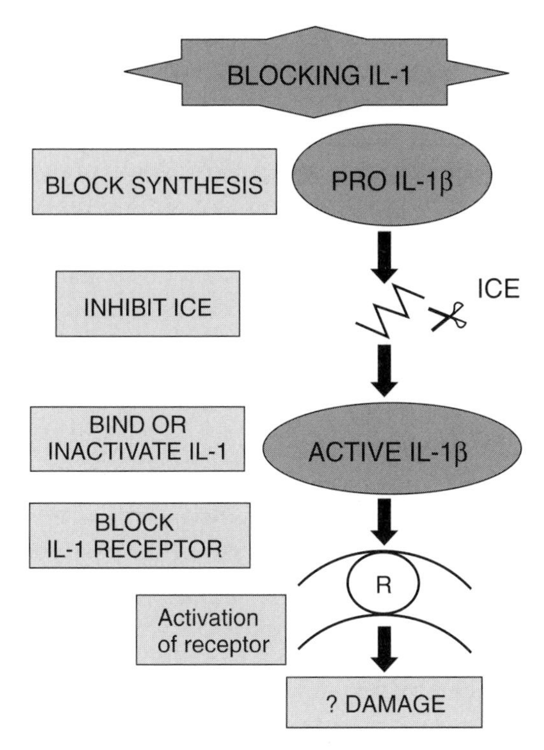


**Fig. 4** Possible strategies to inhibit the synthesis, release, or actions of interleukin-1 (IL-1), which may be beneficial in neurodegenerative disease.

cleave inactive IL-1 to produce the active IL-1. This exciting finding now suggests a direct relationship between apoptosis and IL-1 production. Our recent studies have shown that blocking ICE activity in the mammalian brain reduces the damage caused by stroke.

Therefore, there are already several strategies which may be used to reduce IL-1 activity and which limit neuronal death in experimental animals. These treatments include inhibiting the synthesis of IL-1, blocking its cleavage by ICE, binding or inactivating IL-1, blocking its receptor, or inhibiting the processes beyond activation of its receptor (Fig. 4). As yet, none of these have been tested in humans in either acute or chronic neurodegenerative conditions, but results in stroke and head injury in animals so far are promising.

## Alzheimer's disease

It is clear that there are significant differences between the acute neurodegenerative conditions such as stroke and head injury in which damage can develop over a matter of hours and chronic diseases. However, there are also some common features between these acute conditions and the chronic neurodegenerative diseases such as Alzheimer's disease, which may develop over many years.

There are three primary risk factors for Alzheimer's and related dementias: age, genetics, and head injury. Alzheimer's is a disease primarily of the elderly. Very few cases are seen in patients below the age of 70, and most of these tend to be genetic in nature. A dramatic increase in the incidence of Alzheimer's occurs in people in their seventies, eighties, and nineties, suggesting that either it is a very slowly progressing disease or that there are some factors associated with age which stimulate its onset.

There are several ways in which inherent factors can influence Alzheimer's disease. First, there are a very small number of cases which are clearly inherited and are associated with mutations in specific genes. While these account for only a few per cent of the cases of Alzheimer's, they have been very important in identifying some of the factors which may cause the disease. For example, mutations have been found in a gene which produces a protein known as $\beta$-amyloid precursor protein ($\beta$-APP). $\beta$-APP, is the precursor for a protein known as $\beta$-amyloid, which is believed to be fundamental to Alzheimer's disease. In addition to these genetic causes of Alzheimer's, a significant proportion of cases are familial in nature.

The brains of patients with Alzheimer's show very specific pathology. Certain areas of the brain in particular are affected, and in these regions there is a characteristic pathology. Particularly notable is the presence of plaques dotted around the affected regions. The core of these plaques comprises deposits of a protein known as $\beta$-amyloid, and it is now believed that increased synthesis or reduced breakdown of $\beta$-APP is a key factor leading to the deposition of $\beta$-amyloid which subsequently may contribute to the development of Alzheimer's disease. The way in which amyloid might affect Alzheimer's is not known, but it has been demonstrated that certain types of amyloid protein are toxic to neurones, and may also stimulate an inflammatory response including the activation of glia and the production of immune molecules such as cytokines.

The third significant risk of Alzheimer's disease is serious head injury. This was realized first from studies on the brains of boxers which show in the 'punch-drunk' syndrome, also known as dementia pugilistica. Detailed examination of the brains of boxers revealed that they show many of the features of Alzheimer's disease, including $\beta$-amyloid plaques. Subsequent studies have examined a large number of patients who died after a single head injury, and again start to show the deposition of $\beta$-amyloid. Associated with these amyloid plaques in the brains of many head-injured patients is increased production of the cytokine IL-1. Indeed this is not the only feature of Alzheimer's disease which suggests that it may be an inflammatory condition. Several molecules normally associated with inflammation and activation of the immune system have been found in the brains of patients with Alzheimer's disease. A particularly intriguing observation is that patients taking anti-inflammatory drugs (such as aspirin) for long periods of time, have a dramatically reduced risk of Alzheimer's disease. This suggests that blocking inflammation may reduce either the development or the progression of Alzheimer's.

As it has already been shown that the IL-1 production is increased after head injury or stroke, and that head injury is a risk factor for Alzheimer's, it is possible that IL-1 could be one link between acute brain damage and Alzheimer's disease (Fig. 5). This cytokine is found in increased quantities in the brains of patients with Alzheimer's, and it has been shown from studies in cell culture and in experimental animals, that IL-1 can stimulate the production of $\beta$-APP which may lead to amyloid deposition, and IL-1 also induces many of the features of the brains of patients with Alzheimer's, such as inflammation and activation of glia.

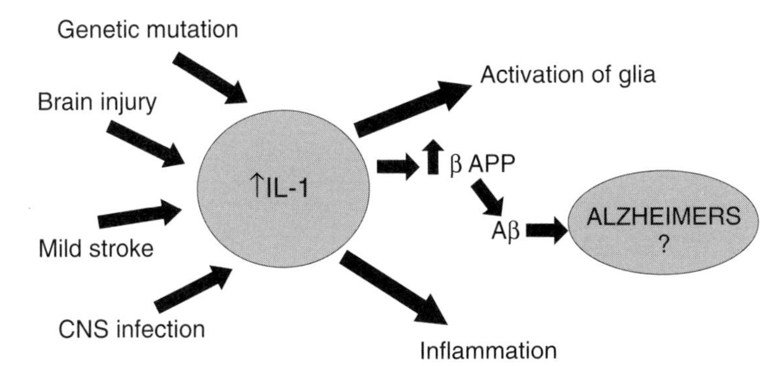


**Fig. 5** Hypothetical scheme depicting how interleukin-1 (IL-1) could contribute to Alzheimer's disease.

## The future?

Clearly, we are a long way from understanding all of the mechanisms which lead to brain damage, and there is much basic research to be undertaken to identify all of the factors. Although there are many common features of neuronal death in a variety of different neurological diseases, there are clearly some reasons why stroke, head injured, and Alzheimer's patients show very different pathologies, behaviours, and progression of the condition. Having identified some of the primary mechanisms of neuronal death, it should soon be possible to develop new drugs to limit brain damage.

## Further reading

N.J. Rothwell (ed.) (1995). *Immune responses in the nervous system.* Bios Scientific, London.

N.J. Rothwell (1996). Death in the brain. *Biomed. Sci. Rev.* **9**: 36–9.

Loddick, S.A. and Rothwell, N.J. (1996). Neuroprotective effects of human recombinant interleukin-1 receptor antagonist in focal cerebral ischaemia in the rat. *J. Cereb. Blood Flow Metab.* **16**: 932–40.

NANCY ROTHWELL

Awarded her PhD and DSc degrees from the University of London studying obesity energy balance and brown fat thermogenesis. In 1987 she moved to the University of Manchester where her research included neuroimmunology, fever, and neurodegeneration. She was appointed as Professor of Physiology in 1994 and Research and Graduate Dean in the School of Biological Sciences in 1996. Her interests include the public understanding of science, interactions between academia and industry, and research training. She delivered the Royal Institution Christmas Lectures in 1998.

# *The science of Murphy's law*

ROBERT A. J. MATTHEWS

Murphy's law states that 'If something can go wrong, it will', and as such has entered popular culture as an expression of the perversity and cussedness of everyday events. While many people jokingly blame their misadventures on the existence of Murphy's law, most scientists appear to regard it as a silly 'urban myth', without basis in fact. In this chapter I will show that, contrary to orthodox opinion, many of the most notorious manifestations of Murphy's law do indeed have a basis in scientific fact.

## Introduction

The suspicion that some things in life are intrinsically likely to go wrong and cause us misery can be traced back centuries. As long ago as 1786, the Scottish poet Robert Burns captured the essence of Murphy's law in his poem *To a mouse*, with the famous lines

> The best laid schemes o' mice an' men
> Gang aft agley [*Tend to go awry*]

A century later, the Victorian satirist James Payn incorporated perhaps the most famous manifestation of Murphy's law in his 1884 parody of Thomas Moore's *The fire worshippers*:

> I had never had a piece of toast
> Particularly long and wide
> But fell upon the sanded floor
> And always on the buttered side

The modern version of the law first emerged during the late 1940s. Like many aspects of popular culture, however, the concept of Murphy's law has accumulated an entire mythology of its own, which has tended to conceal its real origin and meaning. Many more or less humorous articles have appeared over the years claiming to explain its origins or its involvement in some commonly frustrating experience. This has led to the invention of the law being variously attributed to an apocryphal engineer named Edsel Murphy,[1] with even the authoritative *Brewer's Dictionary of Phrase and Fable* suggesting a link with some unidentified Irishman.[2] Yet Murphy was a real person, and both his involvement in the origin of the eponymous law and the aftermath provide a classic demonstration of its validity. Edward A. Murphy was born in Panama in 1918, and graduated from the US Military Academy at West Point in 1940. After serving as a pilot in the Pacific Theatre during the Second World War, he became Research and Development Officer at Wright-Patterson Air Force Base, Dayton, Ohio.

It was in this capacity that Murphy took part in Air Force Project *MX981: Human Deceleration Tests*, performed during the late 1940s. These involved propelling humans at high speeds by rocket sled along a track, rapidly decelerating them, and monitoring the effects. In 1949, Murphy was involved in the operation of a harness equipped with strain gauges designed to measure the forces acting on the volunteers during each run. After he had delivered some load pick-ups to monitor the stresses to the team, a run was carried out. It appeared successful, yet examination of the telemetry recorder revealed that the harness had

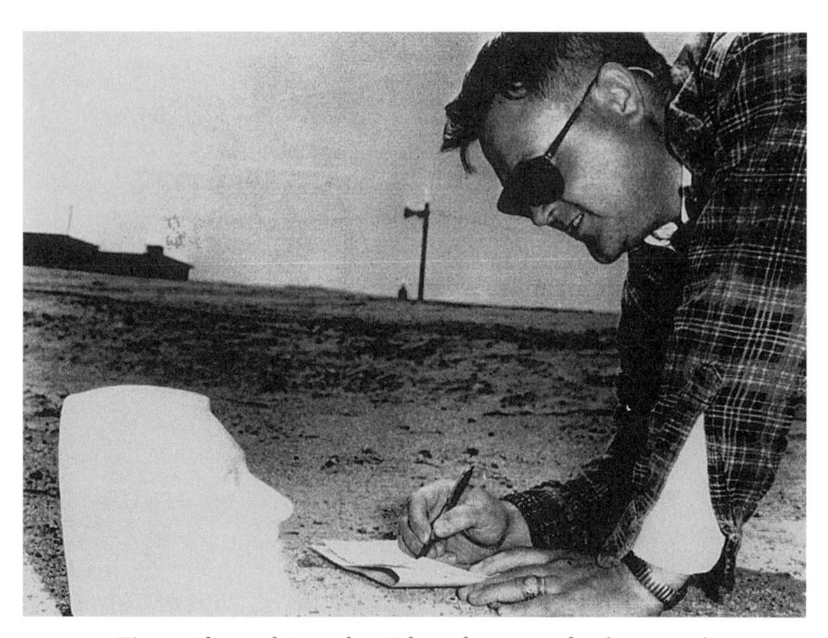

**Fig. 1** The real Murphy: Edward A. Murphy (1918–90).

somehow failed to work properly. Taking the contraption apart, it emerged that all the crucial wiring had been carried out incorrectly. When Murphy learned of the foul-up, he observed that if there was a way for one of the technicians to make a mistake, that would be the way things would be done. This rueful observation was the germ of what eventually became known as Murphy's law (Nichols, G., private communication).

At a subsequent press conference, one member of the project team said that they had become firm believers in Murphy's law, that 'If it can happen, it will happen'. This throwaway, remark was seized upon by the press as a pithy encapsulation of the all-too-familiar cussedness of inanimate objects, and the Law soon took on its classic wording: 'If something can go wrong, it will'.

Murphy himself came to loathe this 'frivolous' interpretation of the law. Following his involvement in the deceleration research, he went on to have a distinguished career that focused on the design of pilot escape systems for high-profile projects such as the X-15 hypersonic rocket plane and the SR-71 'Blackbird' reconnaissance aircraft, and on life support systems for the Apollo missions. In these roles, Murphy came to view the law as an excellent philosophy for safety-critical engineering design: one should always work on the assumption that if something can go wrong, it will. By the time of his death in 1990, the concept of 'defensive' design, in which one tries to foresee and counter the action of human blunders, was widely used in safety-critical applications. Yet Murphy's name seems destined to remain forever associated primarily with the 'urban myth' of the general perversity of the world around us. By failing to have his name associated with his own eminently sensible interpretation of Murphy's law, Murphy himself thus became a victim of his own law.

My own interest in the origins and basis of Murphy's law began in June 1994, after reading a letter in the magazine *New Scientist* from reader Colin Morgan of Warrington, who said that he had discovered why toast normally lands butter-side down:[3] 'The rotation involved in dropping and the distance to the ground combine to allow half a turn to be performed by the toast in mid-air'. He went on to say that if readers doubted his explanation, they should perform an experiment: 'Push a piece of bread or use a biro lengthways to model the bread, slowly over the edge of a table. See?'. In common with, I suspect, many other readers, I initially doubted that so simple and obvious explanation could indeed hold the key to this most famous manifestation of Murphy's law: that if toast or bread could land on the buttered side, it will do. Like many people, I was inclined to resort to the standard explanation for apparent occurrences of Murphy's law: the 'selective memory' effect, in which we are more inclined to remember the few occasions when things go wrong, and forget the myriad times when things go right. However, after a few trials slowly pushing a paperback book over the edge of my desk, I soon found myself agreeing with Morgan's claim, and decided to investigate further. I have since investigated many other supposed manifestations of Murphy's law, and have found that many are both real, and have a solid scientific explanation. In what follows, I outline these manifestations, their explanations—and suggest ways of defeating them.

## Murphy's law of maps

A manifestation of Murphy's law familiar to anyone who regularly uses maps and atlases is that 'If the place you are looking for can lie in an awkward part of the map, it will do'. At first glance, this may appear to be nothing more than a case of selective memory: that we just forget the many times when the place we are seeking is conveniently placed on the map. Some simple mathematics reveals, however, that those who suspect 'enemy action' on the part of the atlas are in fact quite correct.

Suppose the map is rectangular with sides of length $m$ and $n$, ($m \geq n$; see Fig. 2). We can then define a Murphy zone' around the edge of the map, and to either side of the central crease, of width $b$ ($\leq n/2$, to prevent the Murphy zone overlapping itself). The total area of the map that falls into this Murphy zone is then $A$, where

$$A = 2b(2n + m - 4b) \quad (1)$$

As the total area of the map is $m \times n$, the probability that a point picked at random will be in the Murphy zone is $P$, where

$$P = A/mn = 2b(2/m + 1/n - 4b/mn) \quad (2)$$

In the case of a square map, we have $m = n$, so that (2) becomes

$$P = 6(b/m) - 8(b/m)^2 \quad (3)$$

Let us now define the Murphy zone as a thin strip such that $b = m/10$; from (3), we then find that the probability that a point picked at random will lie within this narrow band is $P = 0.52$. In other words, the odds are better than 50 : 50 that our random point will lie in the awkward Murphy zone of a square map. At first sight, the relative narrowness of

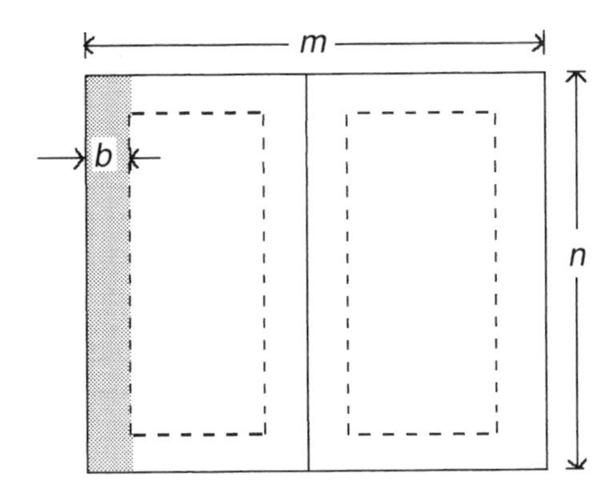


**Fig. 2** The geometry of Murphy's law of maps.

the zone makes this probability seem surprisingly high. The explanation is simply that the zone tracks the outermost—and thus largest—dimensions of the map, so only a relatively narrow width still encloses a comparatively large total area.

Most road atlas pages are, however, not square, but typically have a rectangular 'aspect ratio' $K = m/n$ of about 1.4. Defining the width of the Murphy zone via $r = b/n$, equation (2) now becomes

$$P = [(4/K) + 2]r - (8/K)r^2 \tag{4}$$

Setting $r = 1/10$ and $K = m/n = 1.4$, we now find $P = 43\%$, a somewhat lower probability than with the square map ($K = 1$). Thus, increasing the aspect ratio $K$ reduces the chances of our landing in the Murphy zone; this is essentially because the proportion of it running parallel to the longest sides, $m$, tends to zero.

This result points to a way of combating Murphy's law of maps which does not seem to have been recognized by cartographers: fold-out flaps down each side of the map. The *Reader's Digest Atlas of the British Isles* has such flaps on one edge of its maps, apparently introduced to allow roads to be followed easily from one page to another. This gives the atlas a relatively large aspect ratio of about $K = 1.54$, and thus cuts the chances of landing in the Murphy zone to 41%. If the flaps were on both sides of the each map, $P$ would fall to just 38%.

In summary, Murphy's law is based on a genuine effect, being the result of a kind of optical illusion: a relatively thin band running along the edges and down the central crease of a map mops up a surprisingly large amount of area. As we have seen, however, the frustration caused by Murphy's law of maps could be reduced considerably by relatively simple modifications to the format of atlases.

## Murphy's law of queues

Anyone who has waited at a supermarket checkout or been stuck in traffic jams will have encountered Murphy's law of Queues: If your queue can move slower than the one next to you, it will do'. Again, the most obvious explanation is selective memory, an explanation that gains strength from the queueing theoretic assumption that all the lines in, say, a supermarket checkout will have the same average service speed. That is, each is equally susceptible to random delays due to, for example, bar-code readers breaking down. However, in any *specific* visit to a supermarket, we do not care about long-term averages: we just want to finish first on that visit. Clearly, one can substantially increase one's chances of finishing first by choosing very short queues and avoiding those with families shopping for the winter. But even if the queues are identical in composition and length, the chances that we have picked the one queue out of the $N$ available in the supermarket that will be least affected by the random delays is just 1 in $N$. More importantly, of the three queues we most care about—the one we are in, plus our two neighbouring queues, by which we judge the soundness of our choice—there is just a 1 in 3 chance of our finishing first. In two visits out of three, one or other of our neighbouring queues will be less affected by the random delays, and finish before ours. The situation is akin to that of tossing dice. While it is true that, in the long run, all the numbers are equally likely to come up, the probability that the number we choose on one particular occasion will come up is just 1 in 6.

Can anything be done to mitigate the effects of Murphy's law of queues? While waiting in a queue, one could try indulging in the

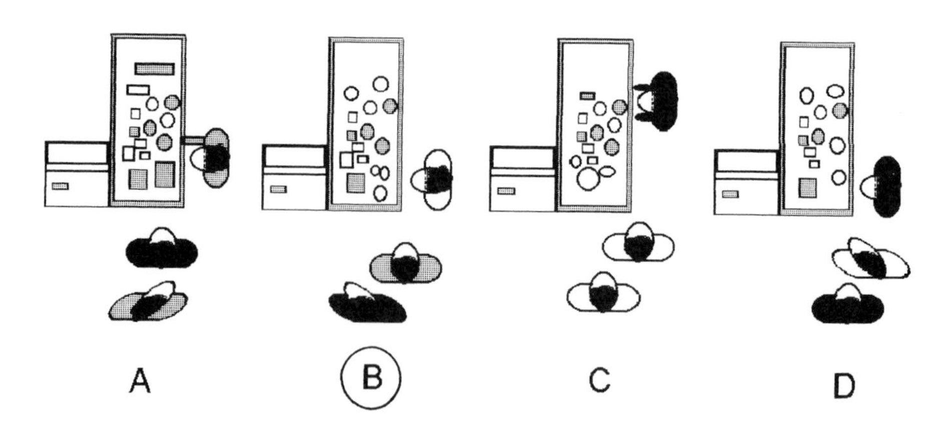


**Fig. 3** Random delays affect all the queues, making the chances of our finishing before both queues to either side of us just 1 in 3.

*schadenfreude* of recognizing that while we may only beat both our neighbouring queues 33% of the time, by the same token one of our neighbours will finish last 67% of the time. Those of an impatient disposition could try choosing queues at either end of the checkout area (A and D in Fig. 3): these have only one neighbour, raising one's chances of finishing first to 50%. One could also try taking a Zen-like approach, spending one's time in the queue pondering the fact that, in the long run, probability theory ensures that every human beings has their share of queueing success and misery.

I conclude this section by pointing out perhaps the most important implication of this analysis of Murphy's law of queues: unless you know the precise cause of a specific delay, jumping between queues makes no sense. Once one recognizes the similarities between picking queues and rolling dice, it becomes obvious that 'jockeying' to avoid delays makes no more sense than continually changing one's mind about which numbers to pick while dice are still tumbling. This should be remembered by all drivers of up-market cars trapped on slow-moving motorways.

## Murphy's law of odd socks

One of the most powerful forms of Murphy's law is experienced by millions of people every day within an hour of waking: 'If odd socks can be created, they will be'. The frustrating penchant of pairs of socks to degenerate into odd socks is, in fact, just one manifestation of a more general phenomenon that affects anything which comes in pairs: gloves, cufflinks, shoes, and saucepans and their lids. The details of the phenomenon emerge from combinatorics, the mathematics of arrangements, but its essence can be understood as follows:

Imagine a drawer containing only complete pairs of socks, all of different designs. Suppose one sock goes missing (do not worry about where or how: our analysis requires only the assumption of random sock loss). Clearly, we have just created an odd sock in the drawer. Now, when the next sock goes missing, it could be either the one odd sock just created, or a sock from a still-complete pair. As the latter will typically outnumber the former, it is clearly more likely that another complete pair will be broken up, leading to the creation of yet another odd sock. We can thus see glimmerings of evidence that Murphy's law can lead to a steady build-up of odd socks. To take the analysis further, we need to turn to combinatorics. Suppose that there are initially $n$ complete, distinct pairs of socks in a drawer, and that $2s$ individual socks are randomly lost. The probability of there being $d$ complete pairs left behind can be shown to be[4]

$$Prob(d, s, n) = \frac{n!(2s)!(2n-2s)!\ 2^{2(n-d-s)}}{d!(2n)!(2n-2s-2d)!(d+2s-n)!} \tag{5}$$

This is the master equation of odd socks, from which many intriguing results flow. For example, one can use it to define a Murphy ratio $M(n)$ giving the relative probabilities of being left with all odd socks and being left free of the things after randomly losing half the socks from our original collection of $n$ pairs:

$$M(n) = (n/2)!^2\ 2^n/n! \tag{6}$$

This shows that when odd socks go missing, they are *always* more likely to leave us with the worst possible outcome than the best. For example, if we began with 10 pairs of socks, by the time we have lost half of them it is four times more likely that the outcome will be the worst possible—a drawer-full of odd socks—than a drawer-full of complete pairs. One can also show that the most likely outcome is for us to be left with just INT[$n$/4] complete pairs, with all the other ($n$—2.INT[$n$/4]) socks being odd (where INT means integer part; e.g. INT[4.6] = 4). Thus, by the time we have lost half the socks from a drawer originally containing 10 complete pairs, we shall most likely find that we are left with just two complete pairs, lost among six odd socks; no wonder complete pairs are so difficult to find in the morning.

Even if we throw out all the odd socks, Murphy's law of odd socks still hinders our attempts to extract a matching pair. One can show that the minimum percentage of socks, $f_{min}$, we still have to extract to stand a reasonable chance of getting just one matching pair is roughly

$$f_{min}(n) \sim 100/\sqrt{n} \qquad \text{per cent} \tag{7}$$

which for $n$ = 10 pairs is about 30%. Thus even if we determinedly weed out the odd socks in our drawers, we will still have to rummage through a substantial fraction of the remaining socks before finding one matching pair.

How can we beat Murphy's law of odd socks? Clearly, keeping pairs together at every stage of the laundering, drying, and storing is crucial in both preventing odd sock creation and speeding matching pair retrieval. Over the years, methods ranging from simply tucking the tops of socks into each other to using patented sock-pinning gadgets have been suggested. One obvious way to avoid this drudgery is simply to replace all our distinct pairs of socks with identical ones. Another (apparently favoured by Einstein, although for reasons concerned with hole generation) is not to wear socks at all.

Happily, however, we *can* allow ourselves a little variety and still avoid much of the pain of Murphy's law of odd socks. The answer lies in restricting ourselves to just two designs of socks. Losing half the socks at random typically still cuts the numbers of both types of half, and thus the number of possible pairs by three-quarters. However, these remaining socks are not lost among a myriad of odd socks, and in general equal numbers of both types of socks will go missing. As a result, one can *guarantee* ending up with a complete pair from any size of two-variety sock collection after drawing out just three socks. The probability of getting a matching pair after drawing out just *two* socks at random is quite high if we restrict ourselves to equal numbers of two varieties. It is easily shown that the probability of getting one complete pair after drawing out just two socks from a collection of $n$ socks of each type is $(n-1)/(2n-1)$, which approaches 50% for large collections.

A word of warning is in order concerning the widespread availability of 'variety packs' of three different designs of sock. These should be avoided at all costs, as one can show that socks making up such packs always disappear in such a way as to maximize the number of odd socks left behind. One cannot help suspecting that these variety packs were invented by sock manufacturers on the advice of a mathematician.

## Murphy's law of rope

Jerome K. Jerome's *Three Men in a Boat* contains several amusing accounts of the cussedness of inanimate objects, one of which highlights a peculiar yet frustrating property of string, flex, rope, and the like:

> *There is something very strange and unaccountable about a tow-line. You roll it up with as much patience and care as you would take to fold up a new pair of trousers, and five minutes afterwards, when you pick it up, it is one ghastly, soul-revolting tangle.*[5]

This appears to be the first extant description of Murphy's law of rope: 'If rope can form a knot, it will do'. However, there is a less amusing side to this wry observation. A single knot in a climbing rope can reduce its breaking strength by up to 50%,[6] a potentially lethal effect that has been recognized since the earliest days of professional mountaineering.[7] This has led to great care being taken in the transport and storage of ropes by all those who professionally rely on them. In addition, forensic scientists are trained to be wary of interpreting the significance of knots found at the scene of crime, following recognition that knots do seem capable of forming 'by themselves' (Budworth, G., private communication).

Our mathematical analysis of this manifestation of Murphy's law[8] begins with a discovery made by research chemists in the early 1960s: as polymer chains grow in length, the probability that they became entangled with themselves rapidly increases. This observation became known as the Frisch–Wasserman–Delbruck ('FWD') conjecture,[9] and it was finally proved in 1988 by Sumners and Whittington using the concept of the self-avoiding random walk.[10]

Many people are familiar with simple random walks, in which the direction of travel along a one-dimensional lattice is chosen by a coin-toss, or in three dimensions by the throw of a die. A self-avoiding random walk (SAW) has the additional feature of forbidding more than one visit to each point on the lattice. Although apparently somewhat abstract, this extra restriction makes SAWs eminently suited to modelling real-life entanglement problems, as it captures both the randomness of the process, and the fact that no two parts of a rope-like object can occupy precisely the same position—as must be the case for any object with finite thickness.

In their proof of the FWD conjecture, Sumners and Whittington set up a specific sequence of positions, {T}, which constitute a simple trefoil knot. If {**i, j, k**} denote the three orthogonal unit vectors of the lattice in three-dimensional space, then

$$\{T\} = \{i, i, j, k, k, -j, -j, -k, -i, -k, -k, j, j, k, k, -j, i, i,\}$$

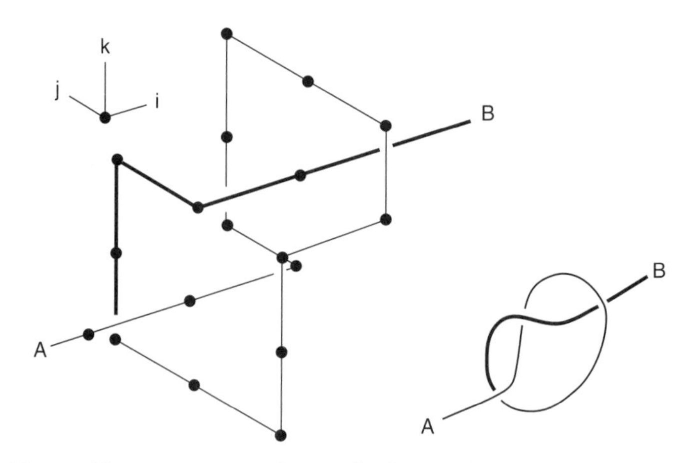


**Fig. 4** The prototypical simple knot, defined on a three-dimensional lattice.

Sumners and Whittington then proved that all except exponentially few sufficiently long SAWs on the simple cubic lattice contain {**T**}. Specifically, by using a theorem due to Kesten,[11] they showed that the probability of a SAW containing at least one copy of $T$ is

$$\text{Prob}(1 \text{ or more } T) \geq 1 - \exp(-k.n + o(n)) \quad k > 0 \tag{8}$$

where $n$ is the number of steps of the SAW and $o(n)$ denotes a rate of increase strictly less than $n$, as $n$ tends to infinity. This result has obvious implications for the knotted rope phenomenon, for if rope, string, flex, or whatever can be regarded as a SAW, then all except exponentially few sufficiently long ropes etc will contain at least one knot. Certainly physical ropes are self-avoiding, but do they perform random walks? If ropes are badly handled and their lateral stiffness per unit weight is not too high, they have ample opportunity to explore randomly the three-dimensional space they inhabit. This can be confirmed by putting some light thread or a jewellery chain into a bag and briefly jumbling it up. Carefully extracting the result—to prevent gravity imposing order on the jumbled thread—and pulling on the two ends does indeed usually reveal a simple trefoil knot. Indeed, this jumbling manoeuvre proves a remarkably effective way of 'forming a knot without really trying'.

It therefore appears that the Sumners–Whittington theorem (8) does provide a mathematical underpinning of Murphy's law of knots. Once again, one can ask whether the worst excesses of the law can be evaded, or at least mitigated. As the knotting probability depends strongly on the length of the rope, one obvious possibility is to store long ropes as halves, re-joining them only when needed. Unfortunately, (8) shows that Murphy's law is not so easily circumvented.

Consider a long length of, say, electric flex modelled as a self-avoiding walk of $2n$ steps. The probability that it is completely free of knots is, by (8)

$$\text{Prob}(0 \text{ knots in length } 2n) = \exp(-\,2n.k + o(2n)) \tag{9}$$

Now, to reduce the probability of knotting, suppose we insert a re-connectable break in the cable at its mid-point. The probability that either of the resulting half-cables are free of knots is, again by (8)

$$\text{Prob}(0 \text{ knots in length } n) = \exp(-\,n.k + o(n)) \tag{10}$$

which is indeed higher than before. However, as the neglect-driven formation of knots in each half-cable is independent, the probability that both half-cables are free of knots is $[\text{Prob}(0 \text{ knots in length } n)]^2$. Thus on concatenating the two half-cables, and assuming that this process typically leaves the number of knots unchanged, we have

$$\text{Prob}(0 \text{ knots} \mid n \text{ joined to } n) = \exp(-n.k + o(n)).\exp(-n.k + o(n)) = \exp(-2n.k + o(2n)) \qquad (11)$$

which is identical to (9). Thus, although each half-cable does have a lower probability of knotting than the length $2n$ original, when one connects up the two halves again, there is just as much chance of finding at least one knot in the result as in the original uncut cable. This result suggests that our best hope of combating Murphy's law of knots is simply to adopt the time-honoured practice of mariners and climbers and handle our ropes with great care, ensuring that they never get the chance to explore randomly the three-dimensional space they inhabit.

## Murphy's law of umbrellas

The weather and the accuracy (or otherwise) of weather forecasts are two frequent topics of conversation among Britons. Being situated in the path of no fewer than five different airstreams, UK weather is notoriously fickle and hard to predict reliably. This can prompt even the most rational of Britons to suspect that behind the overcast skies some malign force is at work. The famous Cambridge number theorist G.H. Hardy devised his own ingenious method for preventing this malign force from ruining a good day's cricket.[12] To ensure that rain did not fall, he would tell an assistant to go outside with an umbrella, and announce 'I am Hardy, and I am going to the British Museum'. This being the ideal rainy-day activity, the perverse nature of the weather would duly ensure that the sun would blaze down—thus giving Hardy himself precisely the weather he required.

Many less distinguished people have suspected that the mere act of carrying an umbrella can be enough to guarantee that it will not be necessary, in other words that 'If an umbrella can be redundant, it will be'. On the face of it, the idea that possession of an umbrella can affect the probability of rain falling seems absurd. However, as I now show, when looked at in the right way, one finds that there is more than a grain of truth behind Murphy's law of umbrellas.

The explanation lies in our reason for deciding to carry an umbrella. Perhaps we have heard a forecast of rain, or we believe the skies seem to foretell a downpour. Either way, our decision is ineluctably tied to the probability of rain falling, via the so-called Base Rate Effect.

Suppose that we are planning to take a lunch-hour walk, and that we have heard a weather forecast predicting rain during that hour. Should

**Table 1.** Outcome of forecasts for 1000 1-h walks

| | Rain | No rain | Sum |
|---|---|---|---|
| Forecast of rain | 80 | 180 | 260 |
| Forecast of no rain | 20 | 720 | 740 |
| Sum | 100 | 900 | 1000 |

we take our umbrella, or not? There are four possible outcomes affecting our decision, shown in Table 1. This has been calculated using real meteorological data showing that, roughly speaking, there is a 10% probability of rain during any given hour in the UK, and Meteorological Office forecasts of rain are now about 80% accurate. The precise meaning of 'accuracy' is somewhat ambiguous,[13] but here I shall take it to mean that eight of 10 occasions of rain are correctly forecast, and similarly for occasions of no rain. The table allows us to read off the key figure of interest when we are trying to decide whether to carry an umbrella or not. Running along the first row, we see that the probability of rain *given* that the Met Office forecasts rain is not 80%, as one might expect. It is 80/260, i.e. about 30%; similarly, the probability that it will not rain, given that the Met Office says it will, is 180/260, or about 70%. In other words, when the Met Office warns of rain falling during our hour-long walk, the 80% accurate forecast is more than twice as likely to prove wrong as right.[14] The reason for this decidedly counterintuitive result is not that the Met Office is being economical with the truth in its claims of accuracy. Rather, it is because the low hourly 'base-rate' for rain of just 10% overwhelms even an apparently impressive level of forecast accuracy.

Thus there is a sense in which 'If an umbrella can be redundant, it will be': if you take an umbrella on your lunchtime walk in response to a Met Office forecast of rain during that hour, then two times out of three the umbrella will indeed be redundant.

That is not to say that you should not take it, of course: if you are going out in your one and only silk suit, then you may well consider that a 1 in 3 conditional probability of a soaking is still too high. Clearly, how one responds to forecasts depends critically on how one views the relative costs of the two possible outcomes of a 'bad' decision: being caught in a downpour without an umbrella, and having to carry an umbrella we don't need. The mathematics of decision theory can be used to identify the optimal decision in such cases, and it can be shown[15] that we should only take a forecast of an event seriously if

$$\text{LR} \times \text{Odds(Event)} > \text{Odds(Concern)} \qquad (12)$$

Here LR is the so-called likelihood ratio, a measure of the accuracy of the forecast, such that

$$\mathrm{LR} = \frac{\text{Prob(Event is forecast } \textit{given} \text{ event does take place)}}{\text{Prob(Event is forecast } \textit{given} \text{ event does not take place)}} \tag{13}$$

Odds(Event) is the base-rate for the phenomenon in question and Odds(Concern) are the odds below which we start to become twitchy about having failed to take the forecast seriously. For the figures used in Table 1, we have an LR of (80/100) divided by (180/900), giving LR = 4 and Odds(Event) = 0.1/0.9 = 0.11. Plugging these into (12) we then find that we should only take an umbrella if our concern about getting wet becomes intolerable at odds of rain below 0.44, or equivalently, a probability below 30%. If it takes more to scare you, then leave your umbrella at home: the base-rate effect will ensure that your decision will prove optimal more often than not.

It must be stressed that this advice depends critically on the base-rate used. For example, if one is planning to go out for the entire day, then the greater base-rate of rain over 24-h periods means it certainly pays to take heed of the weather forecast. For the UK, the 24-h base-rate odds of rain Odds(rain) are approximately 0.7 for the UK, and with an LR of 4, (12) implies that rain forecasts should be taken seriously by anyone who starts to worry about being drenched at probabilities up to about 0.8, which would probably include most people.

In conclusion, it is possible to beat Murphy's law of umbrellas by taking account of the base-rate effect and its pernicious ability to undermine even apparently highly 'accurate' predictions.

## Murphy's law of toast

Undoubtedly the most famous of all manifestations of Murphy's law centres on the fall of toast: 'If toast can land butter-side down, it will do'. As remarked earlier, this propensity has been noted for at least a century, and led to my own involvement in the study of Murphy's law. It was also the centre-piece of a fascinating documentary on BBC-TV in 1991, in which a team led by Professor Ian Fells of Newcastle University investigated various manifestations of Murphy's law experimentally.[16] In the programme, a group of people was supplied with white sliced bread and butter, and instructed to toss the buttered bread into the air and note which side up the bread landed. After 300 trials, the results were statistically indistinguishable from the 50 : 50 split expected from a coin-toss.

However, this apparent refutation of Murphy's law is based on the fundamentally flawed assumption that toast typically reaches the floor after being tossed like a coin into the air. Yet reality is somewhat different, with toast usually heading floorwards as a result of sliding off a plate, or being swiped off a table. Dynamically, this is entirely different from a coin-toss, and as we shall see, leads to an entirely different outcome. The BBC-TV experiment did at least demonstrate the inadequacy of one widely believed explanation for Murphy's law of toast: the presence of butter on one side. Order-of-magnitude estimates show that neither the aerodynamic effect nor mass asymmetry caused by the presence of a thin layer of butter should make any difference to the final state of the toast, and this was comprehensively confirmed by the BBC-TV experiment.

The key to the dynamics and final state of toast lies in what happens as it reaches the edge of the plate or table. Once its centre of gravity has passed over the edge, a gravitational torque is set up, inducing the toast to spin. The final state of the toast is then dictated by whether this torque is large enough to allow the toast to rotate into a butter-up position in the time taken for it to free-fall under gravity to the ground. Thus the fate of toast is controlled by friction, gravity, and the height of the table.

To first approximation, this manifestation of Murphy's law can be modelled as a rigid, rough, thin homogeneous rectangular lamina of mass $m$, side $2a$, falling from a rigid platform set a height $h$ above the ground. Ignoring the process by which the toast arrives at this state and any horizontal velocity, the dynamics of the toast can then be viewed from an initial state where its centre of gravity overhangs the table by a distance $\delta_0$, as shown in Fig. 5.

With these assumptions, the dynamics of the lamina are determined by the forces shown in Fig. 5: the weight, $mg$, acting vertically downward, the frictional force, $F$, parallel to the plane of the lamina and directed against the motion, and the reaction of the table, $R$. The resulting angular velocity about the point of contact, $\omega$, then satisfies the differential equations

$$m\delta\dot{\omega} = R - mg.\cos\theta \tag{14}$$

$$m\delta\omega^2 = F - mg.\sin\theta \tag{15}$$

$$m(k^2 + \delta^2)\dot{\omega} = -mg\delta.\cos\theta \tag{16}$$

where $k$ is the appropriate radius of gyration, such that $k^2 = a^2/3$ for the thin rectangular lamina considered here, and $\delta \equiv \eta a$, with $\eta$ $(0 < \eta \leq 1)$

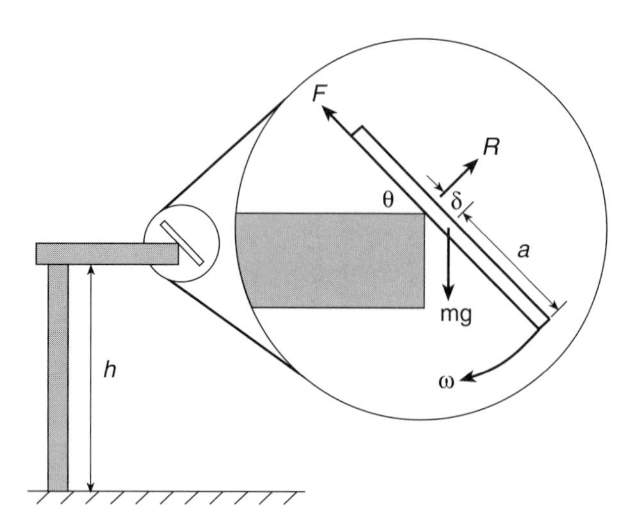


**Fig. 5** The forces on toast about to tumble to the floor.

being the 'overhang parameter'. If the toast begins its descent at angle $\phi$ to the horizontal then for it to land butter-side up again we must have

$$\omega\tau > (3\pi/2) - \phi \tag{17}$$

where $\tau$ is the free-fall time for the height of the table $h$, so that

$$\tau = [2(h - 2a)/g]^{1/2} \tag{18}$$

Solving these equations, making certain simplifying assumptions about the process of detachment from the table or plate-edge and inserting reasonable values of the $\eta$ and $h$, it emerges that toast sliding off a plate or table really does have a bias towards butter-down landings.[17] Furthermore, the low torque acquired by toast sliding off a table or plate leads to the butter-down effect persisting for all tables height below about 2.5–3.0 m.

More sophisticated analysis is possible (see, e.g. ref. 18), but ultimately no amount of mathematics is a substitute for a single practical demonstration. I therefore recommend that anyone who is still not convinced about the reality of Murphy's law of toast simply places some toast (or any similarly sized object, such as a paperback book) on a table or plate held at waist height, and observe what happens as it slides off and on to the floor. The tendency for toast to land butter-side down will become all too obvious.

Publication of my paper in the *European Journal of Physics* triggered a huge amount of media interest from around the world, suggesting that this particular manifestation of Murphy's law is recognized by toast-eating cultures across the planet. While my analysis of the dynamics of tumbling toast was not the first to be published,[19,20] it did go somewhat deeper into the root cause of the phenomenon, with an outcome that many clearly found as intriguing as I did. The essential reason toast lands butter-side down is that the table we eat from (or, equally, the height at which we carry plates from toaster to table or whatever) is insufficiently high to give the toast time to rotate sufficiently to bring the buttered side facing uppermost again. So why are tables the height they are? Clearly, this is because they must be convenient for humans. So why are humans the height they are? As bipedal animals with roughly cylindrical symmetry, we are intrinsically unstable to toppling, and as Press has pointed out[21] if we were a lot taller we would be in danger of severe head injury every time we fell over. With abut 20% of falls among elderly people resulting in serious fractures,[22] it appears that we may already be fairly close to the maximum safe height limit.

At a more fundamental level, this means that there is a limit on human heights set by the relative strengths of the chemical bonds making up our skull, and the strength of gravity pulling us over. And that, in turn, means there is a fundamental limit on the maximum height of tables convenient for bipedal creatures like humans. What is this height? To estimate it, we can use a modified version of Press's original argument.

We begin by considering a humanoid organism to be a cylindrical mass of polymeric material of height $L_H$ whose critical component is a spherical mass $M_C$ (the brain) positioned at the top of the body. Then, by Press's criterion, the maximum size of such an object is dictated by the requirement that the kinetic energy injected into the critical region by a fall will not be sufficient to cause major structural failure—and thus death. Hence we require

$$f.(M_C v_{\text{fall}}{}^2/2) < NE_B \tag{19}$$

where $V_{\text{fall}} \sim \sqrt{(3gL_H)}$ is the fall velocity, $f(\sim 0.1)$ is the fraction of kinetic energy that goes into breaking $N$ polymeric bonds of binding energy $E_B$, and the fracture is assumed to take place across a polymer plane $n(\sim 100)$ atoms thick. A somewhat involved argument[17] then leads to a maximum height on humans of $L_H$ where

$$L_H < 50(\alpha/\alpha_G)^{1/4} \,.\, a_o \tag{20}$$

This makes explicit the dependence of the maximum safe height for humans on $\alpha$, the fine-structure constant that determines the strength of chemical bonds making up our bones, $\alpha_G$, the so-called gravitational fine-structure constant that governs the strength of gravity pulling us over, and $a_0$, the Bohr radius. Inserting the various values, we find that this first-principles argument leads to a maximum safe height for humans of about 3 m. Although fairly rough and ready, this limit has a number of interesting features. First, it agrees well with the empirical observation that a fall on to the skull from a height of 3 m would very likely lead to death. Second, it shows that we are indeed relatively close to our maximum safe height, as suggested by the serious fracture rate following falls among the elderly. Thirdly, even the tallest-ever human, Robert Wadlow (1918–40), was—at 2.72 m—within the bound set by (20).

Most importantly, however, our bound sets an upper limit on the height of a table used by organisms articulated like humans: about $L_H/2$, or 1.5 m. This is barely half the height needed to give toast sufficient time to land butter-side up again. The limit (20) thus implies that all human-like organisms are doomed to experience tumbling toast landing butter-side down—essentially because of the values of certain fundamental constants. As the values of these fundamental constants were 'frozen in' shortly after the Big Bang 15 billion years ago, we are led to the disturbing conclusion that toast lands butter-side down *because our universe is designed that way.*

Happily, as with other manifestations of Murphy's law, it is possible to find ways of ameliorating its effects on toast. Dynamically, we have a choice between either giving toast enough time to complete its fall to the ground, or altering its spin rate. As we have seen, the former leads to somewhat inconveniently tall tables. It is also possible to increase the spin-rate of toast by cutting it into much smaller squares; again, however, at about 20 mm across, these are rather inconvenient. The optimal solution is somewhat counterintuitive: toast seen heading off the table should be given a smart swipe forward with the hand. Similarly, a plate off which toast is sliding should be snatched rapidly downwards and backwards. Both actions have the effect of giving the toast as large a (relative) horizontal velocity as possible, thus minimizing the effect of the rotation-inducing gravitation torque as it passes over the table edge. The toast will then descend to the floor, butter side up.

## Conclusions

The results presented here provide ample evidence that, contrary to orthodox opinion, Murphy's law does indeed have a basis in fact. From the proliferation of odd socks to the fall of buttered toast, a whole range of everyday phenomena do have a bias towards the worst possible outcome. There is, I believe, more to these results than confirmation of a supposed 'urban myth', however. While conducting my investigations into Murphy's law, I have often been struck by the fascination the results presented here hold for many people. I suspect that this is at least partly because questions surrounding such 'trivial' phenomena as tumbling toast are usually airily dismissed by many scientists, who are supposed to spend their time probing the mysteries of the cosmos, not buttered toast. Certainly, in the lexicon of contemporary science, 'triviality' is one of the most pejorative of terms. Yet it is as well to remember that the fall of toast is just as much a demonstration of the laws of physics as the dynamics of distant galaxies: Nature itself does not know the meaning of 'trivial'.

A dismissive attitude towards everyday phenomena also overlooks the fact that the history of science has seen many case of 'trivial' phenomena leading to seminal discoveries. The most famous, of course, is the fall of the apple which by all accounts did prompt Newton to contemplate the concept of universal gravitation.[23] Other examples include Euler's work on hydrodynamics, which sprang from his involvement in the design of the fountains of Frederick the Great of Prussia, Raman's discovery of the eponymous scattering effect after pondering the blueness of the sea, and Feynman's work on quantum electrodynamics, which was partly inspired by his study of a wobbling dinner plate.[24] Who knows: perhaps knot theory, invented in the nineteenth century and now at the forefront of theoretical physics, might have been discovered much earlier if Euler had spent a frustrating day in his garden shed sorting out knotted rope (after all, he did lay the foundations of graph theory after studying the equally 'trivial' matter of how best to tour the city of Konigsberg).

While hardly on a par with, say, universal gravitation, the analysis of one manifestation of Murphy's law has already cast new light on the decidedly non-trivial question of how best to protect people from earthquakes. The geophysics community has long been divided between those who believe that earthquake forecasting—that is, predicting the precise location and timing of major earthquakes—is a realistic possibility, and those who believe it is intrinsically impossible. Using precisely the same mathematics as that used to analyse Murphy's law of umbrellas,[14,15] one can estimate the accuracy required of any earthquake prediction system in which the civil authorities could place their trust.[25] It emerges that the level of accuracy needed to overcome the very low base-rate of major earthquakes is huge—about 10–100 times higher than that achieved by even the very best weather forecasts. For earthquakes, this level of accuracy is simply unattainable: earthquakes are known to be self-organized critical phenomena of far lower intrinsic predictability than weather systems—which may well explain why no one has ever identified a single 'precursor' of an impending earthquake offering even a modicum of reliability. The conclusion is therefore clear: attempts to predict major earthquakes are futile, and the huge sums now spent on such attempts should be spent instead on research into hazard mitigation through, for example, better structural design.

One does not, however, need to live in a region prone to earthquakes to benefit directly from the results presented here. Anyone wanting to minimize the frustration of tangled rope or queueing at check-outs, or to avoid cleaning up after dropping toast should now know just what to do. Perhaps Edward Murphy himself would have enjoyed such a denouement to the story of his law.

## Acknowledgements

One of the pleasures of investigating Murphy's law has been the way in which it has brought me into contact with people from so many disciplines. I would like to thank the following for providing information and advice during my work: Professor David Balding, Professor Ronnie Brown, Geoffrey Budworth, Dr Russ Evans, Professor Ian Fells, Dr Ron Harrison, Professor Dennis Lindley, Irena McCabe, Edward A. Murphy III, George Nichols, Rebecca Nicholson, Neville Snodgrass, Professor De Witt Sumners, and Professor Stu Whittington.

## References

1. Klipstein, D.L. (1967) The contributions of Edsel Murphy to the understanding of the behaviour of inanimate objects, *Electron. Eng.*, **15**, 8.
2. Evans, I. (1995) *Brewer's dictionary of phrase and fable*, Cassell.
3. Morgan, C. (1994) Mystery solved, *New Sci.*, **4 June**.
4. Matthews, R.A.J. (1996) Odd socks: a combinatoric example of Murphy's law, *Math. Today*, **March**, 39–41.
5. Jerome, J.K. (1889) *Three men in a boat*, Chapter 9, Penguin.
6. Warner, C. (1996) Studies of the behaviour of knots, in *History and science of knots* (ed. Turner, J.C. and van de Griend, P.), World Scientific, pp. 181–203.
7. Warner, C. (1996) A history of life support knots, *ibid.* 149–78.
8. Matthews, R.A.J. (1997) Knotted rope: a topological example of Murphy's law, *Math. Today*, **June**, 82–4.
9. Frisch, H.L. and Wasserman, E.J. (1961) *J. Am. Chem. Soc.*, **83**, 3789–95; and Delbruck, M. (1962) Mathematical problems in the biological sciences, *Proc. Symp. Appl. Math.*, **14**, 55.
10. Sumners, D.W. and Whittington, S.G. (1988) *J. Phys. A*, **21**, 1689–94.

11. Kesten, H. (1963) *J. Math. Phys.*, **4**, 960–9.
12. Kanigel, R. (1992) *The man who knew infinity: A life of the genius Ramanujan*, Abacus.
13. Matthews, R.A.J. (1996) Why are weather forecasts still under cloud? *Math. Today*, **32**, 168–71.
14. Matthews, R.A.J. (1996) Base-rate errors and rain forecasts, *Nature*, **382**, 766.
15. Matthews, R.A.J. (1998) The Cassandra Criterion, *Math. Today*, **February**, 7–10.
16. Bootle, R. and Fells, I. (1991) *QED: Murphy's law*, London, BBC.
17. Matthews, R.A.J. (1995) Tumbling toast, Murphy's law and the fundamental constants, *Eur. J. Phys.*, **16**, 172–6.
18. Steinert, D. (1996) *Phys. Teacher*, **34**, 288–9.
19. Held, A. and Yodzis, P. (1981) On the Einstein–Murphy interaction, *Gen. Relativity & Gravitation*, **13**, 873–82.
20. Edge, R.D. (1988) *Phys. Teacher*, **26**, 192–3.
21. Press, W.H. (1980) *Am. J. Phys.*, **48**, 597–8.
22. *Royal Society for the Prevention of Accidents booklet*, 1996.
23. Westfall, R.S. (1993) *The life of Isaac Newton*, Cambridge University Press, p. 52.
24. Matthews, R.A.J. (1996) Think trivial, *New Sci.*, **21/28 December, 32–4.**
25. Matthews, R.A.J. (1997) Decision-theoretic limits on earthquake prediction *Geophys. J. Int.*, **131**, 526–9.

## ROBERT A.J. MATTHEWS

Born 1959, he was raised in Derby and educated at Bemrose Grammar School before reading physics at Corpus Christi College, Oxford. On graduating, he went into science journalism, first with a number of business magazines and then *The Times*. Since 1990 he has worked as a freelance, allowing him to divide his time between writing and academic research. Journalistically, he acts as Science Correspondent to *The Sunday Telegraph* and is a regular contributor to many other publications, including *New Scientist* and *Focus*. Academically, he has published many research papers in areas ranging from celestial mechanics to computer science, with his work using neural computers to probe literary mysteries leading to a visiting fellowship at Aston University in 1993. His current interests lie in applying mathematics—primarily probability theory and Bayesian statistics—to questions the public find interesting but which scientists typically regard with suspicion, e.g. coincidences, parapsychology, and, of course, 'urban myths' such as Murphy's Law. More details about his research can be found at http://ourworld.compuserve.com/homepages/rajm/

# *God, time, and cosmology*

RUSSELL STANNARD

There has in recent years been quite a plethora of popular books on cosmology with attractive titles like *A Brief History of Time*, *The Mind of God*, *Unravelling the Mind of God*, and so on. Some of them suffer from what has come to be known as 'last chapter syndrome'. This is where the authors depart from their specialized scientific field and succumb, towards the end of their book, to the temptation to wax philosophical and theological on the supposed implications of their findings. Thus we find towards the end of *A Brief History of Time*, Stephen Hawking calls into question whether there is any longer a place for a Creator God.

A second example is provided by Steve Weinberg's *The First Three Minutes*. In his epilogue he concludes: 'The more the Universe seems comprehensible, the more it also seems pointless'. A little earlier he dismisses human life as 'a more-or-less farcical outcome of a chain of accidents'.

What are we to make of remarks such as these, which appear at first sight to be damaging to religious belief and to our own sense of worth?

Before going into that, a brief resume of how we think the world originated.

The Sun is a star—one of 100 000 million of them, which together make up a great swirling whirlpool of stars called the Milky Way Galaxy. There are other galaxies besides our own—some 100 000 million of them spread out over vast tracts of space. The furthest lie so far from us that the light we receive from them today has taken 12 000 million years to reach us—even though it has been travelling at a speed of 300 000 km/s. The galaxies are grouped together into clusters of galaxies, our own being a member of a cluster of about 30 galaxies.

When we look at distant galaxy clusters we find that they are all receding from us. The further away they are, the faster they are retreating off into the distance. This is the first clue concerning the violent beginning of the Universe—what we call the Big Bang. According to the Big Bang hypothesis, we envisage that there was a time when all the contents of the Universe were together. They flew apart, and have been doing so ever since. With such a violent occurrence it was anticipated that, like any other violent explosion—a nuclear bomb going off, for instance—it would have been accompanied by a blinding flash of light. That light—albeit in cooled-down form now—must still be about in the Universe today, there being nowhere else for it to be. That radiation has in fact been detected, and has all the characteristics expected of it. So that is a second piece of evidence in favour of the Big Bang theory.

A third comes from a knowledge of the composition of matter in the Universe. Big Bang theory allows us to calculate the relative abundances of the different kinds of chemical elements issuing from the Big Bang. These calculated abundances are in good agreement with what is actually found.

So we know that there definitely was a Big Bang, and from the motion of the galaxies and the distances they have travelled at those speeds, we can conclude that it happened 12 000 million years ago. It was such a cataclysmic event it seems natural to assume that it marked the point at which the Universe came into being. That being so it seems reasonable to ask what *caused* the Big Bang. And that is the first big question we are to address this evening.

The religious response is to say that God created the world. But there are those who do not see God as necessary for this purpose. They argue that the world might have created itself—spontaneously. For this to happen, two requirements must be fulfilled. The first is that they have to get something from nothing. Now that is not as difficult as you might at first think. Let me explain.

When we look at the world we see many properties—electric charge for example. Every atomic nucleus has positive electric charge and every electron negative electric charge. There is an enormous amount of electric charge in the world. But there is no difficulty creating electric charge. In the experiments I used to do at CERN in Geneva with the big particle accelerator there we create electrically charged particles all the time. The trick is that you don't produce them one at a time—you produce several of them so that the new positive electric charge is balanced by an equal amount of negative electric charge—that way there is no *net* increase in charge. So when we look at the Universe, the question we should be asking is not so much 'How much charge is there in the Universe?' as 'How much *net* charge is there in the Universe?' It is then one discovers that the answer—very, very precisely—is *zero*. There is no net charge in the Universe.

Take another property: momentum. This is a property possessed of moving objects and crudely speaking is a measure of the ability of the object to barge other things out of its way. As such, the momentum of the object depends upon the speed with which it is travelling, and how heavy it is. Again there is no difficulty in creating momentum. Get out of your chair and start walking, and you possess momentum that you did not have when seated. This was gained by pushing with your feet against the floor. This action sent the Earth recoiling in the opposite sense with an equal and opposite momentum. Again there has been no *net* change in momentum. When we look at the cosmos, there is as much movement in one direction as there is in the opposite sense, so the *net* amount is again zero.

The same argument holds regarding angular momentum—a property of rotating objects. There are as many stars rotating in one sense as in the opposite sense, as many galaxies rotating in one sense as in the opposite, so again, this yields a nil result for the net total.

You can see how the argument is progressing. You can have as much as you like of any property but it will still add up to nothing if you have an equal amount of its opposite. But, jumping ahead, you might be thinking 'That's all very well, but he can't get rid of *everything* that way—can he? What about the desk he is standing at? Where is he going to find a 'negative desk' to cancel out this positive one?'

Here we must draw on an insight from Einstein's theory of relativity. Energy comes in different guises: heat, electrical energy, gravitational energy, chemical energy, and so on. What Einstein showed was that matter itself is yet another form of energy. An object with a certain mass, denoted by $m$ carries within it a corresponding amount of energy, $E$. All this is embodied in the famous equation $E = mc^2$ ($c$ being the speed of light). For our purposes it is enough for us to recognize that matter—like this desk—is just a form of energy.

Now, an interesting thing about energy is that, like the other properties we were talking about, it comes in negative as well as positive forms. Whenever two bodies are bound together, it takes an input of energy (positive energy) to separate them—to tear them apart. That energy goes towards cancelling out the negative energy associated with the binding. So how much *net* energy is there in the Universe? Obviously there is much positive energy—all that locked up in the mass of the stars to say nothing of that associated with their motions. But we must recall that everything is attracting everything else with gravity, and that introduces negative energy. In fact there is a certain ambiguity as to what to take as the zero level of energy. Certainly a plausible case can be made for claiming that the net amount of energy in the Universe—like the other net quantities we considered—can be taken to be zero.

Summing up this line of argument we can conclude that the resources for making the cosmos could well amount to nothing at all. Thus, there is no difficulty in getting a universe for nothing: the Universe itself amounts to nothing—albeit an ingenious rearrangement of nothing.

Which brings us to the second main problem encountered by those who believe the Universe spontaneously created itself: how to bring about the ingenious rearrangement of the original nothing. This too, so it is argued, need not be an insuperable difficulty—one simply puts it down to a quantum fluctuation. So, what is meant by that?

In the days of classical Newtonian physics, one thought that everything that happens has to have a cause. Cause is followed by effect. Everything is predictable—at least in principle. If one repeatedly sets up the identical causal event, one always gets exactly the same effect.

With the advent of quantum theory all that changed. Now we know that from a given state of affairs, one can predict only the relative *probabilities* of a whole variety of possible later states. For instance, a 60% chance of one thing happening, a 30% chance of another, and a 10% chance of another. There is absolutely no way of determining in advance which way it will go; one just has to wait and see. As I said, all one can deal in are relative probabilities of various possible outcomes.

This element of uncertainty—unpredictability—affects everything happening in the world. The fact that it is not obvious in everyday life is because the effects only become noticeable on the small scale. This generally means one has to be examining the behaviour of individual atoms or of subatomic particles like electrons and protons. But the uncertainty is always there.

So, quantum uncertainty being the rule, some physicists have been led to propose that we can invoke quantum theory to account for the way the Universe came into being. Starting from a state consisting of nothing, it is proposed that there could be a small but finite chance that this will be succeeded by a state consisting of a Universe (the rearranged nothing). One has therefore only to wait around for this quantum fluctuation to occur.

This then is a possible way, so it seems, for having the Universe spontaneously create itself, without the active involvement of any Creator God. The proposal, at first sight at least, appears quite plausible, but it is not without its difficulties. It is all very well talking of a quantum fluctuation, but what exactly is fluctuating if there really is nothing there to begin with? Not only that, but quantum theory was devised to account for the behaviour of the component parts of the Universe; it does not by any means follow that one is justified in applying it to the Universe as a whole. A further difficulty is that if there is a finite probability of this

Universe popping into existence at some point, why not other universes at other points in time? Is one not led to the conclusion of there being universes without number? That seems a rather extravagant claim—a costly way of getting rid of a Creator God.

But setting aside for the moment these various objections, suppose for the sake of argument we were to concede that the world had its beginning in a quantum fluctuation, would that in fact undermine the idea of a Creator God?

I think not. It is all very well putting the Big Bang down to a quantum fluctuation, but why a *quantum fluctuation*? Why was it quantum physics that was in charge of the process rather than some other type of physics? After all, we can all dream up imaginary worlds run according to laws of nature different from our own. Science fiction writers do it all the time. Where is quantum physics supposed to have come from? Would it not have taken a God to have set up the laws of physics in the first place—a God who *chose* the laws for bringing this world (and perhaps others) into existence. This would have put God at one step removed from the origin of the Universe in that, instead of initiating the world by direct intervention, he created the law, it then being the natural outworking of that law that brought the world into existence. And yet responsibility for the existence of the world would ultimately have rested with the creator of the law—with God.

I wish now to turn to what I regard as the most intriguing aspect of the Big Bang. So far we have been speculating on what might have caused the Big Bang—a quantum fluctuation or God. But what I want to talk about now throws doubt on whether there was a cause at all.

In describing the Big Bang I have probably given you the idea that it was an explosion much like any other explosion—bigger, yes, but essentially the same. By that I mean that it takes place at a particular location in space. But this is not how it was with the Big Bang. Not only was all of matter concentrated initially at a point, but also all of space. There was no surrounding space outside the Big Bang.

Perhaps an analogy will help. Imagine a rubber balloon. On to its surface we glue some 5p coins. The coins represent the galaxies. Now we blow air into the balloon. It expands. Suppose you were a fly that has alighted on one of the coins; what do you see? You see all the other coins moving away from you—the further the coin, the faster it is receding into the distance. A coin twice as far away as another is receding twice as fast. But that of course is the observed behaviour of the galaxies—they too are receding from us in exactly that manner.

So far we have thought of the galaxies as speeding away from us as they move through space. But with the balloon analogy in mind, we now have an alternative way of interpreting that motion. It is not so much a case of the galaxy moving *through* space, as the space between us and it *expanding*. The galaxy is being carried away from us on a tide of expanding space. Just as there is no empty stretch of rubber surface 'outside' the region where the coins are to be found (a region into which the coins progressively spread out), so there is no empty three-dimensional space outside where we and the other galaxies are to be found.

It is this interpretation of the recession of the galaxies that leads us to conclude that at the instant of the Big Bang, all the space we observe today was squashed down to an infinitesimal point. Because of this, it becomes natural to suppose that the Big Bang not only marked the origins of the contents of the Universe, it also saw *the coming into existence of space*. Space began as nothing, and has continued to grow ever since.

That in itself is a remarkable thought. But an even more extraordinary conclusion is in store for us—once we acquaint ourselves with Einstein's theory of relativity. What it tells us is that space and time are more alike than one would guess from the very different ways we perceive and measure them. After all, we measure spatial distances with rulers and intervals of time with a watch or clock. Yet despite this, there is an exceedingly close link between the two, to the extent that we speak today of time as the fourth dimension. We are all familiar with the three spatial dimensions. Time is now to be added as the fourth. For our purpose, it is sufficient to accept that space and time are as indissolubly welded together as the three spatial dimensions are to themselves. One cannot have space without time, nor time without space.

The reason why I am telling you this now is because of what I said a little earlier about space itself coming into existence at the instant of the Big Bang. In light of what I have now asserted about the indissoluble link between space and time, we can immediately proceed to the conclusion that the instant of the Big Bang must also have marked the *coming into existence of time*. This in turn means that there was no time before the Big Bang. Indeed, the very phrase '*before* the Big Bang' has no meaning. The world 'before' necessarily implies a pre-existent time—but where the Big Bang was concerned, there was none.

Now, for those who seek a *cause* of the Big Bang—whether a Creator God or some impersonal agency—there is a problem here.

We have already spoken of the causal chain: cause followed by effect. Note the word 'followed': it refers to a sequence of events in time: first the cause, then the effect. But in the present context we are regarding the Big Bang as the effect. For there to have been a cause of the Big Bang, it would have had to have existed prior to the Big Bang. But this we now think of as an impossibility.

It was this lack of time before the Big Bang that prompted Hawking in his book *A Brief History of time* to remark 'What place then for a creator?' Without there being any time, it certainly gets rid of the kind of creator God that most people probably have in mind: a God who at first exists alone. Then at some point in time God decides to create a world. The blue touch paper is lit, there is a Big Bang. But as we have seen, without time before the Big Bang, there could not have been a cause in the usual sense of that word.

It has to be said that exactly the same problem confronts the alternative idea we have been discussing whereby the cause of the Big Bang is thought to have been a quantum fluctuation. According to that scheme, an initial state consisting of nothing was (thanks to the quantum fluctuation) 'followed' by a world that promptly underwent the Big Bang. In the absence of any prior time, there could no more have been that kind of 'initial' state (one that could undergo a quantum fluctuation), than there could have been a God. Indeed, the only kind of quantum fluctuations we know of are those that occur in space as well as in time. Prior to the Big Bang, there was neither.

So, where have we got to? Have these considerations dispensed with a creator God? Before jumping to that conclusion, let us consider the following quotation:

> *It is idle to look for time before creation, as if time can be found before time. If there were no motion of either a spiritual or corporeal creature by which the future, moving through the present, would succeed past, there would be no time at all... We should therefore say that time began with creation, rather than that creation began with time.*

If the archaic expression 'either a spiritual or corporeal creature' had been replaced by a more up to date one—such as 'a physical object'—one could well have thought that the quote came from some modern cosmologist like Hawking, or from Einstein. In fact, those are the words of St Augustine. I think you will agree, they beautifully sum up what I have been trying to say. Modern cosmologists find it hard to come to terms with the fact that, where the beginning of time is concerned, it was a theologian who got there before them—and by 1500 years.

How did he do it—bearing in mind that St Augustine obviously knew nothing about the Big Bang? He argued somewhat along the following lines.

How do we know that there is such a thing as time? It's because things change. Physical objects (for instance, the hands of a clock) occupy certain positions at one point in time, and move to other positions at another. If nothing moves (or in the past had *ever* moved), there would be nothing to distinguish one point in time from another. There would be no way of working out what the word 'time' was supposed to refer to; it would be a meaningless concept. *A fortiori*, if there were no objects at all, moving or stationary (because they had not been created), clearly there could be no such thing as time.

In this way, Augustine cleverly deduced that time was as much a property of the created world as anything else. And being a feature of that world, it needed to be created along with everything else. Thus it makes no sense to think of a time that existed before time began. In particular it makes no sense to think of a God capable of pre-dating the world.

Yet despite all this, Augustine remained one of the greatest Christian teachers of all time. His realization of the lack of time before creation clearly had no adverse effect on his religious beliefs. To understand why this should be so, we have to draw a distinction between the words 'origins' and 'creation'. Whereas in normal everyday conversation we might use them interchangeably, in theology they acquire their own distinctive meanings. So for example, if one has in mind a question along the lines of 'How did the world get started?' that is a question of origins. As such, it is a matter for scientists to decide, their current ideas pointing to the Big Bang description.

The creation question, on the other hand, is quite different. It is not particularly concerned with what happened at the beginning. Rather it is to do with: 'Why is there something rather than nothing?' It is as much concerned with the present instant of time as any other. 'Why are *we* here? To whom or to what do we owe our existence? What is keeping us in existence?' It is an entirely different matter, one not concerned with the mechanics of the origin of the cosmos, but with the underlying ground of all being.

It is for this reason one finds that whenever theologians talk about God the Creator, they usually couple it with the idea of God the Sustainer. His creativity is not especially invested in that first instant of time; it is to be found distributed throughout all time. We exist not because of some instantaneous action of God that happened long ago—an action that set in train all the events that have happened subsequently—an inexorable sequence requiring no further attention by God. We do not deal with a God who lights the blue touch—and *retires*. He is involved at first hand in *everything* that goes on.

A reason why many believers resist the idea of there being no God before the Big Bang, is the thought that this seems to imply that God too must have come into being at that instant. How could God have made himself?

The trouble with this is that again the conception of God is too small. We are assuming that God can exist only *in* time. But again that is not the traditional belief about God. Certainly God is to be found in time; we are interacting with God in time whenever we pray. But he is also *beyond* time—God *transcends* time.

A further concern believers might have as a result of our discussions relates to the Genesis creation story. To put it bluntly, has modern cosmology caught out the Bible with its account of the creation taking place over a period of 6 days, and all this occurring 3974 years, 6 months, and 10 days before Christ—as worked out by Archbishop Ussher on the basis of adding up all the relevant 'begats' of the Old Testament?

This assumes that the Genesis creation account was intended as a literalistic, scientific description of the origins of the world. But was it? This seems most unlikely. For a start we note that there is no such thing as *the* Genesis account of creation. There are two of them, the first commencing at Chapter 1, verse 1; the second at Chapter 2, verse 4. They are quite different from each other; indeed, as literal accounts they would contradict each other. This in itself should be sufficient to assure us that they were never meant to be read in that way. This view is confirmed when we note that people in those days were not particularly interested in scientific questions; our modern way of describing things with literal scientific accuracy had no place in their culture. Instead, much use was made of myths. Here we need to be careful. Words have a nasty habit of changing their meaning over time. Today if we call something a 'myth' we are dismissing it as untrue. In terms of Biblical criticism, the same word acquires a strictly technical meaning. It refers to an ancient narration or story, the purpose of which was to convey to future generations some deep timeless truth or truths. As such, the incidents in the story might not have taken place in exactly the way described; everyone accepted that and it did not matter. The story was merely the vehicle for conveying the truths.

This is not some modern insight. It is *not* the case that there has had to be a retrospective reappraisal of how to read the Bible, the earlier one having been discredited by modern scientific developments. Although admittedly certain people in the distant past regarded the creation myths as literally true, there has always been a strong school that has thought otherwise. For example we have in the fourth and fifth century people like Gregory of Nyssa declaring 'What man of sense would believe that there could have been a first, and a second, and a third day of creation, each with a morning and an evening, before the Sun had been created?' (The Sun supposedly not having been created until the fourth day.)

So much for reflections deriving from Hawking's writings. How about Weinberg's assessment of the Universe as being pointless and human life as being but a farcical outcome of a chain of accidents? Is the nature of the Universe, as revealed by modern astronomy and cosmology, such that one can only conclude that life is a chance by-product?

It is not difficult to appreciate how Weinberg arrives at such a gloomy assessment. Take for example the size of the Universe. Are we really expected to believe that God designed it as a home for humans? If so, it appears to be somewhat excessive—a case of overdesign perhaps. Most places in the Universe are hostile to life. The depths of space are incredibly cold; that is why most planets are freezing. To be warm a planet needs to be close to a star. But get too close—like Mercury and Venus—and they become too hot. And of course, the most prominent objects in the sky, the Sun and the other stars, are in themselves balls of fire and hence not suitable places to find life. Planets tend to be without atmospheres, or if they do have one, it is likely not to be the right sort for sustaining life.

For the great majority of the history of the Universe there was no intelligent life. After a further 5000 million years our Sun will swell up to become a red giant. Though it is unlikely that its fiery surface will reach out far enough to engulf the Earth, our planet will become unbearably hot, and all life will be burned up. This assumes that life has not already been eliminated through the violent impact of a meteorite. We have only to recall the spectacular collision of the Shoemaker–Levy comet with Jupiter to be reminded that an impact of that magnitude here on Earth would be devastating. It is believed that it was the effect of a meteorite impact that eliminated the dinosaurs some 65 million years ago.

And what of the long-term future of the Universe as a whole and of life elsewhere? We have spoken much about the origins of the Universe in the Big Bang, but what of its end? We have seen how the Universe is expanding. The distant galaxies of stars are still receding in the aftermath of the Big Bang. But as they rush off into the distance, they are slowing down. This is due to gravity, each galaxy exerting an attraction on every other one. Keep this up, and eventually the galaxies will be brought to a halt. Except that we have to remember that the force of gravity reduces with increasing distance. So the slowing down force is steadily reducing with time.

The big question is then whether it will have managed to stop the galaxies before its force essentially vanishes to nothing, or whether the speeds of the galaxies are so great they will succeed in escaping the pull of gravity.

If it is the first, then the galaxies will one day come to a halt, and from then on will be drawn back towards each other. All their separations will reduce until eventually everything comes piling back on top of each other in a Big Crunch—with obviously the extinction of all life. So that is one possible scenario.

The alternative is that gravity is too weak to stop the galaxies, and they will continue flying apart for ever. What would be the significance of that for life in the cosmos?

Each star has only a limited amount of fuel. Eventually its fires must go out. For a medium-sized star like the Sun that takes a time of the order of 10 000 million years (the Sun is about halfway through its active life). More massive stars have more fuel, but they achieve higher temperatures and burn their fuel faster—so much faster they might live for only 1 million years. As each star exhausts its fuel, it becomes cold and no longer able to keep companion planets warm enough to sustain life.

Mind you, new stars continue to form. A star is created when the hydrogen and helium gas that was emitted originally from the Big Bang collects together under the influence of its mutual gravity. It squashes down, heating up as it does so (in the same way as air squashed down in a bicycle pump gets hot). If enough gas is collected, the temperature rise becomes sufficient at the centre to ignite nuclear fusion. In a very hot gas, the atomic nuclei are moving about so fast they can fuse together to form heavier nuclei. These heavier nuclei are so efficiently packed together that they are able to release unwanted energy—the energy of nuclear fusion. (The modest heat of the squashed-down gas acts only as a trigger to get the much more energetic nuclear fusion reactions going, in the same way as the lighting of a domestic coal fire involves first setting light to some screwed up paper—the small output of heat from this being the trigger to get the coal burning.)

Not all the gas from the Big Bang was used up in producing the first generation of stars. Our own Sun was one of those formed at a later stage. Still more stars are to be seen today in the very earliest phases of getting underway. But it is clear that this is not something that can go on indefinitely. At some stage, *all* the hydrogen and helium gas will have been drawn together to form stars, or will have been dispersed so thinly as never to be incorporated into a star. From then on the last stars live out there active lives, and die. Everything cools down, and we are left with the Heat Death of the Universe.

So what we find is that if we are dealing with a Universe where the expansion goes on for ever, there will come a point when there can be no further life. One is then left with an ever-dispersing, lifeless Universe for an infinity of time. So Big Crunch or Heat Death, the future is bleak for life either way. Yes, as I said, it is easy to see how Weinberg was led to the conclusion that the Universe seems pointless, and life is but an accidental by-product of no significance.

Or is it? I wish now to change tack and introduce you to some reflections on the cosmos of a very different nature. These have surfaced only comparatively recently—in the last couple of decades. They go under the general heading: *The Anthropic Principle.*

To see what it is about, I want you to imagine that you are going to make a universe. You have freedom to choose the laws of nature and the conditions under which your imaginary universe is to operate. The aim is to produce a universe that is tailor-made for the development life—the kind of universe God presumably *ought* to have created if it were really intended primarily as a home for life.

The first decision is how violent to make your Big Bang. You might feel for example that the actual Big Bang was somewhat excessive if the aim was simply to produce some life-forms. It turns out that if you make the violence of your Big Bang somewhat less—only a little less—then the mutual gravity operating between the galaxies will get such a secure grip that the galaxies will slow down to a halt, and will thereafter be brought together in that Big Crunch I was telling you about. Moreover, all this happens in a shorter time that 12 000 million years—the time needed for evolution to produce us. So, turn the wick down, and you get no intelligent life.

All right you might say, I'll turn the wick up a little. I'll make my Big Bang just a little *more* violent than the actual one. What happens now, is that the gases come out of the Big Bang so fast that they do not have time to collect together to form embryo stars before they are dispersed into the depths of space. There being no stars, you get no life.

In fact it turns out that as far as the Big Bang violence is concerned, the window of opportunity is exceedingly narrow. If you are to get life in your universe, the thrust must be just right—and that is what our actual Universe has managed to do.

The next point to consider is the force of gravity. How strong will you make it in your imaginary universe? If you make it a little weaker than it actually is you will collect gas together after the Big Bang but not enough to produce a temperature rise sufficient to light the nuclear fires. No stars—no life.

On the other hand, you must be careful not to have your gravity too strong. That way you would get only the very massive types of stars. Recall how I said that massive stars can burn themselves out in only 1 million years. For evolution to take place you must have a steady source of energy for 5000 million years—you need a medium sized star like the Sun. Indeed when you come to think of it, the Sun is a remarkable phenomenon. After all, what is a star? It is nuclear bomb going off slowly. Have you any idea how difficult that is to achieve? Yet the amazing thing is that the Sun manages this. The secret is the way the force of gravity in the Sun conspires to feed the new fuel into the nuclear furnace at just the right rate for the nuclear fires (governed by the nuclear force—an entirely different force from that of gravity) to consume it at a steady rate extending over a period of 10 000 million years.

So, in order for there to be life, the force of gravity must lie within a very narrow range of possible values—and the gravity of the actual Universe does just that.

Next we must turn our attention to the materials from which we wish to build the bodies of living creatures. This is no small matter. After all, what do we get coming from the Big Bang? The two lightest gases—hydrogen and helium—and precious little besides. And it *has* to be that way. Remember we need a violent Big Bang to stop the Universe from collapsing back in on itself prematurely. And because of that violence, only the lightest nuclei could survive the collisions occurring at that time—anything bigger getting smashed up again soon after its formation. But you can't make interesting objects like human bodies out of just hydrogen and helium. So the extra nuclei—those that go to make up the 92 different elements found on Earth—must be manufactured somehow *after* the Big Bang. That's where the stars have another important part to play. Not only do they provide a steady source of warmth to energize the processes of evolution, they first serve as furnaces for fusing light nuclei into the heavy ones that will later be needed for producing the bodies of the evolving creatures.

But we are not home and dry yet. Perhaps the most important atom in the making of life is that of carbon. In a sense it is an especially 'sticky' kind of atom very good at cementing together the large molecules of biological interest. But forming a nucleus of carbon is by no means easy. Essentially, it consists in fusing three helium nuclei together. This is as unlikely as to have three moving snooker balls colliding simultaneously. Without me going into any details as to how this comes about—let me just say that it involves something called a nuclear resonance. The occurrence of this resonance is so highly fortuitous, that its discoverer, one-time atheist Fred Hoyle now freely speaks of 'He who fixed it.' In private conversation with him once, it was clear to me that his discovery had had a profound effect on them. While still disavowing any association with 'organized religion', Hoyle nevertheless felt the 'coincidence' was so way out, it could not have been due to mere chance.

So we have our precious carbon. A collision between some of these carbon nuclei and further helium nuclei yields oxygen—another vital ingredient for life—and so on. Thus you must be sure in your imaginary universe to incorporate a fortuitous nuclear resonance.

Does this mean that the stage is now set for evolution to take over, and convert these raw materials into human beings?

Not so. We have our materials, but where are they? They are in the centre of a star at a temperature of about 10 million degrees. Hardly an environment conducive to life. The materials have to be got out. But how, in your imaginary universe, are you going to arrange for that? After all, we know how difficult it is to lift something off the surface of the Earth and out into space—one needs a rocket to do it. Your stars have no rockets and the gravity forces are stronger.

What happens in this actual Universe is that a proportion of the newly synthesized material is ejected by supernova explosions. These occur when massive stars—several times the mass of our Sun—run out of fuel. They suddenly collapse in on themselves. But that raises a problem. How can an *im*plosion produce an *ex*plosion? This was a conundrum that exercised the minds of astrophysicists for many years. In the event the mechanism turned out to be the strangest imaginable. The material is blasted out by neutrinos. Neutrinos are famous for hardly ever interacting with anything. One could pass a neutrino though the centre of the Earth to Australia a 100 000 million times before it had a 50 : 50 chance of hitting anything. They are incredibly slippery. And yet it is neutrinos that are responsible for blasting out the precious stardust. So, my advice to you is that in your imaginary universe make your neutrinos slippery if you wish, but be sure not to overdo it—otherwise you will not get life.

The material is now out among the interstellar gases. In time, this collects together to form a dense cloud, which squashes down to form a star. Outside the star there can be secondary eddies that settle down to form planets. It is now possible to have rocky planets like Earth, Mercury, Venus, and Mars. For the first generation of stars this had not been the case because at that stage there had only been hydrogen and helium around. Given a planet at a reasonable position away from the star for a temperate climate to prevail, one has now at last got a chance of life evolving from the primordial slime.

How likely this is to happen is not known. If one is a physicist one tends to be impressed by the vast number of planets there must be out there—in other words how many attempts one is allowed to produce intelligent life. On this assessment, one is indeed home and dry. If on the other hand you are a biologist, you might be more impressed by the size of the hurdles that have still to be negotiated on the way to intelligent life—like for example the formation of the first cell. You might therefore be inclined to think that there must be some more 'coincidences' to follow—biological ones this time rather than the physical ones we have been considering.

It is impossible to put a hard figure on the likelihood of getting life from simply throwing together a bunch of physical laws at random—laws incorporating arbitrary values for the various physical constants. In talking for example about the strength of gravity having to lie within a narrow range, it is impossible to be more quantitative unless there is some way of specifying a permissable range of values that the strength could conceivably take on. If it could be *any value whatsoever*, then the finite range would be divided by infinity—and the chances would be virtually zero. Whatever the true odds come out to be, it is probably fair to say that to have a universe that is appropriate for the development of life is less likely than winning first prize in the Lottery.

This calls for an explanation. One possible resolution of the mysterious appropriateness of the Universe, is to pin one's faith on science, and to assert that in the end science will come up with a natural explanation of it all. From that vantage point we shall be able to see that there is no mystery; there is no need to invoke coincidences.

The second way of addressing the Anthropic Principle is to assert that our Universe is not alone. There are a great many universes—perhaps an infinite number of them—and they are all run on different lines with their own laws of nature. The vast majority of them have no life in them because one or other of the conditions were not met. In a few, perhaps in only the one, all the conditions happen by chance to be satisfied and there life was able to get a hold. The probability of a universe being of this type is small but because there are so many attempts, it is no longer surprising that it should have happened. We being a form of life ourselves must, of course, find ourselves in one of these freak universes.

This is a suggestion that has been put forward by some scientists, but that does not make it a scientific explanation. For one thing, the other universes are not part of our Universe and so by definition cannot be contacted. There is therefore no way to prove or disprove their existence. Not only that but the suggestion goes against the conventional way scientific development has tended to go. Scientists generally go for the simplest, most economical explanations. It is what we call the application of Occam's Razor. To postulate the existence of an infinite number of universes all run according to their own laws of nature is to go as far in the opposite direction as is imaginable. Which is not in itself to say that the idea of an infinite number of universes is wrong—merely that it does not count as science.

So, an infinite number of universes is the second way of accounting for the Anthropic Principle. The third alternative is simply to accept that the Universe is a put-up job; it was designed for life, and the designer is God.

Now, one always gets a little bit worried over arguments in favour of the existence of God based on design. The original argument from design held that everything about our bodies, and those of other animals, is so beautifully fitted to fulfil its function that it must have been designed that way—the designer being God—and therefore you must believe in God. The rug was pulled from under that argument by Darwin's theory of evolution by natural selection—at least in terms of it being a knock-down proof of God's existence—one aimed at convincing the sceptic.

So it is I would urge caution on those religious believers tempted to make too much of this new argument from design—one based this time on physics and cosmology. It is my contention that one can neither prove nor disprove God on the basis of such reasoning. If one is inclined to reject the idea of God, then one can do so in the expectation that science will one day show how the coincidences are not really coincidences, or it can be done on the grounds of there having probably been many attempts at different universes, so it is again not surprising that the world we know about comes to be the way it is. On the other hand, if one already believes in God on other grounds, say on the basis of religious experience, then the simplest explanation is in terms of a designer God. For religious believers, such an explanation introduces no fresh assumptions at all, over and above what one already accepts as the explanation of other features of one's life.

F. RUSSELL STANNARD

Born in 1931 in London, and educated at University College London. He spent most of his career as a high-energy nuclear physicist working at Berkeley, USA, and CERN in Geneva. One of the first academics to join the Open University, he headed the Physics Department for 21 years. A Reader in the Church of England, and Member of the Center of Theological Inquiry, Princeton, USA, he has always had a strong interest in the relationships between science and religion. He is well-known for his writings and broadcasts on the subject, and is a regular contributor to 'Thought for the Day'. Currently, he is concentrating on writing science books for children. His best-selling 'Uncle Albert' trilogy is in 15 translations.

# *The human singing voice*

DAVID M. HOWARD

## Abstract

Over the centuries, singers have developed vocal techniques to enable themselves to sing with greater acoustic efficiency in order to meet the changing demands of the music they are being asked to perform. Accompanying orchestral forces have increased in size as new instruments have been added, composers have written more challenging scores for singers, and auditoria have been enlarged to admit larger audiences. Laboratory techniques are now available which enable aspects of the vocal techniques employed by singers to be quantified and understood better in relation to their speaking voices. This chapter explores the human singing voice and introduces recent research results that help to explain how singers are able to sustain long notes over a wide pitch range, how they are able to project their sound to avoid being acoustically drowned out by accompanying forces, as well as the acoustic reasons behind some of the difficulties encountered by listeners in distinguishing words when sopranos sing notes towards the high end of their pitch range.

## Introduction

The human singing voice is considered by many to be the most versatile musical instrument with which a human can perform. Singers make use of the same vocal instrument when speaking or singing, but while the ability to speak is something with which almost everyone is endowed, singing requires full conscious control over parts of the vocal instrument for which control can be quite subconscious during speech.

We are able to express our everyday needs as well as communicate complex thoughts, ideas, and emotions by means of our human vocal instruments. It can produce a very wide range of sounds, and it is only a subset of these that are used in any particular language for communication purposes. Children first play their vocal instruments when they cry at birth and they go on to develop a wide range of vocal sounds by imitation and by experimentation. During speech we are not aware of the effort involved or of the transfer of neural messages needed to achieve the highly complex sequence of muscular events which produce the acoustic output. Yet the instrument itself comprises those parts of the body whose prime functions are breathing to sustain life (the nose and airways to the lungs), lung protection (the larynx), and eating (the mouth). Professional singing technique has had to develop over the centuries not only to give singers conscious control over these elements of the vocal instrument, but also to respond to the challenges posed by the historical development of musical ideas, the variety of places where music was to be performed, and the increasing size of audiences.

Early unison plainsong prior to the ninth century was primarily the domain of monks, and it made no great demands on the singers involved as the pitch range used would have been for the most part within that of their habitual speech. In addition, it would have been sung in large resonant abbey churches, often of cathedral proportions, to provide acoustic support. Harmony in its earliest form resulted from the singing of plainsong in parallel octaves when boys and women joined in singing with the men. This practice of singing a parallel part developed into *organum*, where the interval doubled in parallel became the perfect fifth or fourth, perhaps for the comfort of the singers themselves, enabling those with natural tenor and soprano ranges to sing a fourth or fifth above those with bass and alto ranges, respectively.

About the twelfth century, singers added descants above the plainsong tune or *cantus firmus*, which were often of an improvized nature extending the upper bounds of the vocal pitch range heard in music. During the fifteenth century, the *cantus firmus* began to disappear as music was written in which the parts moved more freely away from the bounds of the *cantus firmus* itself and pure polyphony developed. In order to achieve multipart polyphony, composers demanded an increased pitch range from all voices, but without the need for increased loudness as most of the polyphonic choral music was sung unaccompanied, or *a cappella*. When accompaniment was employed, particularly for solo songs, it would have been from a lute, early keyboard instrument, or chamber organ. The music of this period was performed in small rooms in an intimate manner, probably in domestic surroundings.

The late sixteenth century saw solo singers being expected to ornament the music they sang by means of or long phrases of rapid note movement or *divisions*. This decorated or contrapuntal style of vocal writing was used by the composers of early operas at the start of the seventeenth century, and developed further particularly in the music of Bach and Handel in their solo arias in Cantatas and Oratorios. Their scores required a wide vocal pitch range, sustained long phrases, and rapid note movement. In addition, the instrumental forces used for accompaniment were increased and performances were given in large churches or specially constructed buildings for public performances.

During the seventeenth and eighteenth centuries the *Bel Canto* school of singing developed from the Italian singers of the period. These singers developed the tone of their voices to be even throughout their pitch range along with masterly control of rapid passages. The voice was developed and explored fully as an instrument in its own right with particular attention being paid to the tonal qualities and the control of pitch rather than the production of high volume levels. With the building of increasingly larger opera houses and the use of increased orchestral forces during the nineteenth century, singers were called upon to sing louder and louder. Composers demanded wide dynamic ranges for dramatic effect and singers again modified and developed their vocal technique to maintain the instrumental tonal qualities of Bel Canto (or 'beautiful song') but with sufficient power available to fill an opera house in the acoustic presence of a large orchestra.

With the advent of microphones, loudspeakers, and electronic amplification equipment during the early part of the twentieth century, singers were able to perform with loud accompaniment in large venues without the need to produce loud volume levels themselves. Many singers working with microphones are, however, untrained vocally, often due to a belief that singing training can only develop an operatic sound. As a result, singers using amplification therefore often suffer vocal fatigue, sore throats, or more serious voice problems due to a lack of basic knowledge about the human vocal instrument. A modern school of singing known as *belting* has developed around New York's Broadway and London's West End specifically for the healthy development of the style of amplified singing demanded by composers and directors in theatre.

This chapter explores the operation of the singer's instrument with reference to the additional demands placed on singers above those required for normal speech. It is supported by a number of recent research results to illustrate the extent to which the human singing voice is understood from a scientific viewpoint. In many cases, these serve to explain effects quantitatively that have been handed down in a qualitative manner from teacher to singer for generations. In the future, computer-based visual displays may be employed as tools to support the work of teachers in training their pupils during both lessons and the pupil's private practice.

## The singing instrument

Any acoustic instrument can be considered in terms of three main elements: a power source, a sound source, and sound modifiers.[1] In the human vocal instrument, these elements can be related physiologically to the action of: the lungs, the vocal folds, and the vocal tract, respectively. These are illustrated in diagrammatic form in Fig. 1, which also illustrates their acoustic function in terms of an equivalent mechanical model. Those parts of the instrument which can be modified during singing or speech are indicated with double-ended arrows. The action of each element during singing is considered below, and reference to recent research results that provide a basis for enhancing our understanding of the singer's craft are provided as appropriate.

### *The power source*

When any sound is produced, a source of energy is required. In the case of an orchestral instrument such as a violin or an oboe, this might be the forces employed to pluck or bow of a string, or to blow air between a double reed, respectively. In singing, this energy is provided in the form of a flow of air from the lungs following an intake of breath.

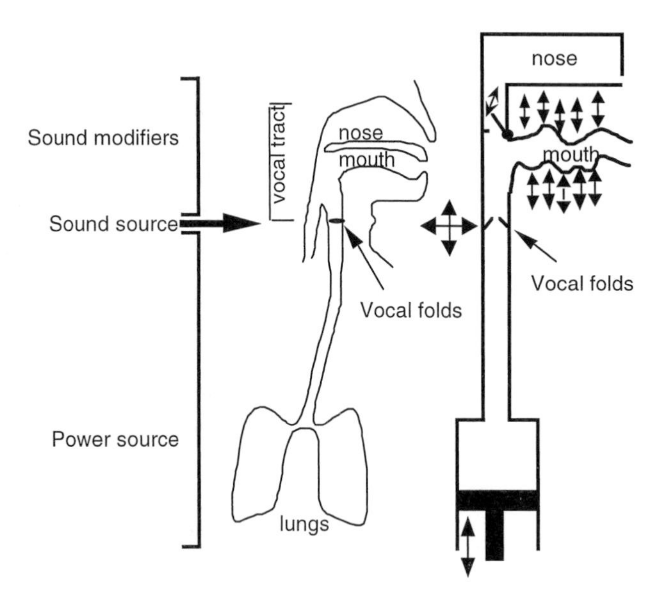


**Fig. 1** The singing instrument in terms of its three main constituent functions (left), physiological equivalents (centre), and mechanical model (right).

Air can flow from the lungs to the outside world while (i) the airways are open, and (ii) the lungs maintain an air pressure that is higher than the atmospheric pressure outside the body. When lung air pressure is lower than atmospheric, air flow is sustained *to* the lungs from the outside world. This provides the basis upon which humans breathe. We draw air in by enlarging the lung spaces through muscular action, thereby creating a lung pressure that is lower than atmospheric, and air flows into the lungs. This is equivalent to pulling the piston in Fig. 1 downwards. We breathe out by muscular contraction, which shrinks the lungs, equivalent to pushing the piston in Fig. 2 upwards, producing a lung pressure that is higher than atmospheric and air flows outwards.

The lungs themselves are made up of a substance that would shrink in size dramatically if the lungs were removed from the body. In this respect each lung compares rather well with a balloon. However, normal lung inflation differs from the manner in which a balloon is normally inflated, in that it requires the lungs to be supported externally in such a manner that they can be physically enlarged to suck air in. This is enabled within the body by means of the ribcage, and they are also supported from below by the diaphragm. Breathing in and out is therefore a result, respectively, of lung expansion and contraction. This is achieved by the actions of: (i) the group of muscles which expands and contract the ribcage, and (ii) the diaphragm and the associated muscles that affect its position. This is illustrated in Fig. 2.

The group of muscles that join and control the size of the ribcage are known as the *intercostals*. There are intercostals used for breathing in by expanding the size of the ribcage (*inspiratory intercostals*) and those for breathing out by contracting the size of the ribcage (*expiratory intercostals*).

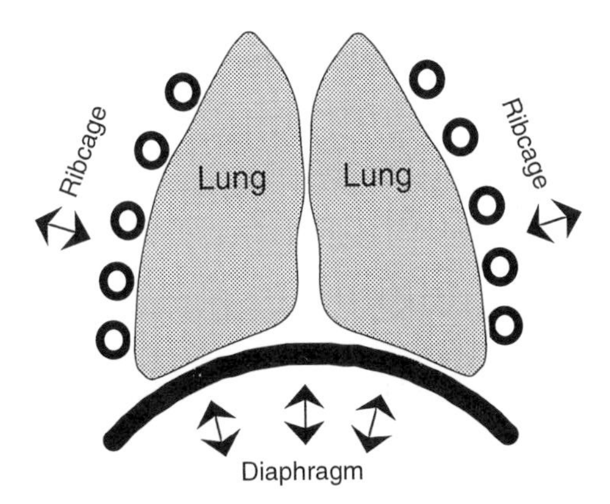


**Fig. 2** Illustration to show how the lungs can be expanded and contracted by the ribcage and/or diaphragm.

The diaphragm itself is a muscle, which is bowed upwards below the lungs as shown in Fig. 2 when it is relaxed. When it is contracted, it becomes shorter and flatter, expanding the lungs by pulling them downwards. The diaphragm sits over the abdominal wall, and as the volume of the abdomen cannot be altered appreciably, any diaphragm contraction serves to push down on the abdomen which causes the abdominal wall to bulge outwards. Abdominal wall expansion and contraction is readily observed externally and is often used by singers and their teachers as an indicator of diaphragm-based breathing. The group of muscles in the abdominal wall are used when breathing out when they raise the abdominal contents and hence lift the diaphragm exerting a force on the lungs which reduces their volume by expelling air.

Singers learn to control their breathing to enable them to sing notes and musical phrases which can be sustained for long periods of time in a controlled manner. In the Bel Canto school of singing, breath control is achieved by concentrating primarily on diaphragmatic breathing, often termed *breath support*, with particular emphasis on keeping the upper chest relaxed and the shoulders lowered. In Bel Canto, the larynx is maintained in a low position compared with that used in speech. Not all professional singers make use of Bel Canto techniques when on stage. For example, singers in Broadway and London's West End musicals make use of a style known as 'belting' which is loud, brassy, and twangy and has its origins in ethnic music world-wide such as the gospel tradition. In order to produce the characteristic belt sound, singers learn to maintain a tense upper chest that pushes down on the lungs to expel air. During belting the larynx is maintained in a high position with respect to that used in speech.

## *The sound source*

The sound source during sung notes is the acoustic result of the vibration of the vocal folds within the larynx. The vocal folds themselves form a valve that can be rapidly closed to protect the lungs from taking in water or food or any air-borne substance which might cause them harm. During singing, the vocal folds open and close many times per second, and the resulting airflow through the glottis (the space between the vocal folds) is of the form shown in Fig. 3. The bottom of the figure illustrates the pattern of vocal fold vibration as if viewed from the front, and the main points to note are that the vocal folds close from their lower edges upward, and part from their lower edges upwards also. In addition, the glottal airflow waveform shows that the vocal folds close

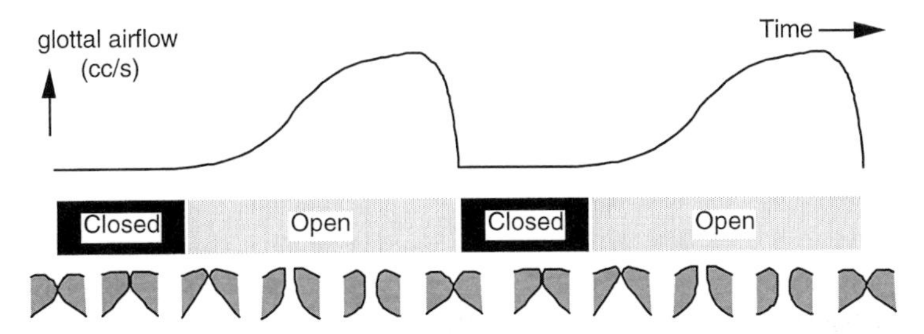


**Fig. 3** Schematized pattern of glottal airflow and cross sections of healthy vocal fold vibration for two cycles, showing closed and open phases. (From ref. 2.)

more rapidly than they open, and it is the instant of vocal fold closure in each cycle that provides the sound source to the vocal tract.

The physical mechanism that is responsible for initiating and maintaining vocal folds vibration can be described with reference to the Bernoulli principle. To initiate vibration the vocal folds are brought closer together than in their rest position or *adducted*. Air from the lungs is expelled through the glottis and this causes vibration of the vocal folds to be initiated and sustained due to the Bernoulli principle, which states that the sum of the kinetic energy (velocity energy by virtue of movement) and potential energy (pressure energy by virtue of position) remains constant (ignoring frictional forces) at any point in a tube through which a fluid is flowing.

The constriction in the airways due to the adducted vocal folds causes the velocity of airflow to increase as it crosses the glottis. This greater air velocity increases the kinetic energy, and therefore by the Bernoulli principle, the potential energy must be reduced at this point, which manifests itself as a reduction in the pressure exerted on the walls of the tube. This pressure drop acts on the folds to pull them towards each other, narrowing the glottis. The air now flowing through this narrowed glottis must increase in velocity, causing the pressure exerted on the tube walls to fall further and the force pulling the folds together to increase further. The vocal folds therefore accelerate towards each other until they meet at the mid-line of the larynx where they snap together at high speed, closing the glottis, and shutting off the flow of air from the lungs, thereby causing a rapid pulse-like drop in pressure immediately above the larynx. This provides a short acoustic pressure pulse to the vocal tract.

Following vocal fold closure, there are two effects which cause the vocal folds to move apart. First, the subglottal pressure is now higher than the superglottal pressure, and this tends to push the folds apart from below. Secondly, the natural pendulum-like action of the folds will tend to return them towards their rest position. The vocal folds therefore part, passing through their rest position to a maximum glottal opening from which their pendulum-like action will cause the glottis closure sequence to begin once again. Both the closing and the opening of the vocal folds is a 'bottom up' process as each begins at the lower edges of the folds as shown in the cross-section in Fig. 3.

When the vocal folds vibrate in a regular fashion during singing, a train of short acoustic pressure pulses of acoustic energy is generated, each resulting from a vocal fold closure. When a soprano sings the note A above middle C, her vocal folds open and close approximately 440 times a second; their fundamental frequency (f0) of vibration is 440 Hz. At the upper extreme of the soprano pitch range the vocal folds may be vibrating with an f0 of well over 1000 Hz. Figure 4 shows typical pitch

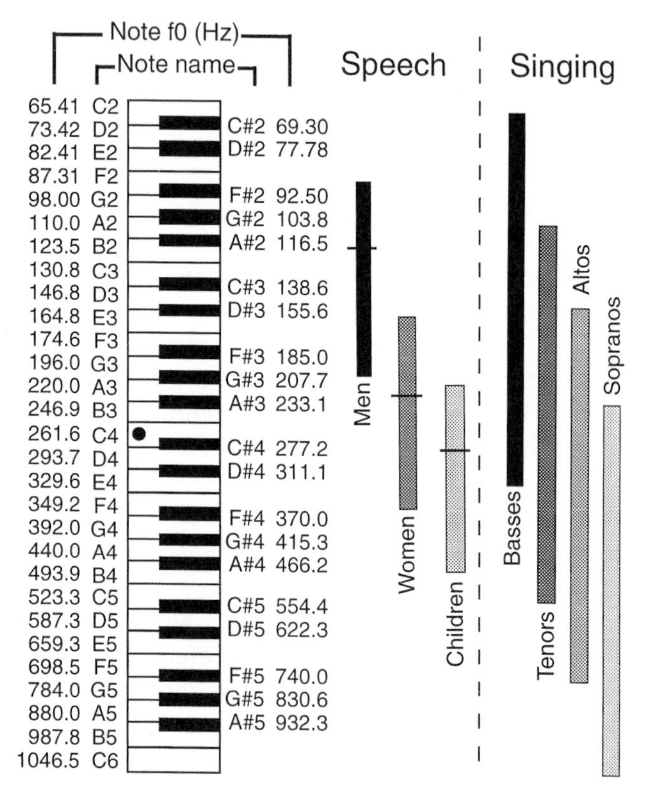


**Fig. 4** Pitch for notes of the musical scale extending two octaves either side of middle C (marked with a black spot). Approximate pitch ranges used in singing and speech are indicated, and average speech pitch values are indicated by horizontal lines. (From ref. 2.)

ranges for sopranos, altos, tenors, and basses, the average speaking pitch ranges for men, women, and children, all set alongside a piano keyboard on which middle C is indicated. Average speech pitches are indicated by the horizontal bars on the speech plots.

The f0 ranges used during singing are considerably greater than those habitually used in speech. The f0 range for a bass extends both above and below the average speaking f0 range for adult males, and that for a tenor extends a considerable way above the speaking range. A similar trend can be noted for the range of the alto and soprano compared with the average speaking pitch ranges for adult females. It is clear from Fig. 4 that all professional singers have to extend their f0 ranges well beyond their habitual speaking ranges.

The f0 of the note produced when the vocal folds vibrate is controlled essentially by the tension and the thickness of the vocal folds themselves; a large proportion of which are muscle. In order for the vocal folds to produce a note at a low *f*0, they have to be relaxed and thick. As the folds are stretched the *f*0 increases up to a point where the folds can be stretched no further, and in order to produce notes at a higher f0 the thickness of the folds has to be reduced. This is achieved by vocal fold muscular action that prevents a significant fraction of the vocal fold tissue from vibrating to leave a thin portion which can vibrate. In practice, the mechanisms by which the vocal folds are adjusted to achieve the extension to the f0 range are somewhat more complicated due to interactions between the muscles involved. The overall effect is, however, that the voice exhibits 'breaks' which can be heard as changes in tone colour as singers make major adjustments to their vocal fold configurations to vary f0 over a wide range, and these are often termed 'register breaks'. One of the key skills that a budding professional singer needs to develop is that of covering over the register break so that they are not audible.

The vibrating vocal folds are observed routinely in the clinic by methods such as an angled mirror placed at the back of the mouth to allow direct observation, or video endoscopy which provides a picture on a television screen by means of a video camera that is focused on the upper surface of the vocal folds. The camera gains access to the larynx either via a stiff tube placed in the mouth (rigid endoscope) or via an optical fibre passed through the nasal cavity (flexible endoscope). While these methods are very valuable for clinical diagnosis of vocal fold problems, they interfere directly with the singer's performance because the means of observation partially obstructs the oral and/or nasal cavity and is thereby invasive. One method which provides rather different but nevertheless useful data with respect to the vibrating vocal folds in a completely non-invasive manner is the electrolaryngograph. This device makes use of two electrodes placed externally on either side of the neck at the level of the larynx. A small high frequency voltage is maintained between the electrodes and the output waveform (Lx) is the resulting current flow. Changes in current flow can be interpreted as variations in the area of vocal fold contact.

Figure 5 shows a few cycles of an idealized Lx waveform. When the vocal folds are in contact a greater current passes than when they are apart, and observation of the detailed shape of the Lx waveform provides useful information in regard to the nature of vocal fold vibration during singing. The fundamental period of vocal fold vibration can be measured from the Lx waveform, marked as Tx in the figure, an the f0 can be calculated as the number of fundamental periods in 1 s:

$$f0 = \left\{\frac{1}{Tx}\right\} \tag{1}$$

The open and closed phases shown in Fig. 3 can be identified from the idealized Lx waveform as shown in Fig. 5.

The closed phase (CP) is more usefully expressed as a closed quotient (CQ), defined as the percentage of the cycle for which the vocal folds are in contact. CQ is measured as follows:

$$CQ = \left\{\frac{CP}{Tx} * 100\right\}\% \tag{2}$$

(For completeness it should be noted that the glottis does not necessarily close completely in the closed phase depending on the voice quality

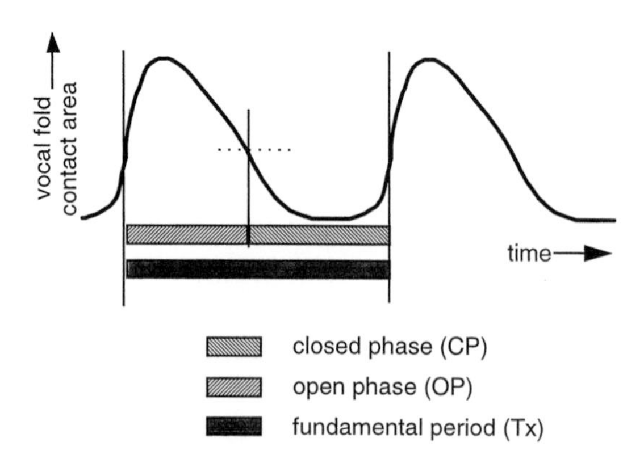


**Fig. 5** A few idealized cycles of electrolaryngograph output waveform (Lx) illustrating the measurement of: fundamental period (Tx), open phase (OP), and closed phase (CP).

and health of the vocal folds; vibration can occur while a proportion of the glottis is held open.)

A real-time computer-based system has been developed to measure f0 and CQ from the Lx waveform in real-time, and Fig. 6 shows the output from this system for a soprano singing a two-octave descending A major scale on the vowel 'ah'. The individual notes of the scale can be identified on the plot of f0 as varying between A5 and A3 when the f0 values for these notes given in Fig. 4 are compared. Highly detailed cycle-by-cycle investigations can be carried out with this system of changes in f0 to study aspects of singing such as vibrato, pitch variations during note onsets and offsets, and fine variation in intonation. The associated CQ values are plotted in the lower part of the figure, and it can be seen that this singer maintains her CQ values between approximately 30 and 45%. It is interesting to note the change in CQ as the lower octave of the scale commences as this is around G4 (392 Hz), which is generally associated with a register break in the female voice.[3]

It has been found that for many subjects CQ varies with f0. Such a variation can be observed for the sung scale plotted in Fig. 6. Another form of plotting CQ and f0 data, which allows the relationship between them to be observed, is shown in Fig. 7 for an ascending and descending two octave A major scale sung by a highly trained professional soprano

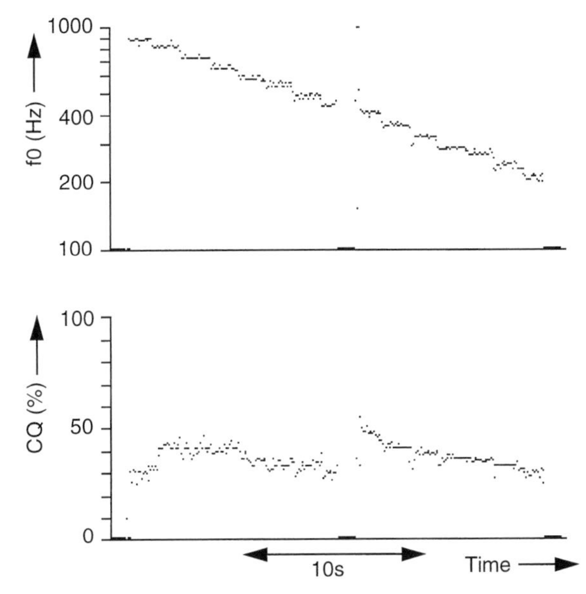


**Fig. 6** A plot of f0 (upper) and CQ (lower) against time for a soprano singing a descending two octave scale of A major.

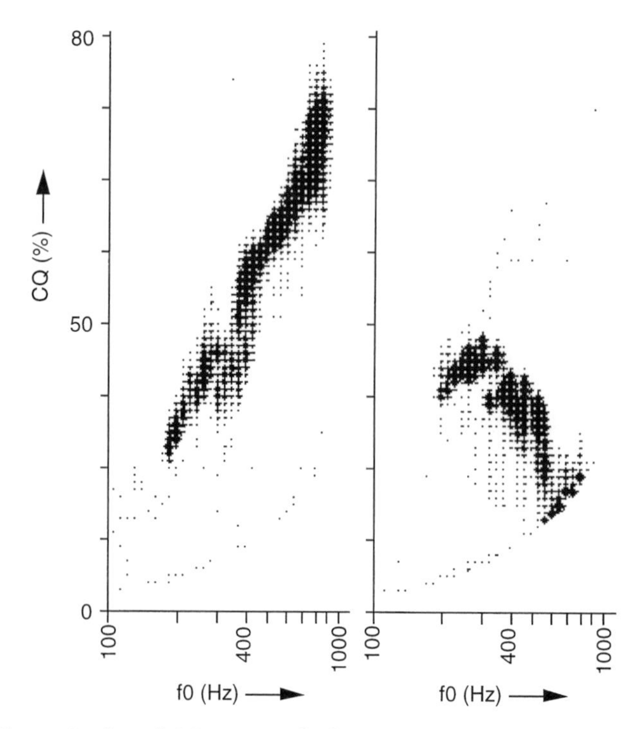


**Fig. 7** A plot of CQ against f0 for a two octave ascending and descending A major scale sung by a highly trained soprano (left) and an untrained adult female (right).

with many years of stage experience and an untrained adult female. A much greater range of CQ values is used by the trained soprano, particularly at higher pitches. Her change of CQ with f0 is essentially linear throughout the scale, which is a very different pattern to that observed for the untrained adult female. Both subjects exhibit clear changes in CQ around the G4 register break as discussed earlier in regard to Fig. 6.

The relationship between CQ and f0 changes when the voice is trained professionally. For adult males CQ remains essentially constant for all values of f0. Following training, CQ values rise and become typically greater than those used in their habitual speech. Untrained adult males use CQ values towards the lower end of those they use during their habitual speech. This is illustrated in Fig. 7. The raising of CQ contributes to improved vocal efficiency as follows:

(1) the time in each cycle for which an acoustic path exists via an open glottis whereby sound energy is lost to the lungs is reduced and therefore greater acoustic energy reaches the ears of the audience;

(2) less lung air is used in each cycle, allowing longer notes and/or musical phrases to be sung; and

(3) the perceived voice quality is less 'breathy' as there is less time during which air flows through the restriction of an open glottis.

Pattern of variation of CQ with f0 are rather different for adult female subjects, and Fig. 7 illustrate the extremes. The nature of this variation changed with training as follows: CQ tends to be reduced for pitches below D4 (see Fig. 2) and increased for pitches above B4, and the CQ/f0 gradient within the pitch ranges G3 to G4 and B4 to G5 tends to correlate positively with singing training/experience.

In effect, there are two pitch regions where different effects manifest themselves, and these are likely to be directly related to the development of inaudible register breaks with training. Suggested summary idealized changes in CQ with f0 shapes with increased singing training/ experience are shown in Fig. 8. At present it is not clear why there is a sex difference between CQ/f0 patterns. However, likely explanations rest with the differences in vocal fold size, physiology, and resulting mechanical properties between the sexes, as well as acoustic differences due to the relationship between the relative pitches of male and female voices (see Fig. 4) and the effect of their sound modifiers.

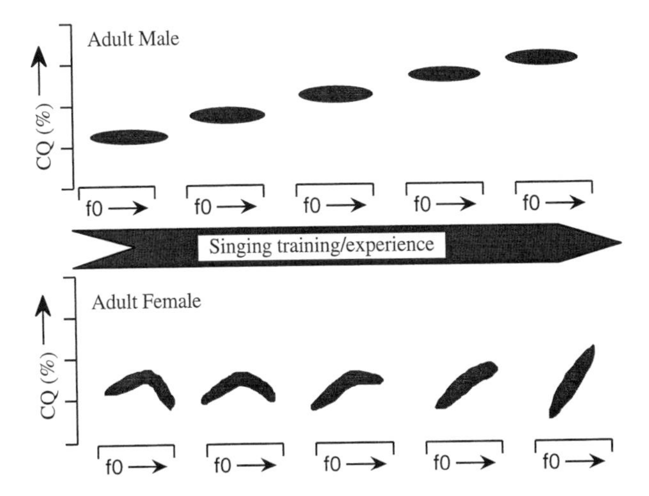


**Fig. 8** Idealized variation in CQ with f0 patterns for adult males and females with singing training/experience. (Adapted from ref. 4.)

## The sound modifiers

The sound modifiers consist of the vocal tract (see Fig. 1), comprising the airways between the larynx and the lips and nostrils. The acoustic excitation from the sound source passes through the vocal tract where it is modified by the characteristics of the acoustic response of the cavities that form the vocal tract. These acoustic properties change as its volume is altered by moving the tongue, jaw, lips, and soft palate (known as the articulators) (Fig. 9). The purpose of the soft palate, often referred to as the velum, is to act as a valve to connect/disconnect the nasal cavity with the airways. The acoustic properties of the vocal tract manifest themselves as a series of resonance peaks in the acoustic frequency response. These peaks are known as formants, and the centre frequencies change as the articulators are moved. Figure 9 shows idealized acoustic frequency response curves for the three vowels in *bee*, *baa*, and *boo*. The lowest three formants are shown in each case as it is only these that change appreciably in frequency during the production of vowels.

When a soprano sings words on high pitches, it is often difficult to understand the words. This is not due (necessarily) to poor singing tech-

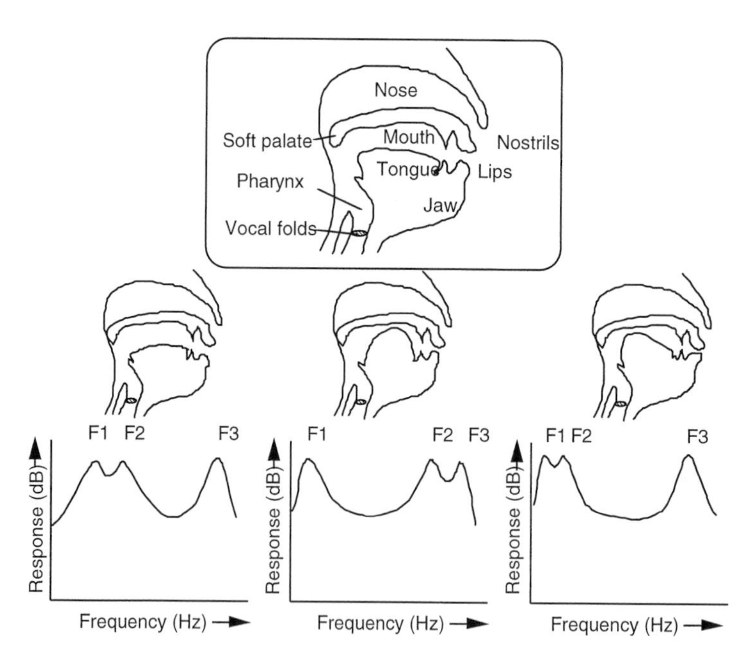


**Fig. 9** The main elements of the human vocal tract including the key articulators as well as the vocal tract frequency response for the vowels in: *baa* (left), *bee* (centre), and *boo* (right). (Adapted from ref. 2.)

nique as there is an acoustic reason why vowels become increasingly indistinct. As the f0 of the voice is raised, the spacing between the harmonics of the acoustic excitation provided by the vibrating vocal folds becomes greater as harmonics lie at integer (1, 2, 3...) multiples of f0. As a result, the formants will be less well defined acoustically at higher pitches as the number of harmonics grouped at or around individual formant peaks is reduced. This effect is illustrated in Fig. 10 which shows the spectra for the sound source, sound modifiers, and acoustic output for the vowel in *bee* sung at two notes an octave apart. The formant positions themselves are the main acoustic cues by which the human hearing system identifies the vowel being uttered. In this example, the positions in frequency of the lowest three formants are clear in the acoustic output sung on G4 but completely obscured when sung on G5. This is entirely a function of the change in the spacing between the harmonics of the sound source. Professional sopranos are trained to take advantage of this effect to help provide more efficient acoustic projection at higher pitches. As the vowels cannot be distinguished, they learn to modify the positions of the lower formants so that they coincide in frequency with individual harmonics of the sound source, ensuring that these harmonics are transmitted preferentially by the sound modifiers to arrive at the ear of the listener with maximum amplitude. In general, vowel distinctions become less clear for note pitches higher than C5. This effect is present in a tenor's voice but to a

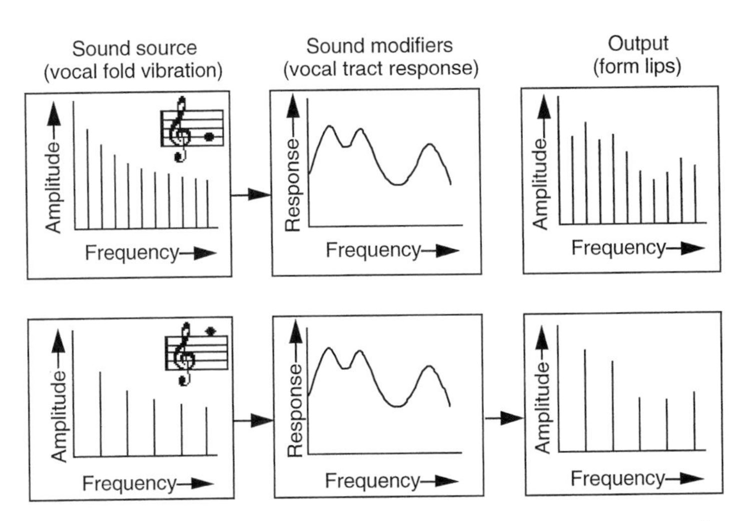


**Fig. 10** Spectral representation of the sound source, sound modifiers, and output for the vowel in *baa* sung on G4 (upper) and G5 (lower). (Adapted from ref. 1.)

lesser degree, because the frequencies of his formants are higher overall with respect to his singing voice pitch range.

Figure 11 from Howard and Collingsworth[5] illustrates this effect by means of spectrograms for the words *bard*, *bead*, and *booed*, which are spoken and sung on G4 and G5. Spectrograms provide an analysis of the energy present (level of blackness of marking) at particular frequencies (*Y* axis) with time (*X* axis). Spectrograms are commonly used in speech research to track the positions of formants and other acoustic cues in different sounds and they were originally designed to provide enhanced understanding of how sounds are perceived by providing a visual indication of the acoustic analysis carried out by the peripheral human

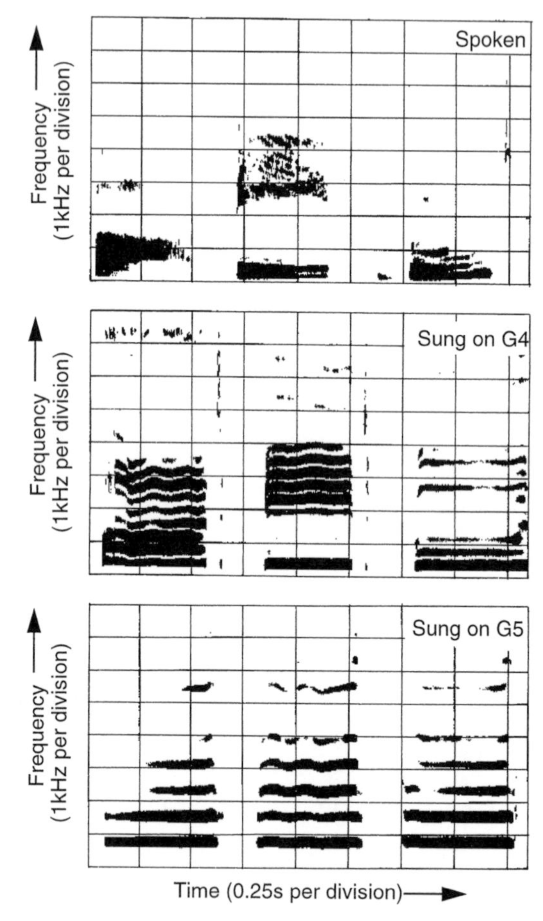


**Fig. 11** Spectrograms for the words *bard, bead* and *booed* spoken (upper), and sung on G4 (centre) and G5 (lower) by a trained soprano. (Adapted from ref. 5.)

hearing system. In these examples, the relative positions of the lowest three formants for these three vowels can be seen in Fig. 9, and it can be seen that these manifest themselves when the words are spoken and sung on G4. However, when she sings the words on G5, the formants are lost in the spectrograms which all look essentially the same because the harmonics of her sound are so widely separated with respect to her vowel formant positions. Vowels sung on G5 cannot be distinguished by the listener.

The knowledge of loss of vowel distinction when they are sung on pitches that are high in the range is demonstrated by composers such as Handel in his choices of note ranges when setting words to music in recitative and aria. An essential difference between these two forms is that recitative is there to provide the narrative with each line of the text appearing once, while in arias the text usually consists of only a few words which are repeated a number of times as the musical ideas are developed. In addition, arias are usually accompanied by a full orchestral sound while recitatives are usually accompanied by the continuo section only (harpsichord and/or chamber organ with a single violoncello). Figure 12 shows histograms of the notes used by Handel in the soprano recitatives 14 and 15 from *Messiah* and the first section of the soprano aria 'Let the bright Seraphim' from his oratorio *Samson*. These histograms represent pitches as members of a note class using the notation given in Fig. 4. There are 238 notes used in the aria to score just 13 words, whereas there are 106 notes used to score 82 words in the recitative. The most commonly used note in the recitative is C5, which is a major third lower than that for the aria (E5). In addition, the first

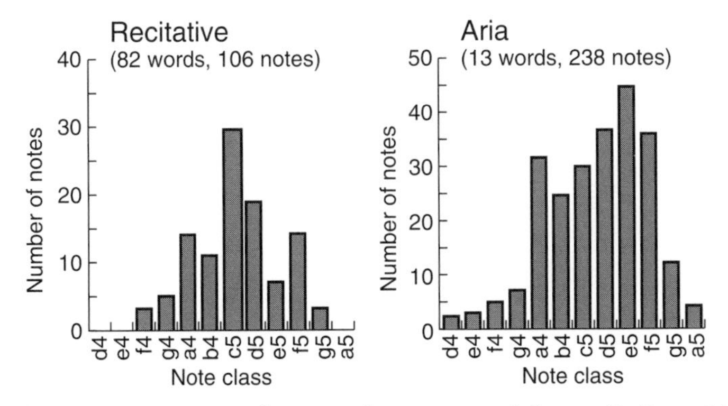


**Fig. 12** Histograms of notes classes scored for recitatives 14 and 15 from Handel's *Messiah* (left), and the first section of 'let the bright Seraphim' from Handel's *Samson* (right). (Adapted from ref. 5.)

presentation of the 13 words of text in the aria is set notes that enable the vowels to be distinguished clearly before the music becomes more florid involving a number of rapid passages of high notes as the text is repeated.

Another important manifestation of the trained singing voice is that it can be heard by the audience when singing operatic arias on a large stage accompanied by large orchestral forces in a grand opera house. In addition, a trained professional singer will often report a feeling of effortlessness during such occasions and certainly no sensation of vocal strain during the louder passages. The average distribution of acoustic energy with frequency during speech, in this case a spoken version of an opera aria, typically takes the form of the plot shown in the left-hand side of Fig. 13. It has a low frequency peak and then it falls away with increasing frequency. The average distribution of acoustic energy with frequency for an orchestra playing the accompaniment to an opera aria has essentially the same shape as shown in the centre plot of Fig. 13, although at a considerably higher intensity level (this difference is not shown in the figure). The average distribution of acoustic energy with frequency for a singer singing the aria when accompanied by the orchestra takes the form shown in the right-hand side of the figure, and an additional resonance peak is apparent between approximately 2.5 and 3.5 kHz. This resonant peak is referred to as the 'singer's formant'.[3] Sundberg suggests that the singer's formant manifests itself due to a clustering together of the third formant with the fourth and fifth formants. The resulting enhanced resonant peak is not generally found in habitual speech, and the amplitude of this peak relative to the lower formants relates to how close together these higher formants are drawn.

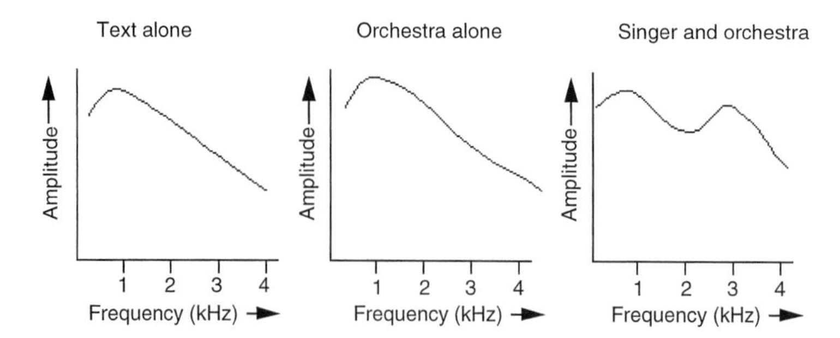


**Fig. 13** Idealized long-term average spectra for a singer speaking the text of an opera aria (left), the orchestra playing the accompaniment to the aria (centre), and aria being sung with orchestral accompaniment (right). (From ref. 1.)

Sundberg further suggests that the singer's formant results from 'a strong dependence on the larynx tube', where 'it is necessary, however, that the pharynx tube be lengthened and that the cross-sectional area in the pharynx at the level of the larynx tube opening be more than six times the area of that opening'. This is achieved in practice by lowering the larynx and enlarging the pharynx.

The use of this particular frequency position for the singer's formant is not pure chance. This frequency region is very close to that where the human hearing system is at its most sensitive.[1] This gives the singer an additional advantage in that they are not only illuminating acoustically a part of the spectrum that is not being employed by the accompanying orchestra, but also they are providing acoustic energy in a frequency region where communication with their listeners is most efficient.

## Summary and conclusions

The human singing voice is one of the most versatile acoustic musical instruments. It consists of the lungs which provide the power source, the vocal folds of the larynx which provide the sound source, and the cavities of the vocal tract whose volume can be altered acting as sound modifiers. The power source provides a steady flow of air through the larynx during pitched notes which sustains vocal fold vibration, providing a sound source at a frequency determined by their masses and tensions. The resulting acoustic excitation is transmitted via the vocal tract which imparts its acoustic resonant properties depending on the positions of the articulators. Important scientific insights are now available to explain the physiological and acoustic bases on which the human vocal instrument works in terms of its sound source and sound modifiers.

Traditional singing voice training will continue to be more of an art than a science with teachers using many different techniques because in most cases, much of their methodology is based on what they themselves have learned from their own teachers and found most useful for developing their own professional singing voices. Singers are given instructions such as: 'sing on the point of the yawn', 'sing as if you have an orange stuck in the throat', 'sing as if through a hole just above the nose', and 'focus the beam of sound on to different positions along the hard palate as you change the pitch'. Rather than actual physical instructions, these instructions can be thought of as psychological hooks which enable singers to play their singing instrument to best possible advantage. Through the use of such psychological hooks, singers work to associate certain articulatory gestures with, for example particular pitches, vowels, consonants, and musical phrasing. Few if any singers could respond directly to an instruction to, for example 'lower your larynx and expand your pharynx' in order to enhance their singer's formant.

It may well be that in the future there are computer-based displays being used during singing lessons and singing practice sessions to indicate levels of parameters such as the relative energy in the singer's formant region and larynx closed quotient. Pilot study work in this area has indicated that changes in these parameters do occur with regular lessons, and that visual displays of such parameters are potentially acceptable to both teacher and pupil. It must, however, be stressed that computer-based displays will never replace professional voice teachers as they could at best only become tools to be employed in support of just a small portion of the teacher's craft. They certainly have the potential to support both the teacher's work and the pupil's practice sessions, giving a quantitative basis upon which progress can be sustained and monitored. Proper application of visual displays could result in better use of time and a more rapid rate of progress on technical matters relating to playing the instrument itself. This would leave the teacher more time during lessons to be devoted to artistic considerations such as: stagecraft, interpretation of the score, musicality, working with an accompanist, and working with an orchestra, as well as the development of confidence.

The art and science of human voice production were just as different 80 years ago. The following is a quote from the introduction to a manual of voice production indicating one reason why a reader might choose to read it:[6] 'A raucous voice and an uncouth tongue are far more alienating to most of us that a fustian coat or a so-called inferior job.'

Sir William Bragg gave the following brief description of human voice production during a series of six lectures on *The World of Sound* at the Royal Institution, which gives a picture of knowledge in the early 1920s:[5]

> *So now we come finally to the most wonderful instrument of all, the organ of the voice, and especially the human voice. In the throat is a reed—the larynx—which may be compared to the reed we have just been using (organ pipe reed); and when we set our lips, teeth, and tongue into various positions we do something which corresponds to the placing of different shaped pipes over the reed. By these changes and subtle variations in the way in which we go from one set position to another, we arrange for a running accompaniment of overtones which are the means of speech, vowels, and consonants of every shade and inflection. And the infinite delicacy and variety of all the tones we produce is matched only by the*

> *wonderful delicacy of the ear which distinguishes them and the brain which interprets them.*

Knowledge of the working of the human voice has change dramatically with the development of electronics, enabling new devices to be implemented with which the human vocal instrument can be explored. Wherever such scientific knowledge leads, it is clear that professional singing is more than simply the production of notes from the human vocal instrument. The physiological and psychological state of the human being in whom the vocal instrument resides must themselves to be fully warmed up, prepared and co-ordinated so that the musical ideas encapsulated in the score can be communicated as completely as possible with the listeners.

## References


1. D.M. Howard and J.A.S. Angus (1996), *Music technology: acoustics and psychoacoustics*, Oxford, Focal Press.
2. D.M. Howard (1998), Practical voice measurement, In: *The voice clinic handbook* (ed. T. Harris, J. Rubin, S. Harris, and D.M. Howard), London, Whurr Publishing Company.
3. J. Sundberg (1987), *The science of the singing voice*, DeKalb, Illinois University Press.
4. D.M. Howard (1995), Variation of electrolaryngographically derived closed quotient for trained and untrained adult singers, *Journal of Voice* **9**(2), 163–72.
5. D.M. Howard, and J. Collingsworth (1992), Voice source and acoustic measures in singing, *Acoustics Bulletin*, **17**(4), 5–12.
6. J.H. Williams (1923), *Voice production and breathing for speakers and singers*, London, Sir Isaac Pitman and Sons Ltd.
7. W.H. Bragg (1920), *The world of sound: six lectures delivered at the Royal Institution*, London, Bell and Sons Ltd.


DAVID M. HOWARD

Born 1956 in Kent, educated in Electrical Engineering at University College London. After completing a PhD at the University of London developing electronic circuits for cochlear implants, he became a lecturer in experimental phonetics at UCL. In 1990 he moved to the Department of Electronics at the University of York, where he was promoted to a Personal Chair in 1996 and became Head of Department. He was awarded the Thorn-EMI Partnership award in 1994 for work on singing pitching development in primary school children, a system which has been featured on Japanese Breakfast television and UK National Radio. He is a Fellow of the Institute of Acoustics and the Institution of Electrical Engineers, the Chairman of the Engineering Professors' Council, and has been President of the British Voice Association. In the last two years he has co-authored a textbook titled *Acoustics and Psychoacoustics* and contributed to *The Penguin Dictionary of Electronics* and other works aimed at communications engineers and health professionals.

# *Asthma and allergy—disorders of civilization*

STEPHEN T. HOLGATE

Allergy is a word that is frequently used loosely to describe human intolerance to environmental factors. This broad definition has led allergy being used to describe such diseases as migraine, irritable bowel syndrome, chronic fatigue syndrome (ME), and, at the extreme end of the spectrum, a 'total allergy syndrome' (allergy to the twentieth century). While a strong case can be made for the human body being intolerant to a wide variety of environmental pathogens and toxins, a more restricted use of the word allergy provides insight into one of the most fascinating and exciting areas of modern medicine. For the purpose of this review, I will narrow C. Von Pirquet's original description of allergy (1906): 'The ability of animals and humans to develop altered responses to foreign substances after repeated exposure' to that more recently described by Gell and Coombes in 1964 of: 'Immune responses which give rise to irritant or harmful reactions'.[1] A key feature of the allergic response is the involvement of the immune system and the exquisite specificity and sensitivity that this is able to impart upon factors encountered in the modern environment.

The classical work of Gell and Coombes divides allergic responses into four categories on the basis of the timing and type of response produced (i.e. antibody or cell mediated). As most of the classical allergic responses fall into the first category of 'immediate' hypersensitivity, which differentiates it from those with a more gradual onset, I will focus on this alone in this review. The range of diseases in which immediate type response is involved is large. Most frequently this type of allergic response occurs at interfaces between the external environment and internal milieux giving rise to such disorders as asthma, perennial and seasonal rhinitis (hay fever), allergic sinusitis, conjunctivitis and, in the gastrointestinal tract, food and drug reactions. In the skin, allergic (atopic) eczema and urticaria (hives) represents two extremes of the type 1 allergic response, the former being a chronic condition with quiet and active periods, the latter being of sudden onset but usually resolving rapidly. In the eye, allergic responses usually lead to an acute superficial conjunctivitis, while a more severe and intractable allergic response, e.g. in vernal and giant papillary conjunctivitis may be sight-threatening through damage to the cornea. One of the most dramatic manifestations of the immediate type allergic response is anaphylaxis in which contact with minute amounts of the offending allergen, e.g. penicillin, bee venom, or peanuts, produces rapid loss of blood pressure, severe narrowing of the airways in the lung, liver, and swelling of the mucous membranes (angioedema) sometimes requiring life-saving measures. In order to place these varied disorders into a mechanistic context, it is necessary to have some understanding of the 'key players' involved.

## Cells and mediators of the allergic response

Irrespective of the organ or organs in which they are manifest, almost all of the diseases referred to above have in common a single triggering mechanism which underpins adverse responses to specific environmental allergens. This factor is present in the serum and was originally named '*reagin*' by Prausnitz and Küstner on the basis that Küstner's allergic response to cod fish could be passively transferred to Prausnitz by injection of serum into the skin followed by local allergen to produce a characteristic weal and flare response.[2] This test for reagin was subsequently named the PK reaction. Forty-five years later Johanssen in Sweden and the Ishizaka's in the United States identified *reagin* as a member of the circulating antibody pool and named it immunoglobulin E (IgE). IgE, along with the other antibody classes, is present in the blood of all humans but, in those who express allergic diseases, it is present in greater amounts and, more importantly, it is directed against specific environmental allergens. The genetic tendency to develop IgE antibodies against common allergens encountered in the environment is referred to as atopy and those clinical disorders that ensure, atopic diseases. One end of the IgE molecule (called the Fab region) is designed to bind to specific components of the offending allergen (epitopes), while the other end (the $F_c$ region) binds with high affinity to specific receptors present on tissue mast cells and circulating basophils (Fig. 1). These cells are important sources of chemical mediators which, when released, lead to the rapid clinical response characteristic of allergy. Fc$\varepsilon$R1 consists of four polypeptide chains $\alpha\beta\gamma_2$. The $\alpha$ chain of the receptor binds to five amino acids at the tail end of the IgE molecule to orientate the IgE

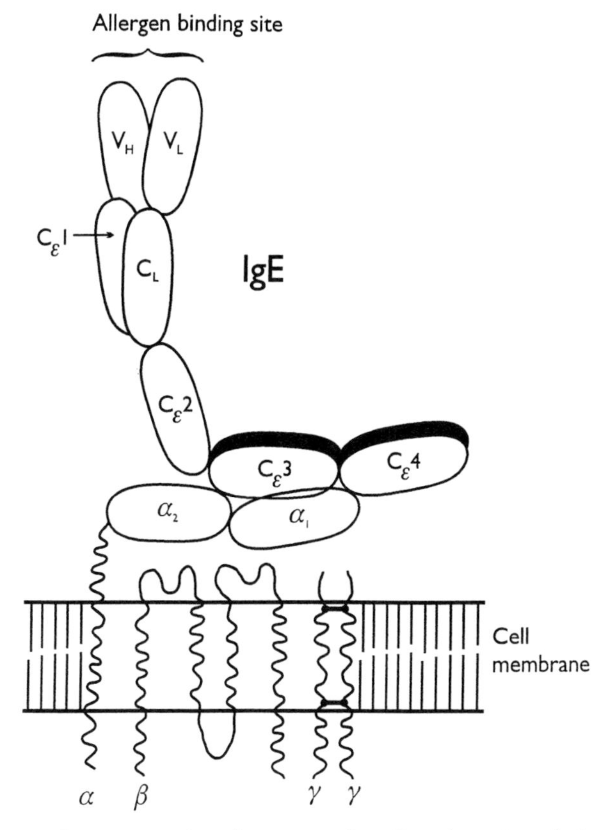


**Fig. 1** The interaction between the $C_{\varepsilon}3$ domain of the Fc portion of IgE and the α-chain of the high affinity $Fc_{\varepsilon}$ receptor ($Fc_{\varepsilon}R1$) expressed on mast cells, basophils and dendritic cells. Cross linkage of two or more receptors by antigen binding to adjacent cell-bound IgE molecules results in cell activation.

molecule so that it lies on its side with the allergen binding site facing outwards (Fig. 1).[3] Binding of allergen to two or more adjacent α chains results in clustering of receptors on the surface of the mast cell or basophil and sets into motion a series of cellular events that trigger mast cell 'activation' with the explosive release of inflammatory mediators. The attachment of IgE to mast cells, a process referred to as sensitization, can be easily demonstrated by pricking a small amount of allergen through the skin of the forearm. A positive response is a characteristic weal and flare identical in appearance to the PK reaction and is due to the release of mast cell mediators, the size of the response being proportional to the amount of IgE directed to the specific allergen.

Within their cytoplasm mast cells and basophils have granules which store a number of chemical substances, including histamine, proteolytic enzymes (tryptase, chymase), complex carbohydrate cleaving enzymes ($\beta$-glucuronidase, hexosaminidase, $\beta$-galactosidase), the anticoagulant heparin, and small proteins (cytokines or interleukins), which serve as messenger between cells. When activated through their IgE receptors, mast cells and basophils produce a variety of newly generated chemical mediators that originate from mobilization of fatty acids from the cell's nuclear and plasma membranes, followed by enzymatic conversion to a create a group of highly potent inflammatory products which include the prostaglandins and leukotriene. The family of three leukotrienes ($LTC_4$, $LTD_4$ and $LTE_4$), previously identified in the 1938s by Kellaway and Trethewie as slow reacting substance (SRS-A) are among the most active chemicals known to produce features of the allergic tissue response. Interaction of these different mediators with specific receptors on target tissues produces the acute allergic symptoms. These include contraction of airways smooth muscle causing wheezing and breathlessness, leakage of small blood vessels causing swelling, stimulation of glands to secrete excess mucus and irritation of nerve endings to create the symptoms of itching, sneezing, and cough. When mast cell and basophil activation occurs in response to circulating allergens, then life-threatening anaphylaxis ensues.

In addition to an immediate component initiated by mast cells, many of the allergic disorders have a chronic component. The cells largely responsible for these are eosinophils (so-called because of their red staining with eosin), which are selectively drawn from the microcirculation into the inflammatory site. These bone marrow-derived cells are selectively removed by the small blood vessels through interactions with a range of molecules expressed on the luminal side of the lining cells that line blood vessels, which provide an adhesive function directed towards the passing leucocytes. Mediators from mast cells, such as histamine and leukotrienes initiate this microvasculature adhesion process by increasing the surface expression of a stick sugar-rich protein called P-selectin, which is stored preformed in granules present in the endothelial cells lining blood vessels. In completing the leucocyte adhesion process and in selecting specific cell types such as eosinophils, is the need for a series of soluble protein signalling molecules (interleukins) which are generated and released upon IgE-dependent mast cell activation. Of particular importance are tumour necrosis factor alpha (TNF$\alpha$), interleukin (IL)-4, IL-5, IL-13, granulocyte–macrophage colony-stimulating factor (GM-CSF) and a series of related molecules, called chemokines, whose specific functions are to direct leucocyte migration. Together these interleukins not only selectively trap circulating eosinophils in microvessels, but also help them migrate through the blood vessel wall into the tissue where, by interacting with the tissue matrix, they secrete their own chemical mediators including the prostaglandins, leukotrienes, and tissue damaging granule proteins and enzymes.

## Orchestration of the allergic inflammatory response

Implicit in the chronic allergic response is a continuous process of IgE generation, mast cell activation and eosinophil recruitment. These processes are orchestrated by principal cells of the immune system T lymphocytes, cells that play a crucial part in protecting against invading organisms both systematically and at mucosal surfaces. In atopic individuals, T lymphocytes receive an allergen signal from highly specialized antigen presenting cells (dendritic cells) located at the interface between the external and internal environments.[5] These 'professional' antigen-processing cells ingest allergen molecules that land on their surface, digest them into small fragment peptides and then present a selected peptide sequence on the surface of the cell held in the cleft of the proteins that make up the repertoire of the human lymphocyte antigens (HLA) alternatively named the major histocompatibility class II (MHC class II) molecules. Presentation of allergen peptides to the T cell occurs in local lymphoid tissue, along with the essential engagement of co-stimulatory molecules, such as B7 on the dendritic cell and CD28 on the lymphocyte, results in the differentiation of the naive T cell to one that generates a range of cytokines which upregulate cells and antibodies involved in the allergic tissue response. The genes for these cytokines are encoded within a small region on the long arm of chromosome 5, a number of which (IL-4, IL-5, and GM-CSF) are co-ordinately controlled. T cells differentiating along this route preferentially release cytokines of the IL-4 gene cluster are called Th2-like. This differentiates them from a second set of helper T lymphocytes (designated Th1-like) which are involved in cell-mediated immunity, a response which is called into play to protect animals against viruses, malignancy, and intracellular pathogens such as tuberculosis and leprosy (Fig. 2). Tissue biopsy studies conducted in patients with allergic diseases in which the cytokine repertoire has been assessed by the presence of the cytokine protein product or the messenger RNA from which it originates demonstrate clear over-representation of Th2-like lymphocytes with their capacity to maintain an ongoing allergic response.

While a number of the Th2-derived products are involved in mast cell, basophil, and eosinophil recruitment and maturation, one particular cytokine IL-4 (and its homologue, IL-13, which shares 30% amino acid

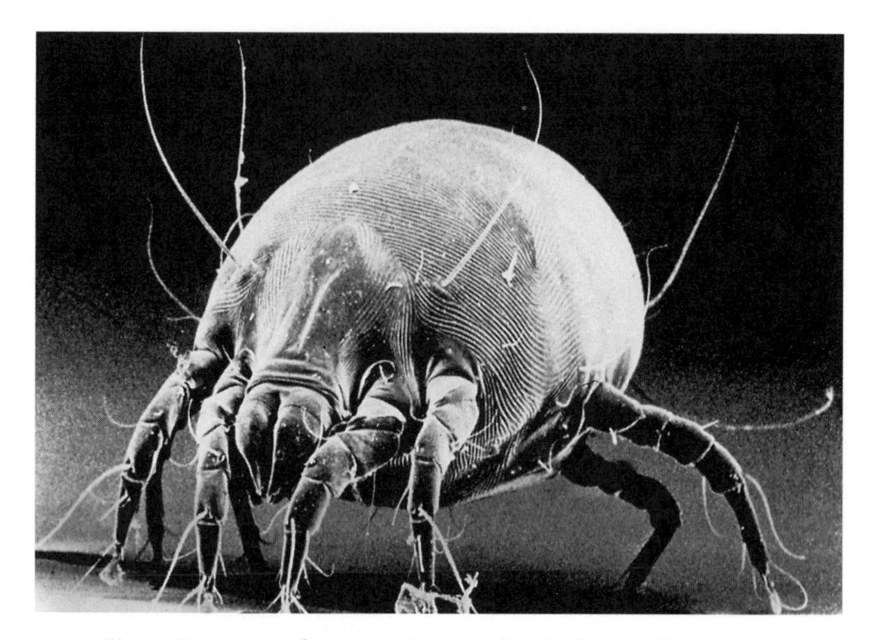

**Fig. 2** Scanning electron micrograph of a house dust mite—*dermatophagoides pteronyssimus*. These and related mites survive off the skin flakes we, as humans, shed. They prefer a damp and warm environment to breed. Under these conditions they excrete faeces in pockets containing the digestive enzymes that we, as humans, become allergic to when we inhale these particles.

homology with IL-4), plays a particularly important part in this arm of the immune response. By interacting with B lymphocytes, these cytokines change the immunoglobulin type being secreted from the short-term protective antibody IgM to the allergic antibody, IgE. As with interactions between the dendritic cell and T-lymphocytes effective signalling involving B lymphocyte requires a close cell–cell interaction with the Th2 cells, antigen presentation, and engagement of a second set of co-stimulatory molecules, called CD40 and its ligand CD40L. Under these cognate conditions and in the presence of IL-4 or IL-13, the B cell starts to synthesize IgE specific to the allergen. If cell–cell contact between the T cell and the B cell is not established but only IL-4 or IL-13 are available, then the results will be the generation of large amounts of IgE which is non-specific. IgE has the important role of linking the recognition of allergen to the signalling of a variety of cells, including mast cells and basophils to stimulate the generation and release of a range of highly active chemical mediators. Inhibition of IgE synthesis can be achieved when IgE becomes bound to a second receptor on B cells (Fc$\varepsilon$R2). The component of the IgE molecule that binds to Fc$\varepsilon$R2 is separate from the

Fc$\varepsilon$R1 high affinity binding site.[6] Of interest is that *Der* $P_1$, the major allergen of the house dust mite through its intrinsic enzymic activity, is capable of selectively cleaving cell-bound Fc$\varepsilon$R2 thereby depriving the IgE secreting B lymphocytes of an inhibitory feedback signal leading to augmented *Der* $P_1$-specific and non-specific IgE synthesis.[7]

## New therapeutic opportunities from understanding IgE-related mechanisms

A particularly exciting new development with therapeutic potential is the opportunity of removing IgE using monoclonal antibodies directed to that part of the IgE molecule that binds to the high affinity receptors (Fc$\varepsilon$R1) on mast cells and basophils.[8] Thus, when IgE is bound to the mast cell via its specified binding site, the antibody is unable to find the part of the IgE to which it was originally directed (epitope) because it is engaged in the $\alpha$ chain of the mast cell IgE receptor and, unlike allergen or anti-IgE molecules that bind to other parts of the IgE molecule, is unable to cross-link the IgE and, as a consequence, fails to activate cells for mediator secretion. However, the antibody is able to bind to IgE both in the circulation and when expressed on the surface of B lymphocytes synthesizing it, to form complexes that the body can easily eliminate. Although these antibodies were originally developed in mice, they have now been 'humanized' so that only a minute proportion of the total monoclonal antibody that is directed to human IgE is of mouse origin, the majority (99%) being human.[9] Phase I clinical studies have already shown that a single injection of a 'humanized' blocking antibody to IgE can remove IgE from the serum within 30 min with a duration of up to 30 days and block the acute asthmatic response to inhaled allergen. Clinical trials are now in progress to see what impact this has on the clinical expression of allergic diseases, including asthma.

Naturally occurring auto-antibodies against IgE and its high and low affinity receptors have also been described. With knowledge of the IgE–Fc$\varepsilon$R1 binding site, it may be possible to develop a peptide vaccine incorporating these amino acids induce artificially an auto-antihuman IgE that would remove IgE without causing mast cell activation. Preliminary clinical studies with such a decapeptide in well defined food allergic subjects looks most promising.

Humanized blocking monoclonal antibodies have also been developed against the key allergic cytokines IL-4 and IL-5. Blockade of IL-4 will not only attenuate IgE production, but will reduce the expression of vascular cell adhesion molecule-1 (an adhesion protein involved in eosinophil recruitment from the circulation) and inhibit development of Th2 cells, which are critically dependent upon this cytokine. Clinical trials of these antibodies are currently in progress and the results eagerly awaited.

## The immune response to gastrointestinal parasites

The observation that IgE, mast cells, eosinophils, and T cells are intimately associated with the pathogenesis of allergic responses, prompts the question as to what the normal function of this system is in the absence of allergic disease. It would appear that all of these components are involved in the protection against the expulsion of parasites. Colonization of the gut with parasites generates a strong Th2 signal with a massive increase in parasite-specific IgE and in the number of mast cells and eosinophils present within the gut lining.[10] The presence of appropriate parasite antigens which interact with IgE, stimulates chemical mediator release from mast cells and eosinophils leading to sloughing of the overlying epithelium, contraction of the gastrointestinal smooth muscle, outpouring of plasma proteins, and hypersecretion of mucus, which are all designed to aid expulsion of the parasite. In addition, the eosinophil is well equipped with parasiticidal molecules e.g. the eosinophil granule proteins, whose job it is to breach the parasite's outer wall and, as a consequence, destroy it.

For reasons which are not understood, the parasite killing and elimination response seems also to be utilized in the development of allergy in tissues distant from the gastrointestinal tract and in response to environment allergens. One possible explanation for this transition is that many of the molecules present in sensitizing allergens, such as house dust mites, cat dander, and fungal spores, have counterparts in parasites, e.g. *Cruzipain* and $\gamma$-gluamyl transpeptidase in *Leishmania cruzii* and *Brugia malayi*, respectively. These enzymes appear to be particularly powerful in driving an IgE response and may in part explain why only some proteins are allergenic whereas others are not.[11] It turns out that the major house dust mite, domestic pet, and fungal allergens have enzymatic functions which may impart particular properties to allergenic molecules enabling them to breach the protective epithelial barriers by enzymatic attack of the cell adhesion molecules that are responsible for maintaining epithelial integrity.

## The genetic basis for allergy and asthma

Allergic diseases have long been known to run in families indicating a strong genetic component. Genetic influences can be divided into two components. The first is the ability of a susceptible individual recognize a common environmental allergen as foreign and initiate an allergic immune response. This operates through the HLA system which provide the mechanism for antigen recognition and presentation to and by T and B lymphocytes.[12] A second set of genetic influences appear to be important to regulating the overall cytokine response. For example, the region on chromosome 5, which contains the IL-4 gene cluster, in which the 'allergic' cytokine IL-3, -4, -5, -9, -13, and GM-CSF are encoded, has been shown to link closely to the inheritance of an increased IgE response and also to bronchial hyper-responsiveness ('twitchiness' of the airways), an important characteristic of asthma. As an example, a structural change (polymorphism) in the gene that controls the secretion of IL-4 from Th2-cells has been linked to a greater IgE response. Conversely, both random and candidate searches of human chromosomal DNA have revealed that the allergic diathesis links with a region on the long arm of chromosome 12 on which there is a gene encoding for a cytokine called interferon-$\gamma$ (IFN-$\gamma$), a powerful suppressor Th2 response.[13] It has been established that there is a reciprocal relationship between Th2 and Th1 responses, a product (IL-10) derived from Th-2 like cells inhibiting the Th1 response, whereas IFN-$\gamma$ generated by Th1-like cells inhibiting the Th2 response. The production of IFN-$\gamma$ is induced by two further cytokines (designated IL-12 and IL-18) which are released from activated dendritic cells, macrophages, monocytes, and epithelial cells and are generated in particularly large amounts during virus infections. Thus, it is possible that, in allergic diseases such as asthma, there is either an increase in the expression of genes which regulate Th2 cytokines or a decrease in expression of genes regulating IFN-$\gamma$, IL-12, or IL-18 production.[13] With the methodology now in place, the next decade is likely to witness an explosion of interest in the genetics of allergy and asthma as, identification of susceptibility genes will not only enable individuals at risk of developing disease to be identified, but will also enable preventive environmental or therapeutic strategies to be directed towards them. In discovering novel disease susceptibility genes, there is the hope that targets for new anti-asthma and anti-allergy drugs will also be found.

## Environmental factors and allergic disease

While $\sim 40\%$ of the clinical expression of an allergic disorder can be accounted for by genetic factors, for these to be manifest there is a need for interactions with environmental factors. The most characteristic feature of the human allergic tissue response is the generation of IgE directed specifically against common environmental allergens derived from indoor and outdoor sources. The most important of these include the dust mites (*Dermatophagoides pteronyssinus* and *D. farinae*), domestic pets (especially cats, rabbits, and rodents) and fungi found in damp housing, e.g. *Cladosporium, Alternaria,* and *Penicillium.*[14] In non-temperate climates other allergens may impose themselves on the allergic response. For example, in the poorer districts of cities in North America and in tropical climates allergens derived from cockroaches seem to be particularly important in initiating and maintaining asthma in these inner city communities.[15] On the other hand, where the atmosphere is dry and therefore not conducive for the survival of dust mites, e.g. Arizona and Saudi Arabia, allergens derived from fungi, e.g. *Alternaria* and domestic pets are more important in driving allergic responses linked to asthma. In the cooler northern climates outdoor allergens, such as birch pollen and in the warmer Mediterranean climates olive and *Parietaria* pollen become important contributors to allergic diseases. With these outdoor sources of allergen (including those derived from grasses and a wide variety of wind pollinated trees) hay fever and allergic conjunctivitis appear to be more closely linked, whereas asthma and eczema, are diseases associated with exposure to indoor allergens such as dust mites, animal danders, and fungi. In the work place, sensitization occurs to certain reactive occupational chemicals, such as isocyanates and acid anhydrides as well as to complex biological molecules.

## Sensitization to aeroallergens and the early life origins of allergic disease

Charles Blackley, a family general practitioner in Manchester, first described the characteristic symptoms of hay fever (*catarrhus aestivus*).[16] Using himself as an experimental model, Blackley was able to show that hay fever and associated symptoms of asthma followed closely the seasonal appearance of grass and other pollens in the air. Henry Hyde-Salter, a physician and later Dean of Charing Cross Hospital in London, wrote a scholarly Treatise on Asthma in 1860 which drew attention to emanations in dusts as a precipitant of asthma.[17] However, it was not until 1967 that Voorhorst first described the domestic house dust mite (*Dermatophagoides pteronyssinus*) as a major source of allergenic material. Rather than the mite itself being the major source of allergens, it is the faecal droppings containing the allergenic digestive enzymes, e.g. $DerP_1$ (a cysteine protease), and $DerP_2$ (lysozyme). To survive and reproduce, the house dust mite has an obligate requirement for a temperature of 25°C and a relative humidity of 80%. The adults feed on the proteins present in human skin scales while the nymphs require hyphae from the fungus *Aspergillus*. In modern housing these conditions are optimally achieved in the bedding and pillows as well as in carpets, soft furniture, and children's soft toys.[18] The faecal particles of the mite contain the highest concentration of allergens. It has been estimated that up to 15% of the contents of vacuum cleaner dust is made up of mites, their body parts and excrement. Contrary to popular opinion, pillows containing artificial fillings provide just as good, if not better, habitats for dust mites than feather pillows. Although various chemical agents that kill dust mites (acaricides) have been developed, the only sure way to kill dust mites is to wash to heat materials to a minimum of 60°C or freeze to –20°C. While freezing will kill mites it will not inactivate allergens.

Sensitization to dust mites and other domestic allergens most likely occurs early in life. In a prospective study of 17 children in Poole, Dorset, we have shown that the level of dust mite allergen present in the home during the first year of life is a major factor in determining whether an infant born of an allergic mother, and therefore genetically at risk of developing allergy or asthma, did in fact do so by the time they reached 11 years of age.[19] Moreover, the amount of allergen in the dust was an important factor in determining the age of onset of first symptoms, with high exposure leading to an earlier disease onset. Although some of the children who were to develop asthma acquired positive skin tests to dust mites before the age of 5 years, the majority did so between the ages of 5 and 11 years. A number of separate studies have shown that in children positive to skin prick tests to mites and cat allergens is the strongest risk factor so far identified for developing asthma. In our Dorset study children who had a positive skin test to mite allergens had a 15-fold greater chance of having asthma at age 11 years when compared with those who were skin test negative.

The mother's influence over the development of allergic disease in the offspring seems especially strong. For example, in the dry climate of Arizona, where sensitization to the fungus *Alternaria*, rather than to dust mites, dominates maternal but not paternal allergy, has been shown to impart a substantially greater influence on the child's development of elevated levels of serum total IgE and the later development of asthma. While it has been suggested that children may preferentially acquire certain genes from the mother (genomic imprinting), this is uncommon in human genetics and a more likely explanation for the strong maternal influence is the intrauterine environment. It is now well established that intrauterine nutrition is an important factor in programming a child for the later development of such adult degeneration diseases as diabetes, hypertension, chronic obstructive pulmonary disease, and osteoporosis. We have some preliminary evidence to suggest that maternal factors also influence the development of allergic responses. In two separate studies (one retrospective in adults and one prospective in 11-year-old children) there exists a positive relationship between greater head circumference at birth and the later development of allergy and high serum IgE levels. At first such an association may sound strange, but those placental and nutritional factors (especially over nutrition) that increase brain growth in the last trimester of pregnancy also influence the maturation of the thymus gland, the site of origin of the immune system. It has been suggested that after 26 weeks gestation, the fetus adopts a Th2-like immune phenotype to prevent maternal rejection and that in the last trimester of pregnancy with increased IFN-$\gamma$ production this converts to a more Th-1-like picture, with a greater capacity to produce IFN-$\gamma$.[22] Interleukin-4 is produced by the human amnion epithelium throughout pregnancy and recently IL-10, the cytokine that inhibits Th-1 responses, has been found in human placenta. If continued IL-4 or IL-10 production continues without check throughout pregnancy then sTh2 cytokine mode will be maintained and an allergic diathesis created. Such a mechanism might also be invoked as a factor in the causation of some cases of the sudden infant death syndrome (SIDS, cot death) in which mast cell tryptase and eosinophils, recognized markers of a Th2 response, are encountered in the lung and circulation.

In babies born of allergic mothers, T lymphocytes removed from the cord blood exhibit enhanced proliferative responses to environmental allergens, such as egg protein (ovalbumin) and milk protein ($\beta$-lactoglobulin) as well as those derived from house dust mite, cats, and tree pollens. The observation that asthma and other allergic diseases appear to be more common in those whose mothers were exposed to high concentrations of allergens, e.g. birch pollen, during the last trimester of their pregnancy. How minute amounts of allergen taken in by the mother can cross the placenta to sensitize the offspring is an intriguing but unresolved question. It is possible that small amounts of antigen are trapped by the placenta and are either presented to the fetus' immune response at this site or that maternal antigen-presenting cells (dendritic cells) or alternatively small amounts of antigen pass across the placenta into the fetal circulation so that antigen presentation occurs in the fetal tissues. A number of groups have now shown that at birth those children who progress to develop allergic diseases (including asthma) have impaired cord blood T-lymphocyte production of IFN-$\gamma$ when their cells are exposed to specific allergen, suggesting an impaired inhibitory mechanism for shutting down a Th2 response. Between the second and third trimesters of pregnancy the T cells from the fetus spontaneously release IFN-$\gamma$, but there are some fetus' whose T cells do not release this Th1 cytokine even when stimulated. It has been hypothesized that the role of this non-specific fetal IFN-$\gamma$ production by circulating mononuclear cells is to counteract the effects of IL-4 and IL-10 produced by the placenta with IFN-$\gamma$ levels reaching maximum at the time of parturition.[23] A mechanism such as this would be needed to prevent an allergic Th2 phenotype from developing in all newborn children and its failure may underlie the development of allergy. Environmental factors in pregnancy may also direct the placental–fetal relationship towards a Th2 like response, including young maternal age and smoking in pregnancy.

## Viruses and the development of allergic diseases

It has long been recognized that virus infections, especially the common cold viruses, can lead to deterioration of asthma lasting for several weeks. It is also known that virus infection in early infancy is responsible for intermittent wheezing illness and is most frequently caused by the respiratory syncitial virus (RSV). Growing from these observations is the view that virus infections are intimately involved in the development of the asthma syndrome and possibly other manifestations of allergic disease. The application of messenger ribonucleic acid (mRNA) detection techniques (polymerase chain reaction) for viruses have paved the way for defining more clearly any link between virus infections and asthma. For example, we have shown that in excess of 80% of acute exacerbations of asthma in school children and ~ 60% in adults are the result of virus infections, the most frequently detected being the common cold viruses (rhinovirus). Mechanisms that may increase the susceptibility of the asthmatic airway to virus driven inflammation are currently being pursued. One attractive possibility is that in the presence of persistent Th2 driven mast cell and eosinophil driven inflammation the release of certain cytokines specifically TNF$\alpha$ derived from these cells leads to an increase in the expression of receptors for human respiratory viruses on the airways lining epithelium. In the case of most rhinoviruses the receptor is an adhesion molecule called intercellular adhesion molecule-1 (ICAM-1), which is normally involved in leucocyte recruitment from the circulation and in cell signalling. ICAM-1 contains an amino acid sequence that binds tightly to a 'cavern region' of the rhinovirus capsid, thereby enabling virus internalization. Once the virus has entered the epithelium it replicates and is also able to generate a wide range of chemical mediators called chemokines (IL-8, RANTES, MCP-1, and eotaxin), which are able to enhance eosinophil and mast cell inflammation.

While viruses can undoubtedly cause deterioration of established asthma, during the first 3 years of life paradoxically there is evidence to indicate that viral or bacterial infection may also serve a protective role against the development of respiratory allergic disease.[25] One of the most consistent risk factors for allergy in children and adults is family size. In studies carried out both in Germany and in national British birth cohorts, the prevalence of mucosal allergy and positive allergy skin test in children have been shown to decline markedly in the last born child with increasing numbers of siblings. A working hypothesis is that over the past 30 years opportunities for acquiring infections from siblings or playmates in early childhood have declined with reduction in the average family size, vaccination programmes and higher standards of personal hygiene. Most viruses and some bacteria are able to evoke a Th1-mediated protective response with the generation of IL-12 and IFN-$\gamma$. Thus, if there are multiple infections during the first few years of life, high concentrations of these Th1 cytokines might be expected to inhibit the release of Th-2 cytokines, thereby biasing the mucosal immune response airways from allergen sensitization.[26] There is some direct evidence to support such an hypothesis. Shaheen and co-workers, have shown that adolescents aged 13–12 years in Guinea-Bissau, Africa, although infected with measles in the first year of life when compared with those vaccinated against measles later, had a 63% lesser chance of developing positive skin test to common aeroallergens.[22] It has also been reported that repeated BCG vaccination of young Japanese children also exerts a protective effect against the development of allergy.[28] If given to animals at the same time as a sensitizing antigen, both BCG vaccination or IL-12 and IFN-$\gamma$ prove to be powerful agents capable of inhibiting subsequent IgE, mast cell and eosinophilic responses. Thus, it is possible that IL-12, IFN-$\gamma$, or vaccines that are able to enhance production of these cytokines preferentially (e.g. *Mycobacterium vaccae*), may form the basis of new preventative or therapeutic strategies for allergy and asthma.

## The changing worldwide trends in asthma and allergy

Epidemiological studies that have used similar methodologies 10–20 years apart have revealed some quite remarkable statistics both in the developed and developing world pointing towards rising trends in allergic diseases, including asthma. While it is recognized that this disorders have an important genetic component, it is only in isolated inbred populations that genetic inbreeding could contribute to a progressive increase in asthma and allergy. One example is the high prevalence of asthma in islanders of Tristan da Cunha all of whom originates from 15 settlers from Scotland, England, North America, Holland, Italy, St Helen, Ireland, and South Africa.[29] However, in outbred populations much more likely causes of these rising trends are changes to our environment.

## The developing world

Studies conducted in Africa, South America, and South-east Asia reveal substantial increases in the prevalence of asthma associated with population shifts from the rural to the urban environment. In 1975 Godfrey investigated the occurrence of allergy and asthma in Gambian school children and showed their association with urban dwelling, higher socio-economic status, and lower total circulating IgE levels.[30] He suggested that in the rural setting parasite infection was protective against the development of allergy and asthma.

Immune defence against invading parasites utilizes the same components as in allergic tissue responses—IgE, mast cells, and eosinophils orchestrated by $Th_2$-like lymphocytes. Because parasitic worms in the gut lumen are too large to be destroyed by conventional white blood cell phagocytic mechanisms, the production of parasite-directed IgE, sensitization of mast cells, and recruitment of eosinophils leads to the release of mediators of inflammation that cause mucosal events designed at expelling the parasite. The antiparasite response in the intestine is almost identical to that of asthma in which excess mucus hypersecretion, contraction of smooth muscle, leakage of small blood vessels, and damage to the epithelial lining underlie the clinical expression of the disease, except that in asthma the immune response is directed against inhaled allergens rather than to intestinal parasites.

To investigate further the relationship between parasite infection and the development of allergy, Lynch and co-workers determined the effect of antihelminthic treatment on the allergic reactivity of children in a slum area of Caracas, Venezuela.[31] The children were divided into two groups, the first being treated for a period of 22 months with the antihelminth drug Quantrel® directed to eliminate the intestinal worms *Ascaris* and *Trichuris* and a second group who declined treatment and used as controls. The antihelminth treatment reduced the gut worm load from 68 to 5% in parallel with a decrease in the total serum IgE level from 2543 to

1124 iu/ml but an *increase* in the occurrence of skin test reactivity to the house dust mite from 17 to 68%. Over th 22-month observation period, in the untreated group parasite colonization of the children further increased from 43 to 70%, serum IgE levels from 1649 to 3697 iu/ml, but dust mite sensitization fell from 26 to 16%. Quantification of specific IgE antibody indicated that the parasite driven stimulation of IgE synthesis resulted in both saturation of mast cell IgE receptors and suppression of specific IgE antibody synthesis. From a public health standpoint high levels of non-specific IgE may protect rural dwellers exposed to parasites from allergy and asthma. It follows that eradication of parasites or reduced opportunities for infection could in part explain the rural to urban gradient in the prevalence of allergic disease.

## Socio-economic factors and the development of allergy and asthma

Population-based surveys in the developed world indicate that the more affluent sections of the community have the highest prevalence of allergic sensitization and diseases associated with it. Of considerable interest have been the recently reported findings of the prevalence of bronchitis and allergic diseases in former East and West Germany. It has been shown that, while bronchitis is much more common in former East Germany, asthma, hay fever and especially allergic sensitization assessed by skin testing was up to threefold more frequent in West Germany. One reason for undertaking these studies was to investigate the influence of industrial- and vehicle-related outdoor air pollution (particulates, $NO_2$ and $SO_2$) on airways disease as this was far worse in former East than in West Germany where government legislation had over the years of separation produced a marked improvement in air quality. Similar findings have been reported when comparing asthma and allergy prevalence in the Baltic States with that in northern Europe. In a further study looking at the influence of age on the apparent East/West difference in the prevalence of clinical allergy and sensitization revealed that it was children and young adults born since the 1960s where the difference was most apparent and was almost absent amongst middle-aged adults. This finding has been interpreted as a cohort effect reflecting economic and cultural differences that followed separation of the two halves of the country as exemplified by the existence of the Berlin Wall. In young British adults, self-reported hay fever and allergy skin prick tests are significantly related to socio-economic class as assessed by the father's occupation, the disorders progressively increasing from social class IV/V to social class I/II. One explanation for the effect of socio-economic factors on the development of allergy is an effect of early maternal programming on allergic sensitization involving dietary, smoking, and other factors.

In Australia and in other developed countries, changes to housing may have produced an important impact. Increased emphasis on energy conservation has resulted in the 'sealing' of homes with tightly fitted windows and doors and, as a consequence, poor air exchange. In older housing burning of fossil fuels in open fireplaces, heating of rooms only when they were being used and an emphasis on 'airing' of rooms may have kept dust mites and other indoor allergens at a low level. One study of housing in Wagga Wagga, South East Australia, where asthma and allergy prevalence has almost doubled over 10 years, has revealed a 15-fold increase in dust mite colonization.[33] Increased use of central heating and air conditioning, as well as soft furnishings and carpets associated with improved socio-economic status, also encourage the habitat for dust mites.

A hypothesis advanced to explain both the past and present epidemiological patterns of clinical allergy is that sensitization might be prevented by infections acquired during early childhood. Thus, in East Germany, in the Baltic States, but also in poorer Western civilisations, lower material standards of living and widespread use of day nurseries from an early age with increased spread of infection, might explain the lower prevalence of allergy in these communities. Linking this with an immunological hypothesis, it is tempting to speculate that reduced childhood infections tilts the mucosal immune response away from a Th1-like to a Th2-like response with the subsequent development of the allergic phenotype.

## Concluding comments

It is most unlikely that the rising trends of allergy and asthma seen worldwide have a single cause. Environmental factors are clearly important with the increased prevalence of asthma largely being accounted for by increased expression of IgE-dependent hypersensitivity. In the developing countries, the rural to urban increase in these disorders can in part be accounted for by changes in parasite infection and increased IgE sensitization to common environmental allergens. In poorer urban developments, increase in the allergen load, e.g. dust mites encouraged by use of Western-style bedding, carpeting, soft furnishing, and cockroach infestation of poor housing has increased the population risk of allergen sensitization. Both in developing and in developed countries environmental factors operating in pregnancy and in the first 3 years of life are most relevant to the observed increase in asthma and allergy. High socio-economic status, possibly operating at the fetal–maternal level, increased exposure to indoor allergens and cigarette smoking are all important factors. In drawing together changes to our environment with the early life development of allergy in asthma, a particularly plausible hypothesis is the 'lazy immune or hygiene hypothesis', i.e. is asthma and allergy an epidemic in the absence of infection?.[34] Reduced exposure of small children to bacterial and infectious agents, by reducing the Th1-like cytokine influence on the development of the immune response at a time when infants are exposed to high concentrations of allergens, provides a particularly attractive hypothesis for explaining the worldwide trends. If this does turn out to be the case, then a vaccine strategy to enhance the production of a Th1 signal in genetically at risk children during a 'window of opportunity' seems attractive.

The introduction and implementation of guidelines for asthma management with emphasis on patient education and effective use of anti-inflammatory drugs ('preventors') rather than relying on bronchodilators ('relievers') for symptom relief alone has resulted in a dramatic reduction in asthma mortality in such countries as New Zealand and the UK. Thus, while understanding of the underlying mechanisms of allergic disease creates new therapeutic opportunities, it is clearly of utmost importance that more effort is made to identify and subsequently remove those environmental factors important in the aetiology of allergic disease and in causing the worldwide increased expression.

## References

1. Gell, P.G.H., Coombs, R.R.A. (eds.), *Clinical aspects of immunology*, Blackwell, Oxford, 1962.
2. Prausnitz, C., Küstner, H. Studien Über Überempfindlichkeit. *Central Bakterol* 1921, **86**, 160.
3. Helm, B.A., Sayers, I., Higginbottom, Machado D.C., Ling Y., Ahmed K, Padlan E.A., Wilson, P.M. Identification of the high-affinity receptor binding region of immunoglobulin E. *J Biol Chem* 1996, **271**, 7494–500.
4. Corrigan, C.J., Kay, A.B. The lymphocyte in asthma. In: *Asthma and Rhinitis*, (eds) (ed. Busse, W.W., Holgate, S.T.) Chapter 34, pp. 450–64, Blackwell Scientific Publications Inc, Boston, 1995.
5. Semper, A.E., Hartley, J.A. Dendritic cells in the lung: what is their relevance to asthma? *Clin Exp Allergy* 1996, **26**, 485–90.
6. Sutton, B.J., Gould, H.J. The human IgE network. *Nature* 1993; **366**, 42108.

7. Hewitt, C, Brown A., Hart B., Prichard, D. A major house dust mite allergen disrupts the immunoglobulin network by selectively cleaving CD23: Innate protection by anti-proteases. *J Exp Med* 1995, **182**, 1–8.
8. Fahy, J.V., Fleming, H.E., Wong, H.H., Liu, J.T., Su, J.Q., Reimann, J., Fick, R.B., Boushey, H.A., The effect of an anti-IgE monoclonal antibody on the early- and late-phase responses to allergen inhalation in asthmatic subjects. *Am J Respir Crit Care Med* 1997, **155**, 1828–34.
9. Corne, J., Djukanovi!c, R., Thomas, L., Warner, J., Botta, L., Grandordy, B., Gygax, D., Heusser, C., Patalano, F., Richardson, W., Kilchherr, E., Staehelin, T., Davis, F., Gordon, W., Sun, L., Louise, R., Wang, G., Chang, T-W., Holgate, S.T. The effect of intravenous administration of a chimaeric anti-IgE antibody on serum levels in atopic subjects: Efficacy, safety and pharmacokinetics. *J Clin Invest* 1997, **99**, 879–87.
10. Moll, H. Immune responses to parasites: the art of distinguishing the good from the bad. *Immunology Today* 1996, **17**, 551–2.
11. Stewart, G.A., Thompson, P.J. The biochemistry of common aeroallergens. *Clin Exp Allergy* 1996, **26**, 1020–44.
12. Howell, W.M., Holgate, S.T. Human leukocyte antigen genes and allergic disease. In: *Genetics of Asthma and Atopy* (ed. IP Hall), Monogr. Allergy, Basel, Karger 1996, **33**, 53–70.
13. Umetsu, D.T., De Kruyff, R.H. $T_{H1}$ and $T_{H2}$ $CD4^+$ cells in human allergic disease. *J Allergy Clin Immunol* 1997, 100, 1–6.
14. Peat, J.K., Woolcock, A.J. Sensitivity to common allergens: relation to respiratory symptoms and bronchial hyper-responsiveness in children from three different climatic areas of Australia. *Clin Exp Allergy* 1991, **21**, 573–81.
15. Chapman, M.D. Cockroach allergens: a common cause of asthma in North American cities. *Insights Allergy* 1993, **8**, 1–8.
16. Blackley, C.H. Experimental researches on the causes and nature of *catarrhus aestivus* (Hay fever or hay asthma), London. Ballière, Tindall, and Cox, 1873.
17. Salter, H.H. *On asthma: its pathology and treatment.* London, Churchill, 1860.
18. Siebers, R.W., Fitzharris, P., Crane, J. Beds, bedrooms and bugs: anything new between the sheets? *Clin Exp Allergy* 1996, **26**, 1225–7.
19. Sporik, R., Holgate, S.T., Platts-Mills, T.A.E., Coggswell, J.J. Exposure to house dust mite allergen (*DerP*$_1$ and the development of asthma in childhood. A prospective study. *N Engl J Med* 1990, **323**, 502–7.
20. Doull, I.J.M. The maternal inheritance of atopy. *Clin Exp Allergy* 1996, **26**, 613–15.
21. Barker, D.J.P. (ed.), Fetal and infant origins of adult disease. Br Med J, 1992.
22. Warner, J.A., Jones, A.C., Miles, E.A., Colwell, B.M., Warner, J.O., Maternofetal interaction and allergy. *Allergy* 1996, **51**, 447–51.
23. Wegmann, T.G., Lin, H., Guilbert, L., Mossman, T.R. Bidirectional cytokine interactions in the maternal-fetal relationship: is successful pregnancy a Th-2 phenomenon? *Immunol Today* 1993, **14**, 353–6.
24. Johnston, S.L., Pattermore, P.K., Sanderson, G., Smith, S., Lampe, F., Josephs, L., Symington, P., O'Toole, S., Myint, S.H., Tyrrell, D.A.J., Holgate, S.T. Community study of role of viral infections in exacerbations of asthma in school children in the community. *Br Med J* 1995, **310**, 1225–9.

25. Strachan, D. Socioeconomic factors and the development of allergy. **Toxicol Letts** 1996, **86**, 199–203.
26. Martinez, F.O. Role of viral infections in the inception of asthma and allergies during childhood: could they be 'protective'? *Thorax* 1994, **49**, 1189–91.
27. Shaheen, S.O., Aaby, P., Hall, A.J., Barker, D.J.P., Heyes, C.B., Sheill, A.W., Goudiaby, A. Measles and atopy in Guinea-Bissau. *Lancet* 1996, **347**, 1792–6.
28. Hopkins, J.M., Enomoto, T., Shimazu, S., Shirakwa, T. Inverse association between tuberculin responses and atopic disorder. *Science*, 1997, **275**, 77–9.
29. Zamel, N., McClean, P.A., Sandell, P.R., Siminovitch, K.A., Slutsky, A.S. and the University of Toronto Genetics of Asthma Research Group. Asthma on Tristan da Cunha: Looking for a genetic link. *Am J Respir Crit Care Med* 1996, **153**m 1902–6.
30. Godfrey, R.C. Asthma and AgE levels in rural and urban communities of the Gambia. *Clin Allergy* 1975, **5**, 201–7.
31. Lynch, N.R., Hagel, I., Perez, M., Di Prisco, M.C., Lopez, R., Alvarez, N. Effect of antihelminitic treatment on the allergic reactivity of children in a tropical *J Allergy Clin Immunol* 1993, **92**, 404–11.
32. Weichmann, H.E. Environment, lifestyle and allergy: the German answer. *Allergology* 1995, **4**, 315–16.
33. Peat, J., Tovey, E., Toelle, B.G., Haby, M.M., Gray, E.J., Mahmil, A., Woolcock, A.J. House dust mite allergens. A major risk factor for childhood asthma in Australia. *Am J Respir Crit Care Med* 1996, **153**, 141–6.
34. Cookson, W.O.C.M., Moffatt, M.F. Asthma: An epidemic in the absence of infection. *Science* 1997, **275**, 41–2.

## STEPHEN T. HOLGATE

Born 1947, received a BSc in Biochemistry in 1968 and completed his medical degree at Charing Cross Medical School in 1971. After completing his training in internal medicine at London Postgraduate hospitals, he moved to Southampton in 1975 where he completed specialist training in respiratory medicine and obtained his MD thesis. In 1980 he completed a 2 year post-doctoral fellowship with Dr K. Frank Austen at Boston, then returned to Southampton where he established his subsequent clinical and research base. He obtained his DSc from the University of Southampton in 1991 on the subject of the Inflammatory Basis of Asthma. He received a personal chair in 1984 and has held an MRC Clinical Research Chair since 1987. He currently runs an interdisciplinary research team of over 50 personnel with research interests in asthma mechanisms. Dr Holgate is a Fellow of the Royal College of Physicians and the Academy of Medical Sciences, a past President of the British Society for Allergy and Clinical Immunology and the 1999 recipient of the King Faisal International Prize for Medicine. He has published over 300 papers, edited a number of textbooks on asthma and allergy and is co-Editor of *Clinical and Experimental Allergy*.

# *El Niño and its significance*

CRISPIN TICKELL

I well remember the first time when I heard the intimidating strike of 9 o'clock in the Royal Institution lecture theatre. In those days the officers of the Society wore white tie and tails, and as now the humbler participants wore black tie or its female equivalent. In strode the palaeontologist Louis Leakey. He wore an open-necked bush shirt. Gazing at the audience like someone who had stumbled into a penguin rookery, he said: 'Animals! That is all you are. Animals!'

We are indeed animals, whether in black tie or bush shirt. Collectively we are threads in an extraordinary tissue of living organisms. This thin blanket covers every part of the earth's surface, itself a combination of rock, water, and air, interacting with, influencing, and being influenced by the life within it. Ed Wilson of Harvard once imagined a journey from the centre of the earth:

> *For the first twelve weeks you travel through furnace-hot rock and magma devoid of life. Three minutes to the surface, 500 metres to go, you encounter the first organisms, bacteria feeding on nutrients that have filtered into the deep-water bearing strata. You breach the surface, and for ten seconds glimpse a dazzling burst of life, tens of thousands of species, plants and animals within a horizontal line of sight. Half a minute later almost all are gone. Two hours later only the faintest traces remain, consisting mainly of people in airliners, who are filled in turn with bacteria.*

You may wonder what this has to do with the perturbation of oceanic and atmospheric behaviour which carries the name of El Niño. In this lecture I want to bring out the smallness and variable conditions in our living space and the enormous effects which even relatively minor and temporary changes can make in it. I want to look not only at the history and science of the El Niño phenomenon, but also at climate change in general and the vulnerability of all species, including our own, to such change, the more so at a time when human activity is seen to be accelerating it.

All ecosystems are sensitive to changes in external circumstances. The closer they are to the limits of what they can adapt to, the more drastic the overall result will be. El Niño events are limited in space and time. But they constitute a marvellous example of what can happen in the event of rapid change.

In the last few months El—or rather the—Niño has entered the currency of household phrases like the ozone hole, or the greenhouse effect. It is a useful piece of shorthand to explain any apparent aberration of the weather, and has acquired a bad name for itself. In the United States today there is a Niño weather radio network which attributes almost any weather to the Niño.

As everyone now knows, it is so called because it happens around the time of the festival of Christmas—the Christ child—along the western coast of South America. It signals the arrival every 4 years or so of warm water from the west which overlies an upwelling current of cool water (first measured by Humboldt in 1802) from the south. It thereby changes weather conditions, first within the region and then in different degrees in other parts of the world. By the following spring the pool of warm water usually disperses, and conditions return to what they were before.

There is little new about the Niño. Living things along the coast must have been aware of it from the earliest times. All societies based on agriculture and water management through irrigation would have had to reckon with it. At sea it causes the virtual disappearance or dispersal of many marine organisms which live near the surface in the cool nutrient-rich waters from the Antarctic. Early written evidence of it in the mid nineteenth century reflected concern about the periodic decline in the droppings—or guano—from sea birds which had by then become a rich source of fertilizer.

In 1891 the President of the Lima Geographical Society drew attention to a change in the pattern of ocean currents which seemed linked to rains in latitudes where rain rarely came. For the desert these were *anos de abundancia*, or years of abundance. In his own words, the effects that year

> *... were so palpable... that large dead alligators and trunks of trees were borne down to Pacasmayo from the north and that the whole temperature of that portion of Peru suffered such a change owing to the hot current that bathed the coast.*

He also drew attention to the name given by local fishermen to the current. He thereby gave the phrase El Niño a new and wider currency.

On the other side of the Pacific, periodic changes in weather conditions had also been observed for a long time. But it was the famine caused by the failure of the Indian monsoon in 1877 which prompted serious efforts to establish the physical mechanisms at work. The search included tracing a possible but, as it turned out, illusory connection with sun spots. But droughts in India and northern Australia, and even temperature fluctuations in south-eastern Africa, south-western Canada and south-eastern United States showed some signs of correlation.

It was the mathematician Sir Gilbert Walker who in 1924, after his retirement from the post of Director General of Observatories in India, identified what he described as the Southern Oscillation, or a see-saw-like variation of atmospheric pressure at sea level at different points across the Pacific. For example when it was low at Darwin in northern Australia, it was usually high in Tahiti, and vice-versa. From these observations was derived the Southern Oscillation Index, which has been elaborated on and is still in use today.

Not until the 1960s was the evidence from the Niño and the Southern Oscillation put together conclusively. Based on information derived from the International Geophysical Year of 1957/58 (by happy coincidence a Niño year), Jacob Bjerknes from the University of California at Los Angeles, in papers of 1966, 1969, and 1972, identified the unitary character of a phenomenon involving large-scale ocean-atmosphere interactions across the Pacific. In the trade it is now known as ENSO (or El Niño Southern Oscillation), but the public, often bewildered by a dense cloud of acronyms, usually prefers to stick to the Niño, and I shall do so tonight. From this has been derived the term La Niña for the more common condition. An alternative term is El Viejo or the old man. Whether such conditions are seen as more female or elderly is a matter of choice.

In one way or another the Niño has probably been sloshing back and forth throughout the Holocene (the 12 000 years or so since the end of the last glacial episode), and possibly for longer still. So far there has been little research into the more distant past, but proxy evidence for it goes back as far as 1525 with consistent irregularity but average frequency of 4 years and intensive events every 10. Recent work on the oxygen isotopic composition of a particular species of coral from the western Indian Ocean over the last 150 years shows a relationship between Niño events and the Indian monsoon. Inter-annual cycles in this coral mesh with Pacific coral and climate records, suggesting a consistent linkage of events across ocean basins in spite of their changing frequency and amplitude.

The Niño must also have some relation to the North Atlantic Oscillation, a more mysterious and longer lived phenomenon with effects on weather on both sides of the Atlantic. Here the complexities are still greater. But all major climatic events affect each other. There are no clear boundaries between them.

What then are the reasons for the Niño? Why does it happen? Until recently, most commentators have described it as anomalous, irregular, or abnormal. Certainly it can vary in character, strength, and timing. But closer scrutiny suggests that it forms part of a natural, regular, and normal oscillation, reaching from the Indian Ocean to the eastern Pacific, operated by understandable physical mechanisms, and modified in its manifestations and impacts by random elements of a chaotic kind. It is difficult to reduce the complexities of such a phenomenon to a few propositions, but at root the story is simple.

The upper layers of the Pacific Ocean in equatorial latitudes are characterized by an imbalance: the surface waters in the west are warm (29–30°C), and those in the east are cool (22–24°C). The gradient between the two is maintained by winds that blow from east to west. The warm water in the west is relatively deep, and, as it warms the air above it, generates intense rainfall. The cool water in the east is the result of upwelling from the south, induced by trade winds that blow surface water towards the warm pool. By contrast it generates relatively little rainfall.

Every few years the regime changes. The trigger is probably unreleased and increasing heat in the warm pool in the west and the air above it. The first sign of change is that the warm winds reverse direction and begin to blow from west to east, thereby pushing warm water eastwards across the Pacific. There it flows over and suppresses the upwelling cool water. This in turn changes the air above it, generating different patterns of rainfall. Sea level in the affected areas rises in a long east-west bulge. Large-scale waves in both ocean and atmosphere transmit the changes far beyond equatorial latitudes, and in different ways and degrees the atmospheric system—the thin film of air around the world—is affected. The Niño has taken over.

The Niño's predominance rarely lasts for more than a year. With the dispersal of the excess heat in the warm pool in the western Pacific, the engine driving the system loses strength. Waters and winds return to the temperature gradient of warmth in the west and cool in the east. The Niña or the Viejo is once more in charge.

Even if the impacts of this see-saw motion, like water moving back and forth in a bath, are not always the same, the main ones are broadly predictable. During a Niño year there are severe droughts in the countries bordering the western Pacific: Indonesia, New Guinea, north-east Australia, and the Philippines, with weakened summer monsoon rainfall over southern Asia generally, including India. In the countries bordering the eastern Pacific it is the reverse: there is heavy rainfall in northern Peru, Ecuador, and Chile with drier spots in southern Peru and Bolivia.

Further away other concurrent changes have been observed. For example north-east South America and southern Africa are drier, and east Africa and the southern United States are wetter. There tend to be fewer Atlantic hurricanes, and more Pacific ones. Such events as volcanic eruptions in low latitudes (El Chichon in 1982 and Pinatubo in 1991) could also effect the working out of the Niño phenomenon.

Such perturbations obviously affect the conditions of life in all its aspects. From micro-organisms through plants and insects to fish and mammals, all have to cope with sudden change. For some it is a disaster, with sharp falls in population density; for others it is an opportunity to be exploited while it lasts; but for most it must be an experience which they are broadly adapted to cope with or at least to recover from. I shall come to the impact of the present Niño in more detail, but three general impacts are worth mentioning now.

Off the coast of South America, primary production at the surface, dependent on such inorganic nutrients as nitrate, phosphate, and silicate, greatly diminishes, and the food chain is damaged if not broken in some places. Pockets of such fish species as anchovy may survive but the natural predators on them, from birds and sea lions to humans, find their catches drastically reduced. By contrast populations of scallop and shrimp seem to increase in the warmth, and other warm water fish—for example skipjack tuna—move in. They move out again with the return of the Niña.

Droughts in the Indonesian archipelago and neighbouring countries subject forests and plants to extreme stress, and forest fires, with destruction of peat as well as mature trees, affect all parts of the food chain. The animals that live in the leaf litter of the forest floor (insects, reptiles, amphibians, and so on), those in the trees and canopy (birds, and such primates as orang-utans), and those in mountain or savanna, decline sharply in numbers. Obviously, the speed of recovery depends on the intensity and distribution of deluge or drought, but renewal, when it begins, may (as elsewhere) serve eventually to revitalize the landscape, even if some of the losers are lost for ever.

A particularly interesting feature of Niño events is change in the distribution of micro-organisms, and thus in the susceptibility of such creatures as ourselves to disease. Floods bring opportunities for the vectors of such diseases as malaria, dengue and yellow fever, encephalitis, and schistosomiasis, and for the agents of such diseases as hepatitis, dysentery, typhoid, and cholera. Recent research has suggested mechanisms for the well attested spread of cholera in South America during Niño events. Warm or brackish water combines with run-off from the land to produce local plankton blooms in which the cholera bacillus flourishes. The bacilli are ingested by copepods of all kinds, and in poor sanitary conditions are soon conveyed to humans. Similar conditions have produced similar results in other parts of the world.

Of recent Niño events, the present one is probably the most serious yet recorded. But the 1982/83 event ran it fairly close. I saw some of it for myself. For nearly everyone it came as a surprise, and we only knew it was happening after it had begun. I had a privileged ride in *Britannia* from Acapulco to La Paz in Baja California along the western coast of Mexico. No bluer skies or more sparkling seas could have been imagined. But when the Queen went further north into US waters, she was greeted with torrential rain. President Reagan's helicopter could not land on *Britannia* and a royal visit was condemned to yellow oil skins throughout. A weaker Nino took place between 1986 and 1987, and a still weaker but prolonged one between 1991 and 1994.

1997 was special. By that time some of the lessons of previous Niños had been learnt. Comprehensive ocean–atmosphere models were in place. The results of an international Tropical Ocean Global Atmosphere (TOGA) research programme were also available, together with a network of buoys monitoring ocean and atmospheric conditions across the tropical Pacific. Thus an on-coming Niño was confidently predicted from late 1996 onwards.

By the spring of 1997 the sea surface temperature in the eastern Pacific and subtropical areas to the north had risen, with strong east-blowing winds from the western Pacific adding to the effect, and by last October the sea surface temperature was as high as 5°C above the average. Subsurface temperatures rose even more sharply, and in December it was 9°C above the average. The bulge in sea levels could even be seen from space. If previous patterns are followed, the warm water should soon begin to disperse, and there are predictions that the opposite—Niña—phenomenon will develop later this year.

The consequences reach far and wide. As always they are affected by other events, and the Niño label on them must be attached with caution. Even so the direct effects are impressive: typically increased rainfall in some areas, equally typical droughts in others but in most cases pushed to unwelcome extremes. Rainfall in Ecuador and northern Peru has reached record levels, and I believe the Ecuadorian economy is in a state of collapse. Elsewhere, through so-called teleconnections, events have followed familiar paths with a high measure of local variability.

Thus the Indian monsoon was weak but sufficient. Central China, South-east Asia and west Africa have endured droughts and southern Africa is having a poor rainy season. There have been deluges in east Africa, California, Florida, the southern parts of the United States, and southern Brazil, northern Argentina, and central Chile. Whether the recent devastating ice storms in eastern Canada and the warm wet January weather in western Europe are also connected is a matter of debate. Certainly the European weather was predicted by the European Centre for Medium Range Forecasts at Reading as long ago as early December.

As usual ecosystems have been placed under heavy strain. A particularly sad case is what has happened in the Galapagos Islands, that treasure house of evolutionary distinctness. Although on the equator, the islands are usually kept cool and dry by the Humboldt current, and many species found nowhere else are adapted to these special conditions: green algae off shore, marine iguanas, tortoises, and a wide range of birds, including flightless cormorants, Galapagos penguins, finches, and blue-footed boobies. As on previous occasions, water temperatures have risen, and there has been heavy rainfall. Corals have been bleached, here as elsewhere. In the past the indigenous species have survived even big population crashes. Now the future is less certain. The introduction by humans of alien organisms, from fire ants, flies, and other insects to rats, cats, and goats has compounded the Niño effect, and Galapagos ecology, already under threat, may never be the same again.

The costs of the present Niño to human affairs are literally immeasurable. In some cases the Niño was the predominant effect. In others it mixed in with other phenomena. In yet others it had a minor role. A good example are the recent forest fires in Indonesia in which land clearance practices combined with Niño-induced drought to cause profound and lasting destruction of ecosystems. A pall of ash and smoke overhung South-east Asia with widespread effects on human life and economic activity.

After the 1982/83 Niño efforts were made to assess overall costs: in terms of human life rather than livelihood there were over a thousand deaths; and in financial terms the price was between $8 and $9 billion. But these figures are no more than guesses. We have only to look at the range of human activity affected: habitats generally, water supplies, fisheries, agriculture (wheat, maize, cocoa, coffee, tea, and livestock), forestry, banking, insurance, even the operation of the Panama Canal, and, as we have seen, the character of the micro-organisms which prey upon us.

In such circumstances and in view of the chaotic element in regional impacts, it may be questioned if detailed prediction is really worthwhile. I suppose that we could simply sit back and let it all happen. But Niño risks, like other risks, can be calculated. The impact on water management, agriculture, forestry, and economic activity may vary in severity, but are much the same each time in the main areas under threat. This time the world had ample notice of what was likely to happen. The risks at the two ends of the geographical see-saw are pretty high. Others further afield, as in Africa, are lower and more chancy. But in most cases warnings could be given and precautions of a kind could be taken.

The UN Food and Agriculture Organization drew attention to the threat to world food supplies as long ago as last September, and commodity markets have reflected the uncertainty. The World Bank also issued early warnings. On 19 December the UN General Assembly adopted a resolution calling for 'an internationally concerted and comprehensive strategy towards the integration of the prevention, mitigation and rehabilitation of the damages caused by the El Niño phenomenon.'

I am afraid that the prose style has not improved since I left New York in 1990, and I suspect that this was a veiled request for more economic aid. Even so the costs will be high. When the Niño has finally given way to the Niña, I think it will be more important to assess the value of the warnings given and the precautions taken than to worry about inevitably artificial balance sheets.

Even if some Niños are stronger and longer lasting than others, they are no more than temporary manifestations of an underlying oscillation, and previous conditions will eventually return, or have always done so up to now. They are part of a global climatic system which is itself subject to constant change. I look now at that wider system, and the significance of the Niño within it.

Most climatic change is slow by human standards. Climate seems to have ridden a slow roller coaster for the past 60 million years, before entering a long undulating downward slide towards the ice age world of the last two and a half million years. Within the ice ages there has been a broad rhythm, with over 20 glacial periods interspersed with 10 000–15 000 year interglacials like the present. 125 000 years ago there were hippopotamuses in Trafalgar Square. Eighteen thousand years ago tundra in the same place was trodden by mammoth, woolly rhinoceros, and reindeer. Even in the last relatively stable 10 000 years there have been important wobbles, with farming in Greenland in the tenth century, vineyards in Salisbury plain in the twelfth century, and ice fairs on the River Thames in the seventeenth.

The reasons for these variations, large or small, are still not fully understood, but most see a combination of three main factors. There is the impact of changes in the earth's orbital relationship with the sun known as the Milankovitch effect: eccentricity of orbit (100 000 years), tilt (41 000 years), and wobble like the spin of a top (21 000 years). Secondly, there is the particular arrangement of land and sea produced by the continuing movement of tectonic plates. Last there is the rise and erosion of mountains: for example, the rise of the Himalayas in Miocene times probably changed weather patterns by creating deserts, intensifying the monsoon, and possibly helping to cool the earth by drawing carbon dioxide out of the atmosphere through soil weathering.

Other factors should also be taken into account: variations in solar radiation (there was a remarkable absence of sunspots—the so-called Maunder minimum—during the cold spell of the sixteenth and seventeenth centuries), and the effects of volcanic materials which reach up into the stratosphere (the explosion of Mount Toba in Indonesia some 75 000 years ago may have been a trigger for renewed glaciation).

Until recently it was generally believed that all climatic change was slow. Now we are less sure. The evidence of cores drilled through several parts of the Greenland and Antarctic ice caps suggests a series of cold and warm spells that could have raised or lowered the average winter temperature in northern Europe by as much as 10°C over the course of only a few years.

A prime example is the Younger Dryas event, a cold episode of about 800 years after the end of the last glaciation when a rapid cooling of surface waters in the north Atlantic led to a southward re-advance of the ice and a return to glacial conditions in western Europe. It may have been triggered by the discharge of large quantities of melt water from a glacial lake in the area of the Great Lakes into the Atlantic through the Gulf of St Lawrence. Icebergs floating southwards from the melting Greenland ice cap would have had the same effect.

More important it was also associated with a sudden change in the pattern of ocean circulation. In the Atlantic, the upper waters of the ocean, warmed in the tropics, flow north-eastwards to the vicinity of Greenland where the Arctic air cools them. Anyway they are saltier and thus denser than those in the Pacific, and sink before beginning the return journey southwards at depth. In turn they draw the warmer surface water northwards. The so-called Atlantic Conveyor is the engine of a global circulation system of immense power. The flow of the Gulf Stream north eastwards across the Atlantic is roughly equal to that of a hundred river Amazons, and results in an enormous transport of heat. Without it winter temperatures in western Europe would fall by more than 5°C. Liverpool or Dublin would have the climate of Spitzbergen.

This system is critically vulnerable to changes in temperature and chemistry at the surface. It faltered, changed direction, or shut down many times during the oscillations or flickerings associated with the ice ages. There could have been as many as two dozen such between 100 000 and 20 000 years ago. There are suggestions that shutting off the flow of warm water into the north Atlantic may increase the flow of warm water into the south Atlantic so that there is an alternation in glacial conditions in the Arctic and Antarctic.

How long it may take for the climate to jump from one mode of operation to another is still unclear. Current chronometric methods are not sufficiently precise to yield a firm result. But changes could certainly have been within a human lifetime, if not within a decade. No wonder that we need to watch the behaviour of ocean currents south of Greenland. No wonder also that we should watch the continuing build up of greenhouse gases which could, as part of a process of global warming, set in motion the melting of the Arctic ice cap, and thereafter another reorganization of the ocean circulation and the weather patterns which depend upon it. The Niño oscillation arising in the Pacific is a matter of months. A major oscillation arising in the Atlantic could be a matter of decades, centuries, or thousands of years.

There is now a new factor at work. Environmental change is accelerating because the human foot is on the accelerator. A periodical visitor from outer space would find more change in the surface of the earth in the last 20 years than he would have found in the last 200, and in the last 200 more than in the last 12 000.

There are five main aspects. Like any other animal species on a bonanza, the human species has multiplied its numbers at a giddy-making rate. There were perhaps about 10 million of us around the end of the last ice age. The introduction of agriculture, the specialization of human function and the growth of cities caused rapid proliferation. By the time of Thomas Malthus, when the industrial revolution had barely started, our numbers stood at about 1 billion. By 1930 they had risen to 2 billion. There are now almost 6 billion, and by 2025, short of some catastrophe, there will be 8.5 billion. At present there are about 90 million new humans every year. Since the Rio Conference on Environment and Development in 1992, 450 million new people (more than the whole world population at the time of the Roman Empire) have come to inhabit the earth.

Partly as a result in this dizzy increase in our numbers, we have done lasting damage to the earth's green covering. The 1993/94 UN Environmental Data Report showed that 17% of the world's soils had been damaged to a greater or lesser extent since 1945. Over the last century the effects of industrialization have become a global rather than a local problem. Even within the enormous land mass of the former Soviet Union, some 16% was judged an ecological disaster by the Academy of Sciences in 1995. Waste disposal may soon become a bigger problem than consumption of resources. No part of the world is exempt from the waste produced by human activity.

We have polluted both salt and fresh water. The oceans may seem vast, but a sea lion in the Antarctic which has never seen a human being will probably carry human-made chemicals in his blubber. Coastal areas are particularly at risk from the toxic materials brought down by rivers to the sea. There is widespread pollution of rivers and underground aquifers. At the same time demand for fresh water has doubled every 21 years while supply is the same as it has been for thousands of years.

We have depleted the diversity of life. Mass extinctions have occurred before in the history of life, most famously at the end of the Cretaceous period 65 million years ago when the long dominance of the dinosaur family came to an end. What is happening today is comparable. Current rates of extinction could be up to 1000 times what they would be under natural conditions. It is a crisis with two aspects: mass extinctions of species and their habitats, and gross depletion of genetic variability within species as a large proportion of their numbers is wiped out. When the archaeologists of the future look at the deposits of the last quarter millennium, they will find a biological discontinuity as big as any in the past. They will expose a richness not of fossils but of plastic bags, discarded products, and, if they are unlucky, radioactive or other toxic waste.

Finally, there are the changes we have brought about in the chemistry of the atmosphere. Acid precipitation is a problem for those down wind of industry, but it is manageable if the political will is there to solve it. Depletion of the stratospheric ozone layer is more serious. Damage to the human metabolism may seem alarming to us, but a more fundamental problem could be the effects on other organisms, including phytoplankton at the base of the food chain.

Then there is climate. Since the industrial revolution, we have been using the sky as a waste unit by enhancing the natural greenhouse effect with emissions of carbon dioxide, methane, nitrous oxide, chlorofluorocarbons, and related molecules into the atmosphere. Apart from water vapour, carbon dioxide accounts for the largest proportion of greenhouse gases, and we know from ice cores that during the last ice age its concentration in the atmosphere was on average between 180 and 210 parts per million . The interglacial average was 280 parts per million, the level before the beginning of the industrial revolution. It is now over 360 parts per million and rising. According to the most recent models, its effects on warming are mitigated by aerosols, or particulate matter, mostly originating in industrial activity in the northern hemisphere.

There has been controversy about the consequences of the increases in atmospheric carbon dioxide but more about the degree of change than about change itself. In its most recent assessment, the Intergovernmental Panel on Climate Change suggested rises in average global temperature of between 1°C and 3.5°C by the end of the next century (an average rate of warming greater than any in the last 10 000 years), and an average rise in sea level of up to half a metre in the same period (an average rate three to six times faster than that of the last 100 years). It concluded that '......the balance of evidence suggests a discernible human influence on global climate'.

Although confidence in the modelling has increased, there remain many uncertainties which make it difficult to quantify the risks involved, or the regions most likely to be affected. But in general terms there is likely to be more precipitation, and more extreme and irregular rainfall, particularly in areas subject to the monsoon. Areas that are already wet are likely to get wetter, and arid regions may see more prolonged and severe droughts. Whatever the human contribution to global climate change, the world is at present becoming warmer, and 1997 was the warmest year on record. The figures speak for themselves.

Inevitably there have been suggestions that the frequency and intensity of Niño events may be linked to global warming. There has been work on the subject at the National Center for Atmospheric Research in Boulder, Colorado, and at the Commonwealth Scientific and Industrial Research Organization (CSIRO) in Australia. The evidence is inconclusive. But the Boulder team concluded:

> *... that both the recent trend for more ENSO events since 1976 and the prolonged 1990–95 ENSO event are unexpected given the previous record, with a probability of occurrence of about once in 2,000 years. This opens up the possibility that the ENSO changes may be partly caused by the observed increase in greenhouse gases.*

Even the possibility is significant. It reminds us of the numberless interconnections which characterize the world climate system. The point was well made recently by Professor Fairbanks of the Lamont-Doherty Earth Observatory at Columbia University in New York (and quoted in a perceptive article by Richard Lloyd-Parry in the *Independent on Sunday* on 7 December):

> *... The best analogy is to the cogs in a watch, all different sizes, turning at different speeds, some of them directly connected, others not. People are distracted by the predictions of gentle, long term global warming, a few degrees spread over centuries. But small perturbations in the climate can lead to large consequences, and they do not necessarily have to be gradual changes.*

As I remarked at the beginning, we are all animals, adapted like the ecosystems of which we are part to relatively stable living conditions. Changes in those conditions whether abrupt or gradual require changes in us. One response is to move to where conditions, whether warmer or cooler, are more congenial. When there were fewer people in the world, that was always an option. But with the steep rise in human population and the other trends I have described, that option is no longer open to us. One of the forecasts for the next century is a strong increase in the number of refugees and of their accompanying pathogens, generated by changing conditions of temperature and moisture.

Looking back into the past we should not forget that of the 30 or more urban societies which have ever existed none but our own has survived. The reasons may be various. Population pressure, degradation of the resource base in its different aspects, and environmental change, whether natural or human-induced, all played a part. For a society living at its ecological limits, any change is perilous, and represents a challenge which most cannot meet. The Niño is a not—so—gentle reminder of our vulnerability. It should make us look again at the actions we are taking, most of them unwitting, which cumulatively and in ways beyond current knowledge are changing the life system on the surface of the planet, and increasing our own vulnerability within it.

Public awareness of the perils has greatly increased over the last quarter century. There are several international landmarks: from the publication of the first report of the Club of Rome in 1972 and the UN Conference on the Environment in Stockholm in the same year, to the UN Conference on Environment and Development at Rio 20 years later, and a succession of other UN conferences on population, human settlements and food.

On issues of climate, the Intergovernmental Panel on Climate Change was set up after the report of the Brundtland Commission on Environment and Development in 1987, and has since produced two major assessments. Another is in preparation. In this process science has come into its own as never before. We have a Framework Convention on Climate Change, and there have been three meetings of the parties to it, most recently at Kyoto last December. Another is due at Buenos Aires at the end of 1998.

Whatever the reluctance of governments to do anything about greenhouse gas emissions, and the controversy surrounding responsibility about who should do what, at least the problems have been recognized, international mechanisms for coping with them have, however tentatively, been set in place, and people the world over are beginning to think differently about what faces them in the next century.

Here is the rub. There is nothing more difficult than thinking differently. New ideas, paradigm shifts and new models of behaviour are usually painful. They need time to sink in. As Lord Keynes once remarked, '...the difficulty lies not in new ideas but in escaping from old ones.' There is a powerful inertia built not only into our mental habits but also into the management of our society, which is after all the only one we know. We often use a wrong-headed vocabulary and wrong-headed systems of measurement. Our educational system is incomplete, especially at higher levels. We cherish out of date values, and look on the world in a resolutely short-term perspective. Perhaps it is our genes, adapted to a brief life span, which are partly to blame. As has been well said, we are Stone-Agers displaced in time.

Yet we can and do change. If we could recall our intellectual frame of reference of even 10 years ago, we would be startled by the difference between then and now. How much greater is the difference between generations, and between generations over centuries. I remember a conversation I once had with a monk on Mount Athos. For him the world had begun 4000 or so years ago, and the hands of God and the Devil could be seen manipulating the events of daily life. After a few moments the talk between us on all but trivialities became meaningless. Yet I suspect that, for many of us, bits of that monk's frame of things lie uneasily juxtaposed with those of Darwin, Freud, or Marx: or last week's book; or yesterday's television programme; or today's press headlines.

How then does change come about? I suppose there are three main mechanisms. First is leadership from above. Intellectual conviction joined to persuasive exposition joined to political weight can divert the course of history. We can all think of examples for good or ill.

Second is pressure from below, at least in democratically organized countries. In matters of the environment, popular feeling, expressed through the media, non-governmental organizations, and the electoral process, particularly at local level, is a potent force for change. It may not always be right, and is sometimes misdirected. But it is nearly always a corrective to the conventional wisdom and, as in other matters, can correct itself if it gets thing wrong. Look at the positive outcome of the row over the Brent Spar.

Last is the role of catastrophes. Too often they have done useful service by jerking us out of dangerous inertia. As Johnson said: 'Depend upon it, Sir, when a man knows he is to be hanged in a fortnight it concentrates his mind wonderfully.'

Again we can all think of examples. The Niño stands out as a recurring illustration of what can happen if we come to depend overmuch on one set of circumstances. I doubt if the United States would have changed position on climate change, in time for the Second World Climate Conference in 1990, without the crippling drought that struck the Middle West in 1988. But if there has to be a catastrophe, let it be big enough, but not too big; small enough, but not too small; quick enough but not too quick; slow enough but not too slow; and preferably affecting no one reading this today.

I sometimes wonder what would happen if we, like nearly all species that have ever lived, were simply to be eliminated. Consider a major asteroid impact. If we perished more or less together—say over five years rather than 50 million years—what would become of the earth? How long would it take for the injury to heal, for our cities to fall apart, for the earth to regenerate, for the animals and plants we have chosen for ourselves to find themselves a more normal place in nature, for the waters and the seas to become clean, for the chemistry of the air to return to what it was before we disturbed it? Niños and Niñas would come and go, the ice would stretch back and forth from the poles, and living organisms would adapt or transform themselves as they have for more than three and a half billion years. Whatever may happen, we are no more than a tiny if precious episode in the vast procession of life. Let us cherish it while we have the chance.

## Background references

1. S.G. Philander. *El Niño, La Niña and the Southern Oscillation.* Academic Press.
2. E.A. Laws. *El Niño and the Peruvian anchovy fishery:* University Science Books, 1997.
3. A. Rob. *El Niño Southern Oscillation and climate variability.* CSIRO Publishing, 1996.
4. J. Webster and T.N. Palmer. *The past and the future El Niño. Nature,* 11 December 1997.
5. M.H. Glantz: *Climates of changes: El Niño's impact on climate and society.* Cambridge University Press, 1997.
6. R.R. Colwell. *Global climate and infectious diseases: the Cholera paradigm. Science,* 20 December 1996.
7. W.S. Broeker. *Thermohaline circulation, the Achilles heel of our climate system: will man-made* $CO_2$ upset the climate balance? *Science,* 28 November 1997.
8 C.D. Charles, D.E. Hunter and R.G. Fairbanks. *Interactions between the ENSO and the Asian monsoon in a coral record of tropical climate. Science,* 15 August 1997.
9. P. Lehodey, M. Bertingnac, J. Hampton, A. Laws, and J. Picaut. *El Niño Southern Oscillation and tuna in the western Pacific. Science,* 16 October 1997.

10. R.B. Alley and M.L. Bender. *Greenland ice cores: frozen in time. Scientific American*, February 1998.
11. UCAR. *El Niño and global warming: what's the connection?. UCAR Quarterly* Winter 1997.

## Acknowledgements

My particular thanks are due to Dr Michael Davey of the Hadley Centre for Climatic Prediction at the Meteorological Office at Bracknell.

SIR CRISPIN TICKELL

Sir Crispin Tickell is Chancellor of the University of Kent at Canterbury; Chairman of the Climate Institute of Washington, DC; Director of the Green College Centre for Environmental Policy and Understanding; Chairman of the Government's Advisory Committee on the Darwin Initiative; and Convenor of the Government Panel on Sustainable Development. He was a member of the Diplomatic Service, his final appointment being British Permanent Representative to the United Nations (1987–90). He was also President of the Royal Geographical Society (1990–93), Warden of Green College, Oxford (1990–97), and Chairman of the International Institute for Environment and Development (1990–94). He is the author of Climatic Change and World Affairs (1977 and 1986), and Mary Anning of Lyme Regis (1996). He has contributed to many other books on environmental issues including human population increase and biodiversity.

# *Lasers—making light work*

COLIN WEBB

## Introduction

It is now over 35 years since the first laser was demonstrated. Since that time, lasers have found applications in fields as diverse as eye surgery, telecommunications, data storage, target designation, cutting, and welding, as well as providing entertainment at rock concerts. They have established themselves as indispensable laboratory tools for the research chemist and physicist. Even in the home the familiar CD player, which uses a laser to read the information stored on a compact disc, provides evidence of the maturity of this very elegant technology. On a grander scale, lasers can be used to create artificial stars in the night sky. These beacons enable the images captured by large astronomical telescopes to be corrected for the effects of blurring due to the Earth's atmosphere so that photographs of a quality previously only available from the Hubble space telescope can now be made from telescopes on the ground.

As someone involved in the development and applications of lasers from the very beginning, I will attempt to trace the evolution of the current range of laser devices employed for these tasks. In the early days they were called 'optical masers' rather than lasers. In 1960 I was the first student at Oxford to be engaged on optical maser research—in fact the world population of graduate students in this field (including me) numbered just two.

I was asked by the other graduate students who started on the same day in October 1960, 'What use would this optical maser thing be used for, even if it works?' I remember mumbling something about 'Well perhaps they might be used for communications, or for wavelength standards or metrology.' Actually all of these have indeed become real applications for lasers but I couldn't have imagined then the range of uses to which these very versatile devices have been put in the 36 years since then.

## Einstein's theory of radiation

To understand why there was excitement in 1960 about the prospects for making an optical maser it is necessary to retrace the history of the subject up to that time. It all starts with a famous paper in 1916 by Einstein. He pointed out that there was something wrong with the then existing theory of the way radiation interacts with atoms or molecules. There was something definitely missing—the process of stimulated emission. It was known pre-1916 that if you raise the energy of one of the outer electrons of an atom—that is, put the atom into one of its allowed excited energy states—it will after a time revert to a lower energy state, giving out a photon or quantum of radiation in the process. Just like radioactive decay of nuclei the process is totally spontaneous.

As shown in Fig. 1 the atom has a natural tendency to lose energy by making transitions from a state of high energy $E_2$ to a state of lower energy $E_1$ and in doing so it gives out a photon of radiation in the form of light whose frequency $\nu_{12}$ is proportional to the energy gap between $E_2$ and $E_1$ with a constant of proportionality known as Planck's constant $h$ such that

$$E_2 - E_1 = h\nu_{12}$$

Now it was also known prior to 1916 that if you took an atom initially in the lower state $E_1$ and shine on to it radiation of exactly the right frequency $\nu_{12}$ (so that the photons have exactly the right energy to match the energy gap $E_2 - E_1$ between the two states) you can excite the atom from $E_1$ up to $E_2$.

As shown in Fig. 2, after such an absorption event the photon has disappeared, the atom has taken up the photon's energy as internal energy. The process of absorption will only occur if there is resonance between the radiation frequency and the energy gap of the atom.

What Einstein pointed out was that there should be a third process in which radiation could interact with atoms. As shown in Fig. 3 this process—stimulated emission—is one in which you start off with an atom already in an the excited state, and now shine on a beam of radiation containing photons of the characteristic frequency $\nu_{12}$. These shake up the atom in a resonant way such as to make it give up another photon exactly identical to the one you put in. As a result we have two photons coming out. Stimulated emission is an amplification process because if you take those two photons coming out they go on further and each one of those can stimulate further photons to make a cascade.

In his classic paper of 1916 Einstein worked out the relative probabilities of all these processes to happen. He showed that the probability

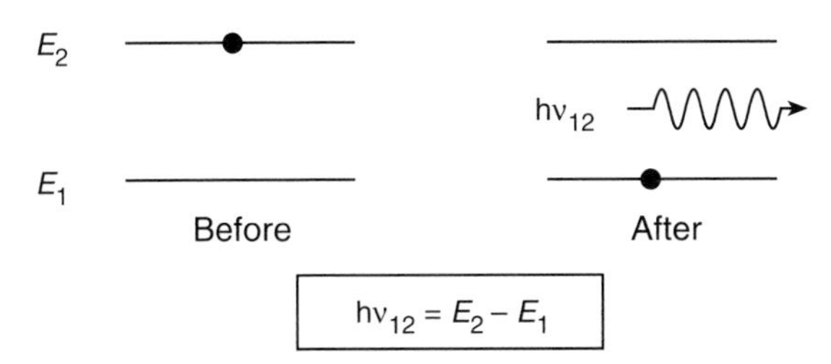


**Fig. 1** Random process. Rate does not depend on number of photons already present.

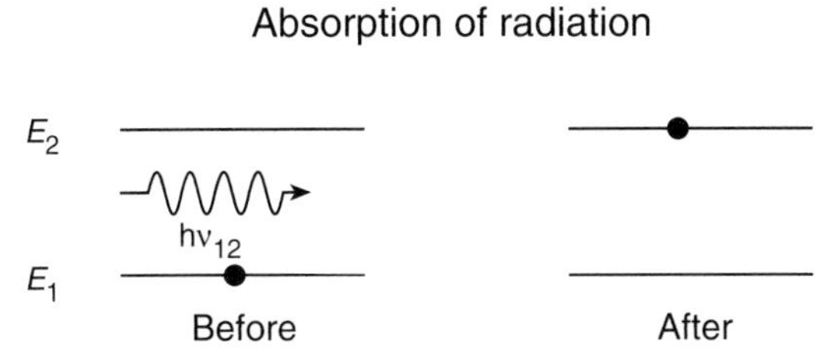


**Fig. 2** Rate depends on flux of $h\nu_{12}$ photons and on number of atoms in $E_1$ (capable of undergoing absorption).

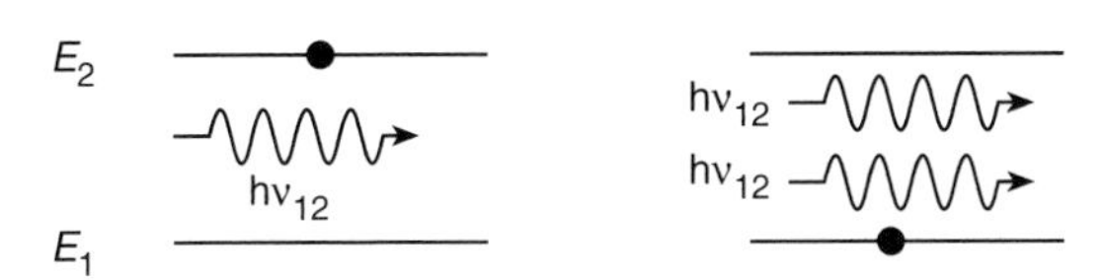


**Fig. 3** Rate depends on flux of $h\nu_{12}$ photons and on number of atoms in $E_2$ (i.e. number of atoms capable of undergoing stimulated emission).

per $E_2$ atom to undergo stimulated emission is exactly equal to the probability of an atom of the same kind in $E_1$ to undergo an absorption event in a given beam of radiation. Herein lies a snag if you want to make an amplifier using the process of stimulated emission, because absorption and stimulated emission go on at the same time and compete with one another—one trying to add photons to a beam of radiation and the other trying to remove them.

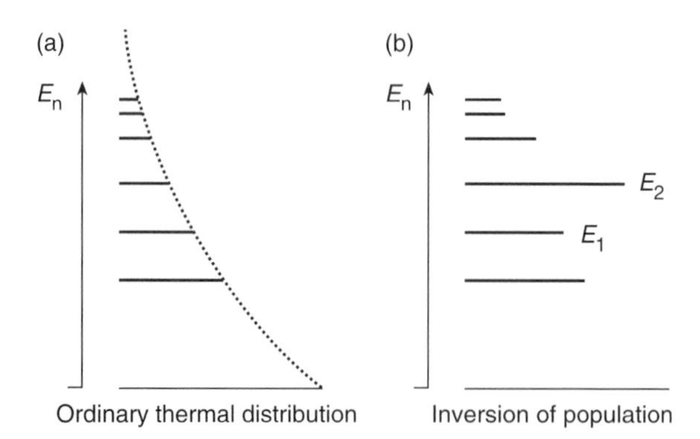


**Fig. 4** Condition for amplification.

Figure 4(a) shows what happens in ordinary situations if you put atoms in a box and allow them to come to thermal equilibrium. The various energy states of the atom are shown as a ladder of (not necessarily equally spaced) bars with the number of atoms occupying each energy state being indicated by the horizontal length of the bar. The atoms distribute themselves across the available energy states according to Boltzmann distribution formula, which simply implies the higher the energy state, the fewer the number of atoms which occupy it. So, for any pair of energy states there are always fewer in the upper state than in the lower state of the pair. That means that, as a beam of photons goes through the box, more absorptions will occur than stimulated emissions and so the net effect is overall absorption.

What took so long was to learn how to do was to somehow bring about the situation shown in Fig. 4(b) in which, for pair of energy states, there are more atoms in the upper energy state $E_2$ than in the lower energy state $E_1$. When such a situation exists—a population inversion—then you have the possibility of making an amplifier.

## The MASER

In 1954 Gordon, Zeiger and Townes invented the MASER, a device for amplifying radiation in the microwave band (a band of frequencies used for radar and microwave cookers) by creating an inversion of population in a stream of ammonia molecules. By passing them through an inhomogeneous magnetic field the stream of molecules was separated into two—those in the excited state and those not in the excited state. By collecting those in the excited state in a box whose sides were exactly a small number of half-wavelengths long they created inside a resonant structure (a resonant cavity) an inversion of population and in this way they made an amplifier for microwaves of the correct frequency. The device was called a MASER, where the letters of the acronym stand for Microwave Amplifier using Stimulated Emission of Radiation.

## The LASER concept

Throughout the later 1950s scientists were arguing whether it would be possible to extend the MASER idea upwards in frequency into the visible region to make an optical MASER. One major objection raised against this possibility comes from the fact that the microwave cavity used in the MASER is a box with sides only a few half-wavelengths long. If you tried to make an *optical* cavity only a few half-wavelengths long on each side it would be so small that it would be difficult to put much power into it. Further, if the cavity were that small the atoms would collide with the walls very frequently which would simply de-excite them wastefully. In any case, the objectors asked, how could the atoms be put up into the excited state fast enough to compete with the rate at which they fall down naturally by the process of spontaneous emission which in the visible region goes much faster than in the microwave region? But in a famous paper in 1958 Schawlow and Townes, pointed to ways around these objections and predicted that not only should it be possible to build an optical MASER but that it would have very interesting properties. They showed that the resonant cavity could be tens of thousands of wavelengths long. It could be made with two highly reflecting mirrors facing one another, just so long as they were exactly an integral number (possibly several million) of half wavelengths apart. Neither did the cavity have to be only a few wavelengths wide, in fact it didn't even need reflecting side walls at all.

The race was then on to demonstrate optical maser action by putting gas or solid media inside optical cavities of this type and exciting the atoms by running a discharge in the gas or by subjecting the solid to the intense light of a flashlamp. If by some such means it would be possible to create an inversion in the medium then somewhere in the medium an atom would sooner or later spontaneously emit a photon travelling along the length of the cavity as indicated diagrammatically in Fig. 5(a). As the photon goes through medium it will cause an avalanche of new photons which will be travelling towards one end mirror of the cavity. Most of the beam of such photons will be reflected at the mirror and double back on itself, traversing the length of the cavity again, and causing further

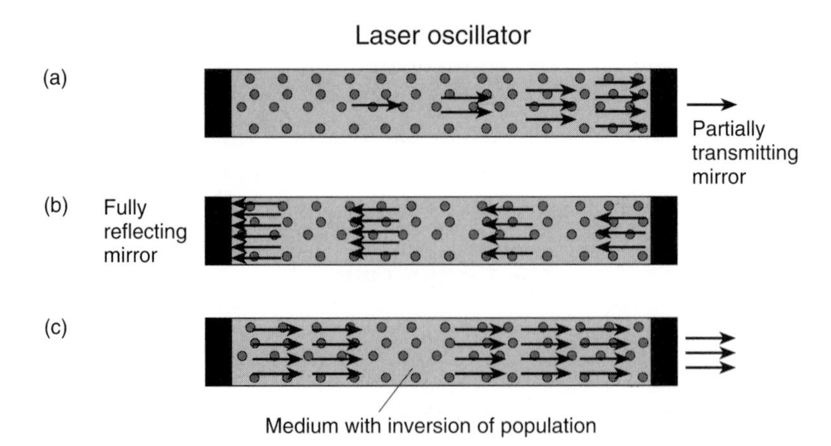


**Fig. 5** The build-up of laser oscillation within the laser cavity.

photons to join in the cascade as in Fig. 5(b). The beam as it bounces back and forth between the two end mirrors of the cavity will build up a standing wave of light and make an optical MASER. Some of this energy is allowed to pass through into the outside world by the partially transmitting coating of the mirror at the output end to form the laser output beam as indicated in Fig. 5(c).

As the device we are now talking about operates with light rather than microwaves it didn't take long before the clumsy description 'Optical MASER' was replaced with the simpler acronym LASER. The initials stand for Light Amplification by Stimulated Emission of Radiation. Actually, as I have described it, the device is not so much an amplifier as an oscillator and so really it ought to be called LOSER, but nobody particularly wanted to invest in that concept.

## The ruby laser

In May 1960 came the first report of successful laser operation. Maiman of Hughes Research Laboratories in Malibu California, observed optical maser oscillation in a rod of pink ruby. Ruby is a crystalline material much studied by microwave spectroscopists. The active ingredients are chromium ions $Cr^{+++}$ embedded in a lattice of aluminium oxide $Al_2O_3$. Maiman's laser, shown in Fig. 6 employed a high power flash lamp—rather like a photographic flash lamp—coiled in helix around the ruby rod to excite the chromium ions from their unexcited state (the ground state) up to two broad bands of energy states labelled $^4F_1$ and $^4F_2$ in Fig. 7.

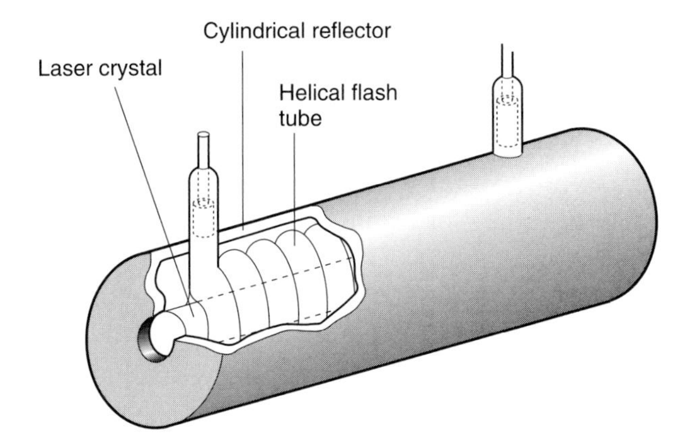


**Fig. 6** The first laser: schematic.

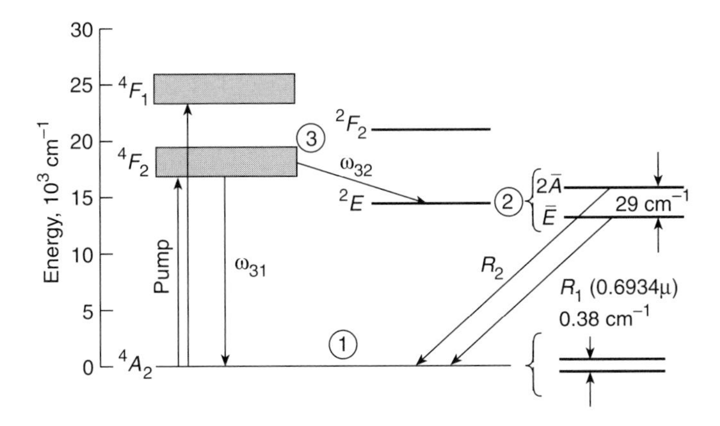


**Fig. 7** Energy levels of $Cr^{+++}$ ions in the ruby crystal.

The $Cr^{+++}$ ions quickly decay to an intermediate energy state (labelled $^2E$ in Fig. 7) and as the excitation continues eventually enough of them are transferred out of the ground state and into the intermediate energy level $^2E$ where they pile up until eventually there are more of them in the intermediate state $^2E$ than are left in the ground state labelled $^4A_2$. This works because the absorption bands from the ground state up to $^4F_1$ and $^4F_2$ are relatively broad and a flashlamp emits light of a broad range of wavelengths (i.e. white light) so that a good proportion of the flashlamp's output is usefully absorbed. Once $Cr^{+++}$ ions in the ruby rod reach inversion the laser goes off in oscillation on the red (694.3 nm) transition connecting $^2E$ back to the ground state. In Maiman's original laser the cavity was formed by plating reflective coatings directly on to the ends of the ruby rod, which was polished with its flat ends accurately parallel to one another, although later versions employed mirrors separate from the ruby rod.

Ruby lasers are still around—they have proved to be an amazingly enduring variety of laser—for example they are still used for removing tattoos and for making holograms. However, to go back to the 1960s, immediately the laser came along the military took an early interest because it was felt that one day this technology could lead to development of a battlefield weapon—the sort of death ray that had features in science fiction since HG Wells or before. There was even a comic strip in the early days called *The Adventures of Ruby Laser* featuring the adventures of a wonder-woman character equipped with this death ray technology.

## Modern solid state lasers

In many ways the ruby laser represents something of a dinosaur in the history of the laser because in order to make it work you have to get half the total number of $Cr^{+++}$ ions in the entire ruby crystal out of the ground state and into the upper laser state $^2E$.

A refinement on the ruby idea which came along slightly later was the Nd : YAG laser. The active species is triply ionized neodymium $Nd^{+++}$ embedded in a host lattice of yttrium aluminium garnet (YAG). This type of solid-state laser involves four rather than three energy states. As shown in Fig. 8, the laser transition at 1064 nm is between an excited state $E_2$ but the lower state $E_1$ is not the ground state $E_0$, but one sufficiently far in energy above the ground state that it is effectively not populated at all at room temperature. Like ruby, we have to pump the atoms from the ground state to a high lying broad energy band $E_3$ from which they relax to the intermediate energy $E_2$ state which forms the upper state of the laser transition but, unlike the case of ruby, the transition out of this state does not lead back to the ground state but to the lower lying intermediate level $E_1$. In order to get an inversion we no longer have to get more than the half the number of atoms out of the ground state, we merely have to get sufficiently more of them into $E_2$ than in $E_1$ so that the amplification on the $E_2$ to $E_1$ transition is enough to overcome the loss from the end mirror. Nd : YAG lasers are therefore much easier to bring to the threshold for lasing than ruby lasers.

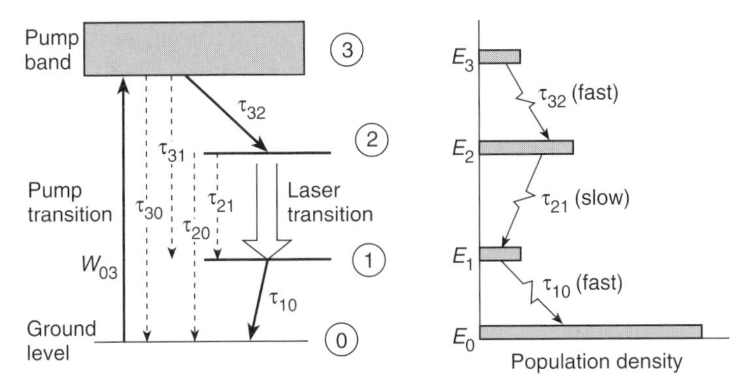


**Fig. 8** Simplified energy level diagram of a four-level laser.

Nowadays, they form the basis of most of the solid state lasers which have taken over from ruby lasers for many practical applications.

For low to medium power applications it is possible to use as a pump to excite the active ions up to the upper laser state the output of an array of semiconductor laser diodes instead of a flashlamp. A modern diode-pumped solid state laser derived from the Nd:YAG laser is shown in Fig. 9. This laser, manufactured by *Schwartz Electro-Optics Inc.* is actually a holmium:YAG laser which puts out 42 W at a wavelength of 2.1 $\mu$m which is very suitable for certain medical applications. The operating head of this laser is small enough to be held in the palm of a hand.

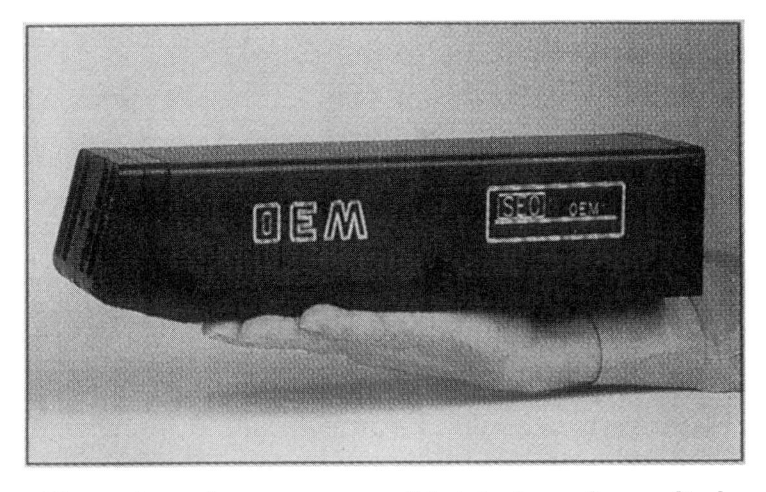


**Fig. 9** A modern compact solid-state laser for medical applications.

Nowadays, in high power solid-state lasers, the active medium of is often made in the form of a rectangular slab of glass in which the appropriate rare earth ions, usually neodymium, are embedded. As shown in Fig. 10(a) the pump light is provided by an array of some 10 or more flashtubes which irradiate the large rectangular faces of the slabs. The laser beam inside the cavity makes a zigzag path within the slab, being internally reflected several times along the length of the side walls of the rectangular slab. With such devices nowadays it is possible to achieve average output powers of several thousands of watts which makes them suitable for welding and cutting metals on an industrial scale.

Another descendant of the early ruby solid state laser is the titanium doped sapphire laser in which the active medium comprises $Ti^{+++}$ ions embedded in a crystal lattice of aluminium oxide instead of the $Cr^{+++}$ ions in the case of ruby. Unlike the lasers we have discussed so far, titanium sapphire lasers can actually be tuned to operate at different wavelengths over a broad band. In these lasers, the final level is not a simple sharp level but a band of levels which are coupled to the vibrations of the crystal lattice and can lose energy to the lattice. By supplying a cavity containing a tunable element (prisms or filters) a laser which can be tuned from 650 nm at the red end of the spectrum to about 1000 nm in the infra-red can be made. Further, by frequency doubling its output, it can provide coverage of the range from 325 to 500 nm. These and similar devices employing non-linear optics have enabled laser radiation to be generated over the region from near ultraviolet, throughout the visible and on into the mid infra-red.

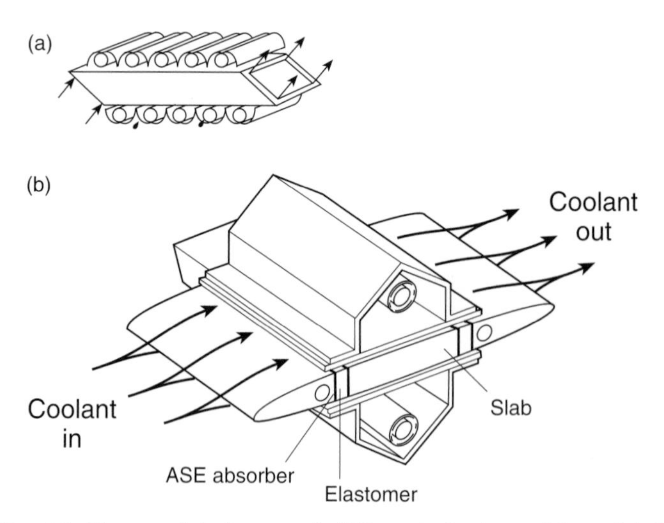


**Fig. 10** Zigzag slab lasers of different designs: (a) multiple flashlamp and (b) double flashlamp pumped configurations.

## The helium–neon laser

Returning again to the early 1960s, an independent line of research by Javan and co-workers at Bell Labs led to the development in 1961 of the first gas laser. It comprised a silica glass tube containing a mixture of neon and helium gases excited by coupling radiofrequency power from a small ham radio transmitter into the gas filled tube via external electrodes. The laser cavity was formed by two flat highly reflective mirrors mounted inside the gas envelope. The radiofrequency discharge set up in the tube caused the gas inside the tube to glow, with the familiar red colour of a neon sign. The laser transition, which has a wavelength of 1.15 $\mu$m, connects a high lying excited state of the neon atom and one somewhat lower in energy. The helium is there to help to excite the upper laser state of neon in a more selective way than would happen in a pure neon gas discharge. One feature in which the first gas laser differed strongly from the early solid state lasers was in its ability to generate output in the form of a steady beam rather than in short pulses, although more modern solid state lasers are also capable of continuous operation.

Shortly afterwards, another type of gas laser was developed by White and Rigden also at Bell Labs. Like the first gas laser it used a neon and helium gas mixture, but this time it operated on a transition of the neon atom in the visible region of the spectrum, at the familiar red wavelength of 633 nm. From 1962 until the present time the red He–Ne laser has been the most common form of laser, familiar by its use at supermarket checkouts for reading bar codes. Its design has been simplified to that shown in Fig. 11—a single gas tube which not only encapsulates the discharge tube but acts a gas reservoir. The cavity mirrors are permanently aligned and incorporated as part of the tube structure itself. The gas discharge is powered by a simple DC circuit, and runs via the narrow bore section of tubing between a pin anode and a hollow cathode.

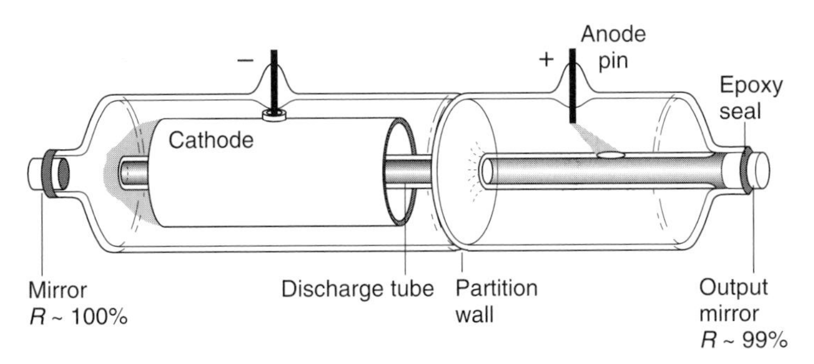


**Fig. 11** An early form of helium-neon laser.

He–Ne lasers have also found their way into teaching laboratories in schools and universities and provide dramatic and highly visible demonstrations of diffraction and interference effects. Such lasers are used in civil engineering construction sites to provide a reference datum ensuring the straightness of tunnels and bridge construction. They are also used in the military sphere as gun sighting devices. The threat that a 100 mm artillery shell can be placed accurately and centred on the projected spot is a powerful argument, and concentrates the mind wonderfully.

## The argon ion laser

The next important development in the field of gas lasers took place in 1964 with the discovery of the argon ion laser—independently and virtually simultaneously by three groups. Bennett, working at Yale had predicted theoretically new laser transitions in singly ionized argon. He was searching for them by pulsing extremely high current discharges through a tube containing low pressure argon gas and duly found a number of laser transitions in the blue green region of the spectrum. A group in France led by Convert, discovered independently the same group of laser transitions in the argon ion and at Hughes Research Labs, Bridges also discovered them. At the time, Bridges had been investigating some rather weak laser transitions of ionized mercury vapour and his tube accidentally became contaminated with argon. He communicated his results to Gordon of Bell Labs who speculated that it would be possible to make these transitions (which had hitherto only shown pulsed laser operation) oscillate continuously. The continuous wave (CW) argon ion laser evolved from the helium neon laser by making the tube of silica rather than glass, water-cooling the narrow bore section to prevent the high current density discharge from melting it, and by providing a coiled return tube from anode to the cathode ends of the tube so that the strong pumping action of the discharge would not cause all the gas to accumulate in the anode bulb. In 1964 the first CW argon ion lasers, giving about half a watt of blue-green light became available at Bell Labs. Further development quickly led in 1965 to lasers capable of 5–10 W continuous output—at the time, the highest laser power achieved anywhere.

Even in these very early days there were attempts to use this laser for medical and biological applications despite the fact that there was no means of moving the beam around but rather the patient or experimental animal had to be moved within the beam. Developments of this laser over the intervening years have made it into a very useful device for eye surgery and a number of other medical applications. These lasers have

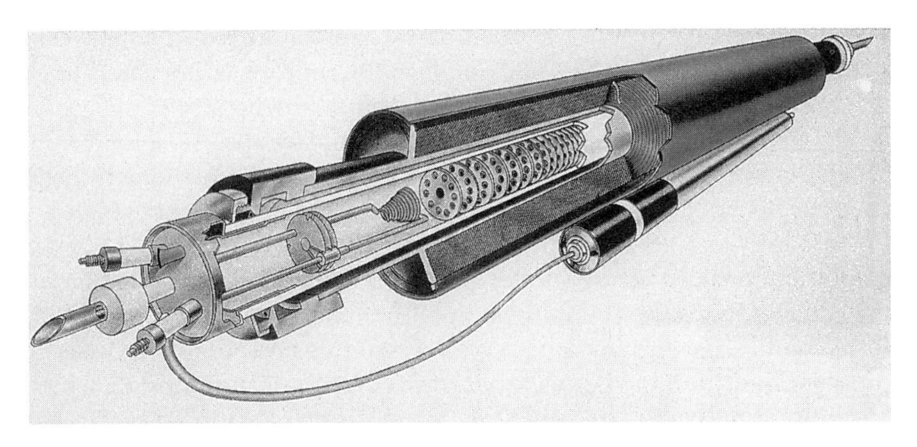

**Fig. 12** The discharge tube of a modern argon-ion laser.

also been used quite widely for laser shows and displays associated with pop and rock music and indeed in one year the Christmas decorations in Regent Street.

Argon ion laser technology has proved remarkably durable. Over the years there have been many predictions that within a year argon ion lasers would be totally replaced by all-solid state devices employing semiconductor technology but this hasn't happened yet, although there are signs that it is about to happen at last.

In modern argon ion laser designs, as shown in Fig. 12, the gas discharge is constrained by a number of metal discs to run down the axis of a ceramic tube and there are elaborate precautions to prevent the internal surfaces of the end windows of the tube becoming contaminated with metal particles and means to top up the gas pressure when gas atoms become buried in the electrode or other structures. Devices giving up to 25 W continuously are available and in the laboratory devices providing up to more than 100 W have been demonstrated. They are big devices consuming a lot of power—they are not efficient and require huge water cooling to get rid of the unused input power. Part of the problem is that the input energy to the discharge is not channelled in any selected way to the upper laser level. You have to excite a great many states of the argon ion just to access the few that lase.

## Dye lasers

One important application of the argon ion laser in scientific exploration of the detailed properties of atoms and molecules is via its use as a pump for tunable dye lasers which led to a renaissance in atomic and molecular physics in the 1970s and 1980s. In dye lasers the energy levels are those of a complex organic molecule in which the upper laser level is excited from the molecular ground state by absorbing light from a flash lamp or a fixed wavelength laser such as the argon ion laser. Once in the excited state which belongs to a group of levels which are vibrationally as well as electronically excited, the vibrational part of the energy can relax as the dye molecule makes collisions with molecules of the surrounding solvent liquid medium (frequently ethanol or methanol) in which the dye species is dissolved. From this level they can decay in the laser transition to various levels of the electronic ground state which are vibrationally excited and from which they can relax back to the overall ground state again by collisions with solvent molecules. Because the transitions cover many states which are spread into one another by these collisions, they are widely tunable. These lasers were the forerunners of the titanium-doped sapphire systems as tunable lasers for spectroscopy. Indeed, because they cover the tuning gap between the fundamental and second harmonic tuning ranges of titanium sapphire for yellow-orange lasers they still have some importance.

Tunable lasers, particularly dye lasers pumped by argon ion lasers and perhaps more recently titanium-doped sapphire lasers, have so taken over the fields of atomic physics that indeed it is difficult to find an experiment nowadays in an atomic physics laboratory that does not somewhere use a laser.

## The carbon dioxide laser

Perhaps the most impressive development of gas laser technology dating from the 1960s was the invention by Patel of the carbon dioxide laser, which almost immediately gave continuous output powers of 100 W and soon afterwards was demonstrated at power levels of 1 kW and more. The James Bond film *Goldfinger* was made in 1964 and showed James Bond being threatened with bisection by a laser beam. In those days, the only lasers around would have hardly caused him any worry as the helium–neon lasers and early argon ion lasers would not have done much more than singe his trousers. However, within a year science fiction had become science fact and the $CO_2$ laser would have indeed given him something to worry about.

The output wavelength of the $CO_2$ laser is at 10.6 $\mu$m in the mid-infrared region of the spectrum. The energy levels involved in the case of the $CO_2$ molecule all belong to the electronic ground state in which the

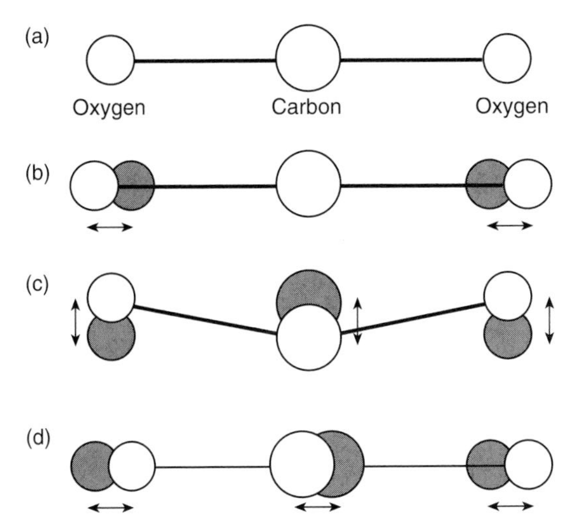


**Fig. 13** Vibrational modes of the $CO_2$ molecule (a) the equilibrium molecular configuration, (b) the symmetric mode of vibration, (c) the bending mode of vibration, and (d) the asymmetric stretch mode of vibration.

two oxygen nuclei and the carbon nucleus, which in equilibrium are arranged in a straight line with one another, execute different types of vibration as indicated in Fig. 13.

Because all the levels involved in laser action in the $CO_2$ molecule are contained within a very small spread of energy—even the upper laser level is only about a quarter of an electron volt (eV) above the ground state—the energy investment needed to excite the molecule to the upper laser level is very small compared with the 35 eV typically needed in the case of a visible gas laser such as argon ion. Furthermore, almost half of the energy invested in getting to the upper laser level is recovered as a laser photon when the $CO_2$ molecule is stimulated to emit. It is a much more efficient type of laser than the argon ion because it avoids most of the unnecessary waste of energy in going around the excitation loop, from ground state to upper laser level and from the lower laser level back to the ground state.

In Patel's original $CO_2$ laser the $CO_2$ was excited by collisions with molecules of nitrogen gas which had been excited in a radio-frequency discharge much like the ones used for the early helium–neon lasers. In this case, however, the discharge excited the $N_2$ molecules (which are simple dumbbells in their nuclei arrangement) into the first excited state of their vibrational energy level ladder which happens to be at about

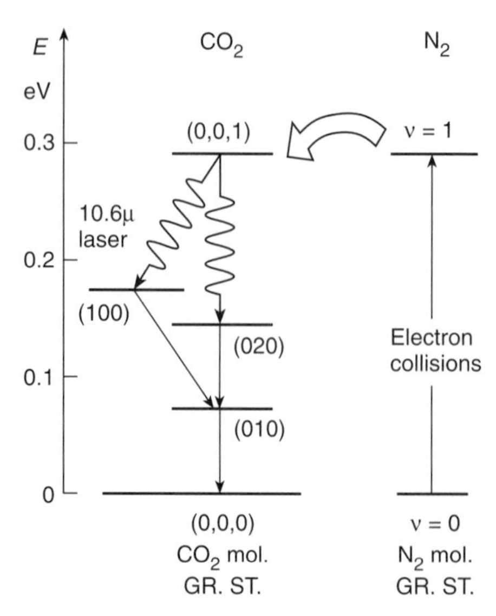


**Fig. 14** Energy levels of the $CO_2$ and $N_2$ molecules important in laser operation; the numbers in brackets beside the $CO_2$ levels denote the number of vibrational quanta in the symmetric, bending, and asymmetric modes, respectively.

2.5eV—exactly the right energy to excite resonantly the $CO_2$ molecule in collisions as shown in Fig. 14.

The next development in the case $CO_2$ laser was the addition of large quantities of helium gas to the $CO_2$–$N_2$ mixture. The lightweight helium atoms are very good at conducting away the heat locked up in the form of molecules of $CO_2$ in the lower laser level and the level belonging to the first excited state of the bending mode ladder which happens to lie almost exactly halfway between the energy of the lower laser level and the unexcited ground state. The existence of this intermediate level is very important for the ability of the $CO_2$ laser to reach high powers as the collision between a $CO_2$ molecule in the lower laser level (labelled 010 in Fig. 14) and a ground state $CO_2$ molecule leads to two molecules in this intermediate level and provides a way out of the lower laser level which prevents bottle-necking of molecules in this level which would otherwise spoil the inversion and restrict the $CO_2$ laser to low power operation.

$CO_2$ lasers have found applications in industry for cutting and welding—even wood working, decorative scroll work for furniture nowadays is often made using $CO_2$ lasers, and provided one is not interested in feature sizes much smaller than 1/10 mm, the $CO_2$ laser is an effective way of cutting many materials otherwise difficult to machine such as ceramics and refractory metals. $CO_2$ lasers can also be used to surface treat metals to provide localized case hardening, etc. Medium power $CO_2$ lasers have also found many applications in surgery, especially those where the laser radiation can be delivered in a direct line to the treatment site as a flexible optical fibre transmitting at the mid-infrared wavelength of 10.6 $\mu$m is still not available.

## Semiconductor diode lasers

Another line of research which also started in the 1960s has led to the development of lasers in which the active medium is a semiconductor, for example: an alloy of gallium and arsenic–gallium arsenide. The gallium arsenide laser was demonstrated first as a device that worked for only a few seconds when cooled to the very low temperature of liquid nitrogen (–196°C) developed in 1962 at the laboratories of the General Electric Co. in the USA.

The chip which is the key component of the semiconductor diode laser illustrated in Fig. 15 is very small, about the size of a grain of salt. Excitation is supplied by direct current across the junction between p-type and n-type material, which is biased in the forward direction. The combination of electrons and holes which occurs in the junction region is responsible for the formation of a population inversion in the gap between the conduction and valence bands and causes light in the red or near infra-red region of the spectrum to be amplified in a very narrow channel. Cleaving the face of the gallium arsenide crystal forms a naturally flat surface which even without coating has enough reflectivity

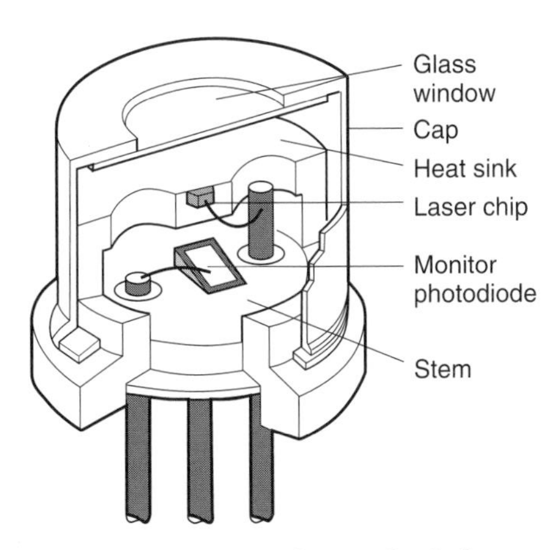


**Fig. 15** A semiconductor diode laser.

to act as an end mirror of the laser cavity so that the crystal itself forms its own cavity. In fact, sometimes the high reflection coefficient which naturally exists between the gallium arsenide/air interface is something of a nuisance if we want to operate the semiconductor laser in an external cavity with a wavelength selective element in order to tune it. In that case it is necessary to coat the cleaved face of the crystal with a special coating in order to reduce its reflectivity so that the external cavity can take control.

Semiconductor diode lasers have found widespread applications, not least as low power laser pointers as a lecture demonstration aid. Lasers giving a few milliwatts of power are quite cheap and widely available and operate typically at 670 nm—a little too far off the peak of the eye response curve for maximum visibility.

The next step in this development was quite a long time in coming. To make these lasers operate continuously over a long period and at room temperature took a huge amount of development in many laboratories throughout the world. It was not until 1970 that room temperature operation for periods of more than a few seconds was reported. A great effort on the part of the material scientists went into growing highly purified materials and fabricating them in the forms required without cracks and defects. The semiconductor laser has now achieved dominance over all other types of lasers, certainly in terms of number of units sold. They have found their way in such everyday devices as CD players and in telecommunication much of modern information networks depend on the ability to produce coherent light in a device that is small and cheap. With semiconductor fabrication technology it is possible to make many thousands of such devices on the same chip.

## Lasers in telecommunications and data storage

It is with the coming to fruition of semiconductor laser technology that the original motivation for developing the laser has been fulfilled. Fibre-optics communications depends on the ability of the laser to beam its energy into a narrow cone and hence into the tiny end face of a glass fibre, much smaller in diameter than a human hair. In the UK many thousands of miles of trunk fibre network are already installed and in daily use with the promise in the near future of fibre-optics network extending to the home, bringing a giant step in the information carrying capability which will be accessible to the individual subscriber.

The first optical fibres that I remember from the 1960s were very poor. Indeed, it would have been necessary to boost the signal back to its original value after only a few tens of metres of transmission path. Now with improvements in fibre fabrication technology many tens to hundreds of kilometres can be spanned before the signal needs boosting. Recent advances in fibre technology made at the University of Southampton by Gambling and Payne have resulted in fibres implanted with ions of rare earth element erbium. These fibres can act as laser amplifiers so that the signal can be boosted without even having to go from fibre-optics into electronics and back again. The erbium amplifiers which operate on a three-level energy scheme analogous to the ruby scheme can be pumped by light from semiconductor diode lasers so that the staging amplifiers along this long path communication network can be made entirely from solid-state semiconductor devices all operating at low voltages. With this technology it is possible to make fibre-optic links which can span even the large distances between continents. Across both the Atlantic and Pacific oceans the information carrying capacity of fibre-optic links will increase enormously in the coming decade.

In the 1960s it seemed that intercontinental communications would be dominated by satellite technology in which microwave signals would be beamed from the earth to satellites in geosyncrynous orbits and from there down to the fixed receiver stations. In fact, for sender to receiver station communications between stations at fixed locations the cost of fibre-optic communication is now far less than that of satellites. Satellites do, however, retain their attractiveness for transmission originating at remote locations or situations such broadcasts from war zones.

Another technology which has become ubiquitous in recent years is that of the compact disc (CD) recording. The information is converted to a digital form and registered in the form of indentations on a highly reflective disc. As shown in Fig. 16 in reading the information from the disc, a semiconductor laser illuminates the disc from below on which side the indentations appear in the form of a series of bumps and gaps along a spiral track on the surface of the disc.

By some ingenious optical tricks, including for example making the height of the bump equal to a quarter of the wavelength of the light of the laser, the reflection of the laser light back along its path is either high from when the laser light illuminates a flat surface or low when it encounters a bump. The back reflected light is monitored on a detector which converts it back to a digital signal once again and from then on electronics can be used to convert it to mimic the original sound wave. The diameter of the laser spot as focused on the surface of the disc is about twice the width of one of the individual bumps. This means that about half the light is reflected from the top of the bump and half from the surrounding flat area but this light has travelled an extra half wavelength and is therefore

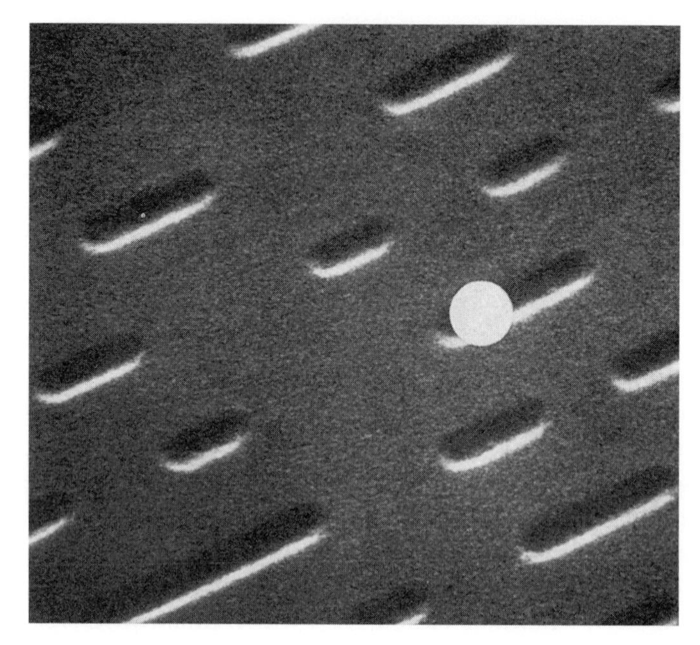

**Fig. 16** Scanning electron microscope image of the information layer of a CD disc. The optical effects, 'pits' with a depth ~ 100 nm are arranged in more or less parallel tracks. The track pitch is 1.6 $\mu$m. The 'pits' and 'lands' between the pits have lengths of $0.9 + n \times 0.3$ $\mu$m, where $n = 0, 1, \ldots, 8$. An indication of the size of the scanning spot is given.

exactly out of phase with the light reflected from the top of the bump with which it interferes destructively, helping to provide the cancellation of the beam reflected from the spot as a whole. The digits are actually coded in as the change in reflectivity in going from flat to bump area.

Without semiconductor lasers this technology would not be possible and without the low manufacturing cost which semiconductor fabrication technology allows, the cost would be prohibitive. Production of semiconductor lasers exceeds billions per year.

While each semiconductor laser is limited to a maximum output power of a few milliwatts, they can be fabricated in rows to form bars and the bars can be stacked vertically into two-dimensional arrays. In this way arrays of diode bars can be made which put out 20 or even 100 W of average power and recently Livermore Labs have demonstrated an array capable of putting 4000 W of power. That is an amazing achievement. We have a semiconductor device whose average output power is equal to four electric fire bars but now in the form of coherent light. The cooling arrangements for this device which are necessary to prevent it simply melting down are also fabricated by elegant lithographic techniques.

In the past few years, it has become possible to use such arrays of semiconductor lasers as the starting point for pumping conventional solid-state lasers such as neodymium YAG and the output of such a laser which is at 1.06 $\mu$m can be frequency doubled to 532 nm in the green region of the spectrum and so output from visible solid state lasers pumped by semiconductor diode lasers is now a reality. These form very compact all solid-state devices which operate quite efficiently and their proponents claim that within a few years they will have taken over all the functions currently the realm of argon ion and other gas lasers.

## Copper vapour lasers

Since the late 1970s a particular research interest of my group at the Clarendon Laboratory has been high power pulsed metal vapour lasers, especially the copper vapour laser.

The copper laser produces output at 578 nm in the yellow region and 511 nm in the green region of the spectrum. There is also its analogue, the gold vapour laser, which produces output on the corresponding transitions at 628 nm in the red and 312 nm in the near ultraviolet. As far as its atomic structure is concerned, the gold atom is very similar to the copper atom. Of course, bankers have a very different view of the similarity of these metals. By heating a ceramic discharge tube containing a few lumps of copper and a buffer gas of helium and pulsing a high current discharge through this mixture, Walter was able to demonstrate laser action on the green and yellow transitions but with very low average power.

The next step forward was taken by a group at the Lebedev Institute in Moscow under Petrash and Isaev who showed that it was possible to heat the copper laser up to its working temperature and to maintain it there by simply pulsing the discharge sufficiently rapidly that the waste heat from the discharge itself took over the role of heating the tube, retaining the heat by a suitable thermal insulation.

Another step forward was taken in 1978 when Smilansky at the Nuclear Research Centre in Israel demonstrated a practical self-heated laser in which the increase of output power was obtained by increasing the tube diameter well beyond the range of earlier devices. Previously, it had been thought that it was necessary to keep the tube diameter small in order to maintain good discharge uniformity but this demonstration showed that it was possible at the elevated temperatures at which the copper laser operates to have good discharge uniformity even in tubes of diameter 40 mm and more.

A lot of the early development of the copper laser was funded by the requirements of the laser isotope enrichment programme at Lawrence Livermore Labs in the USA. Copper lasers are used as the pump lasers for dye lasers, which are tuned to the individual absorption lines of the uranium isotope $^{235}U$ which, in natural uranium, is present at a concentration of only 0.7%. The vapour of natural uranium, entrained in a flow of rare gas, is irradiated by lasers beams tuned to the absorption lines of this one isotope, which is, therefore, selectively excited from its ground state and up through its various excited states until it becomes ionized. Electric fields are used to extract the ions from the gas flow but as ideally the ions are all of one isotope, the metallic uranium that deposits on the collector plates at the edge of the gas flow is therefore enriched in the wanted isotope for fabricating reactor fuel rods for which the concentration of $^{235}U$ needs to be raised to 3.2% relative to the common isotope $^{238}U$.

The copper laser is chosen as the starting point for the isotope enrichment process because it is, as gas lasers go, very efficient. The reason for its good efficiency can be seen from the energy level diagram of the copper atom (Fig. 17). If now an electron in the discharge collides with

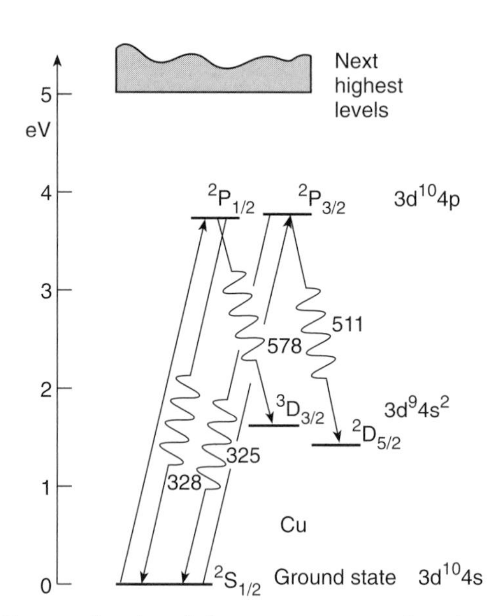


**Fig. 17** Energy levels of the copper atom involved in laser operation.

an unexcited Cu atom, the overwhelming probability, if the free electron is fast enough, is that the 4s electron will be excited upwards to the 4p level.

There is in the copper atom another pair of states, lying only 1.5 eV or so above the ground state. These arise from an electron arrangement in which there is a vacancy of one electron in the 3d shell and two 4s electrons outside. The rules of quantum mechanics forbid transitions from these states back to the ground state and as to get to this state from the ground state involves the movement of an inner electron, it is not very probable that it will be excited by the collision of a free plasma electron with a ground state copper atom even if such an excitation is energetically allowed by the energy available from the kinetic energy of the electron.

If you take a discharge containing neon with some copper vapour and pulse a high current through it, the free electrons in the discharge preferentially excite the copper atoms to the states belonging to the $3d^{10}$ 4s arrangement rather than the $3d^9$ $4s^2$, so creating a population inversion transiently between the corresponding energy levels. Of course, then, once the laser oscillation has happened, producing a short but intense pulse of laser output lasting maybe only 20–40 ns, the population builds up in the D levels and laser action can only recommence when these atoms have been re-excited back to the ground state.

If it were necessary to wait until the atoms in the D levels hit the wall of the tube and got de-excited that way before we could start the laser pumping cycle over again, the repetition frequency of the laser pulses would be limited to a rather low value and the average output power of the laser would be correspondingly very limited. Fortunately, however, once the discharge current pulse is over the free electrons which carried the current rapidly cool down to fairly low energies and these slow electrons are quite effective in de-exciting copper atoms in the D levels back to the ground state. This means that the copper laser is capable of being operated at pulse repetition frequencies in the range 5–40 kHz and it is this ability to operate at high pulse repetition frequency which is responsible for its high average power capability and also for making it a very suitable laser for many specialist applications in which pulse repetition frequencies of this order are needed. So a typical copper vapour laser, indicated schematically in Fig. 18, operates in a refractory ceramic discharge tube maintained by the self-heating effect of the discharge at temperatures in the range of 1500°C and pulses of the order of 1000 amps of duration 250 ns repeated at frequencies of 10 kHz.

To achieve reliable operation in such a combination of conditions has required several technological problems to be solved but they have now been solved and commercial copper lasers with good efficiency, high

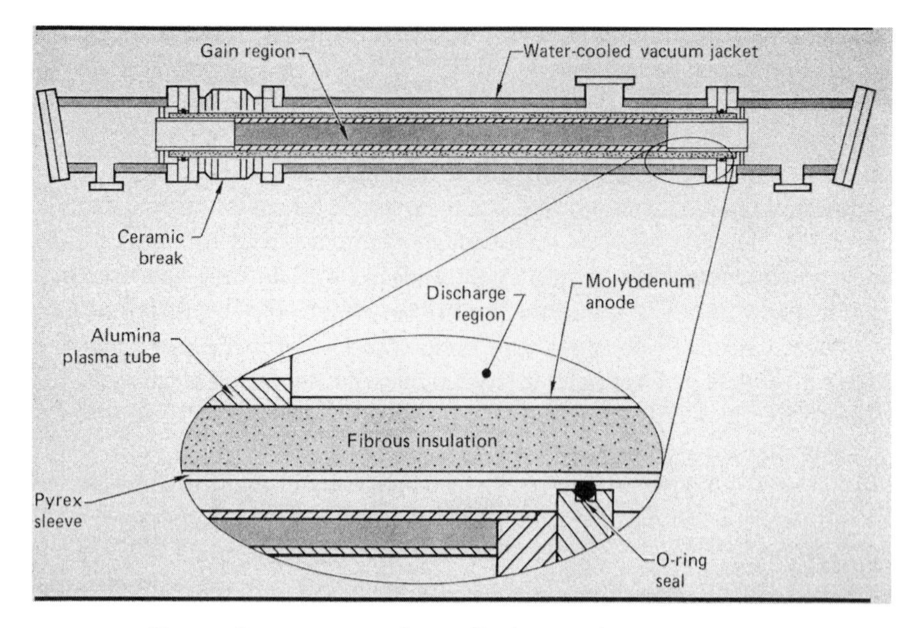


**Fig. 18** Copper vapour laser: discharge tube construction.

average power and long operating lifetime are now available and one of the companies offering such devices is Oxford Lasers Ltd, a company spun off from my research group in 1977.

The copper laser operates well as an oscillator needing only quite low reflectivity mirrors on its cavity as the amplification factor which light experiences traversing the tube is very high. In fact, the device operates very well as an amplifier so that chains of copper laser amplifiers, each adding energy to the initial input pulse provided by a single copper laser oscillator, make the realization of very high average power systems possible. At the Lawrence Livermore Labs such oscillator/amplifier chains exist, each capable of providing several kilowatts of average power in a beam which can be used to pump the dye laser arrays used for the isotope enrichment process.

There are, however, many other applications which the copper lasers have found, including medical applications, high speed photography, and ultra-precision machining of metals. The yellow output at 578 nm can be used for dermatology, in particular for treating skin blemishes and vascular lesions. Remarkable results have been obtained in improving the cosmetic appearance of patients with port-wine stains. Another medical application is in treating cancer by photodynamic therapy (PDT). In this procedure the patient is injected with a drug called Haemato-porphoryn derivative (HPD) which circulates in the blood stream and has the property of attaching itself to tissue connected with tumours so that after a few days, the tumour regions contain a significant quantity of this chemical. The patient is then recalled and the regions where tumour is suspected are irradiated with laser light over a broad area. The light is not concentrated so as to cut or break the skin but it does have the action of penetrating deeply into the tissue and activating the drug deep inside the tumour. This activates the cytoxic action by exciting the drug molecules to a state from which they transfer their energy to oxygen molecules in the blood stream which are themselves promoted to a highly active and chemically aggressive state. The active oxygen molecules attack everything in their vicinity and, in particular, they shut down the blood supply to the tumour which then dies off leaving a wound which heals in the normal way. In order to be able to penetrate the light needs to have a wavelength longer than about 610 nm otherwise it would be strongly absorbed by blood. On the other hand it must have a wavelength shorter than 640 nm otherwise it would be insufficiently energetic to excite the HPD. Laser light in the region 610–640 nm is therefore required at powers in the order of 3–10 W. This can conveniently be obtained from the compact dye laser pumped by a copper vapour laser illustrated in Fig. 19, although gold vapour lasers

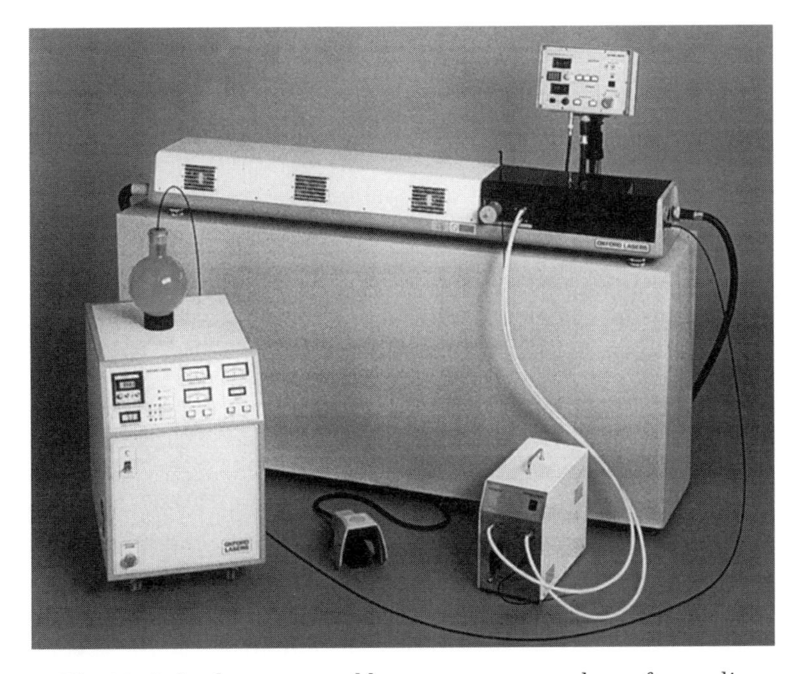

**Fig. 19** A dye laser pumped by a copper vapour laser, for application to the treatment of cancer by photodynamic therapy.

operating at 628 nm have been used. The copper/dye combination is, however, somewhat more efficient and has a better record for reliability.

Because the copper vapour laser can be synchronized with a high speed motion picture camera, the combination provides an ideal way of studying high speed events such as fluid flow and the break up of aerosol jets into individual droplets as well as events within jet engines and diesel engines under operating conditions.

With recent advances in controlling the quality of the copper laser beam in terms of its ability to be focused to a fine spot, nearly equal to the limiting value imposed by the laws of physics, the power density at a fine focus which can nowadays be obtained from even quite modestly sized copper vapour lasers is sufficient to cut and drill practically any material. Slots of widths 10–50 $\mu$m can be cut in materials such as stainless steel, titanium, and alumina ceramic. For example, it is possible to create artificial cracks in pipes to simulate the ageing effects experienced by nuclear reactor vessels.

With this technology we can also cut neat holes in diamond and also to drill holes which make a very shallow angle of entry to the surface, previously impossible with any other technology.

One of the most exciting recent developments, however, has been in the field of astronomy and the use of copper vapour lasers to make artificial guide stars in the earth's atmosphere to allow a system of adaptive optics to correct the image forming system for the effects of distortion due to atmospheric turbulence.

For centuries this problem has prevented astronomical telescopes from forming images of the sharpness of which their optics, operating in an airless environment, would be capable.

All ground-based optical and infra-red telescopes of more than a few centimetres aperture suffer from this limitation since the turbulence of the Earth's atmosphere distorts the wavefronts arriving from stars or other astronomical objects, causing what should be a point-like image to wander erratically and hence degrade into an extended blur on any photographic or video recording medium exposed over the necessary period. The same effect is responsible for the characteristic twinkling of stars viewed by the naked eye.

With the type of artificial guidestar shown in Fig. 20 Rayleigh scattering of laser light from air molecules provides a steerable reference object for an adaptive optical system fitted to an astronomical telescope. This allows atmospheric distortion to be nullified by reflecting the wavefronts from a deformable mirror whose instantaneous shape is computer controlled so as to produce continuously a sharp image of the artificial star. By 1983 Fugate at the USAF Phillips Labs in the USA had achieved

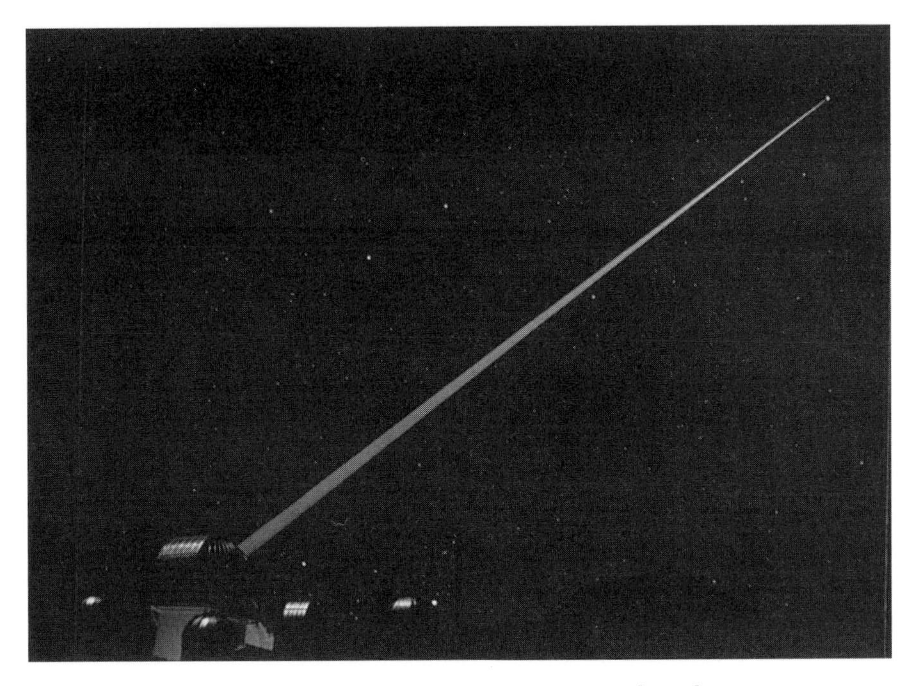

**Fig. 20** Artificial star created by copper vapour laser beam projected from an astronomical telescope. Courtesy Dr R.Q. Fugate, Phillips Laboratory, Kirtland AFB, USA.

proof of principle in 'open loop' operation in which wavefront distortion was measured but not actively corrected. In 1989 Fugate's group achieved the first truly continuous 'closed loop' operation of an astronomical telescope fitted with adaptive optics in which the deformable mirror was used to correct atmospheric distortion on a moment by moment basis.

Because the work was initially carried out under a classified programme none of the ideas or results could be disseminated to the international community of astronomers. When analogous concepts independently proposed by astronomers in France and the USA began to appear in the open literature the case for declassification became overwhelming.

With declassification in 1991 the stunning images obtained by Fugate and his colleagues at the Phillips Laboratory could be published. These immediately sparked off such interest in the technique that several European countries including the UK, France, Spain, and Germany have embarked on programmes to upgrade their national observatories by retrofitting existing telescopes with laser guidestars and adaptive optics.

One leading astronomer has said that it is certain now that all new telescopes constructed in the future will include provision for adaptive optical systems in their design.

At the moment my group at the Clarendon Lab is actively engaged in the effort to provide a high power laser system for the next generation of research in this field. The objective is to create a bright artificial star in the layer of atomic sodium which exists naturally in the earth's atmosphere at a height of 100 km above ground level. To do this we will have to solve many problems including how to make a laser tuned exactly to the centre of the 589 nm transition of atomic sodium (the familiar colour of sodium street lamps. This must be powerful enough to project several hundred watts in an optical beam of perfect quality and yet be light enough to be mounted on the gantry which moves with the primary mirror of a big astronomical telescope such as the William Herschel telescope. Further it must not dissipate any excess heat inside the telescope dome. This is a challenge which we find exciting and worthwhile.

Lasers have come a very long way in the 35 years since the first demonstration. The applications they have found include not only those which were foreseen from the very beginning but also very many which could not possibly have been foreseen and which have led our research into exciting and unexpected areas. That is one of the major rewards of doing the research.

## COLIN WEBB

Born 1937 in Erith, Kent and educated at the Universities of Nottingham and Oxford. From 1964 he was a member of the technical staff at Bell Laboratories, New Jersey, USA. In 1968 he returned to Oxford to take up a Research Fellowship and to head the laser research group at the Clarendon Laboratory. In the 1970s he became a University Lecturer and Tutor in Physics at Jesus College, Oxford. In 1977, with some of his colleagues, he set up Oxford Lasers Ltd. to exploit commercially the technology generated by their research. Today he continues to be Chairman of Oxford Lasers as well as Professor of Laser Physics and Head of Atomic and Laser Physics at Oxford. His research interests are in the scientific and industrial applications of lasers, particularly for environmental monitoring and as laser beacons for astronomy.

# *Fuel cells—fuelling the future*

A. HAMNETT

## Introduction

As a nation, we are living on the accumulated reserves of the past; in energy terms, the long past. The oil reserves that continue to fund our life-style were laid down many millions of years ago, but are now being used up far more rapidly than they can ever be replaced even though energy is historically cheaper at the moment than it has been in the memory of anyone in northern Europe. Deregulation of the energy market is likely to reinforce this trend: we expect the price of energy to stay low for the foreseeable future; our oil companies, like the lilies of the field, 'sow not', although they do reap with ever-increasing efficiency. It is a truism to say that we cannot continue in this vein: *tempus edax rerum*, and our reserves of fossil fuel must eventually become exhausted, or at least so depleted that their price becomes unattainable. We must, in other words, find sustainable cycles of energy usage and generation which reduce, and perhaps eventually eliminate, our dependence on the irreversible extraction of the earth's bounty. The purpose of this lecture is to examine one such cycle, in which solar energy as the ultimate fuel source, is coupled to traction systems that will enable us to maintain both a constant level of $CO_2$ in the atmosphere and at the same time to minimize the changes we will need to make both to our habits and to our expensively provided infrastructure.

Solar energy is, in fact, the origin of all our fossil fuels; the energy harvested by plants; their growth in the mighty forests of the carboniferous era and their subsequent anaerobic decay to form oils and coals has left us a vast reservoir of stored chemical energy. However, the brightness of today's sun, and the energy conveyed by its rays differs but little from Palaeozoic times, and much effort and ingenuity has gone in recent years into finding ways of harvesting solar irradiation and converting the light

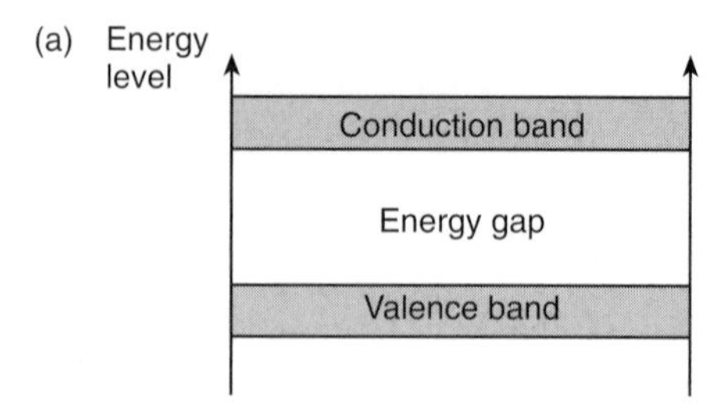


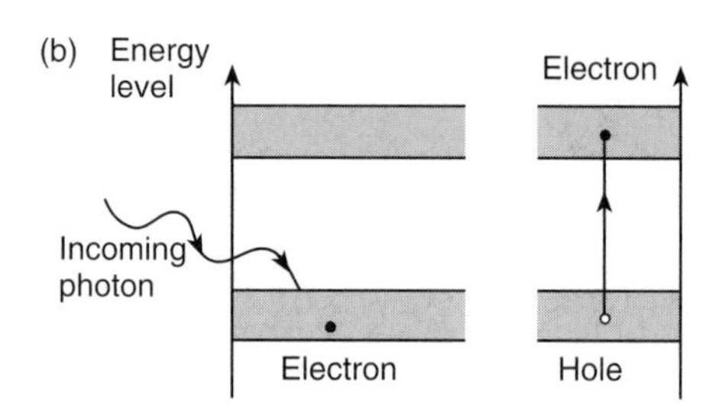


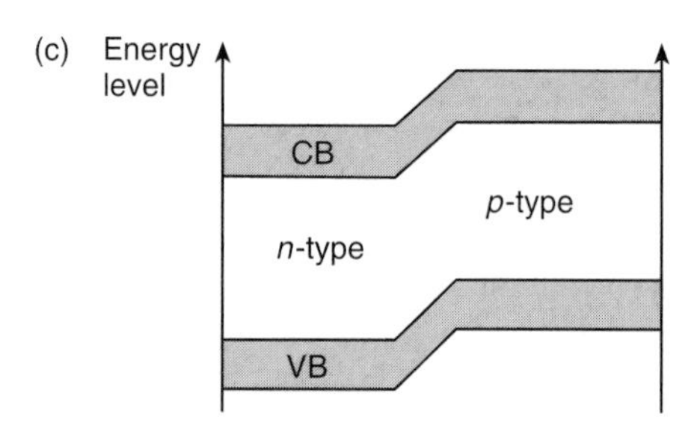


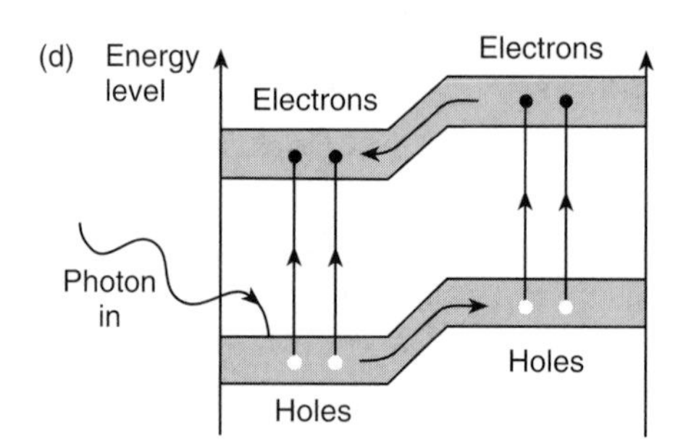


**Fig. 1** (a) Energy bands in a normal intrinsic semiconductor, (b) promotion of electron from the valence to the conduction band on illumination; (c) internal electric field formation at a *p*-*n* junction; (d) operation of a *p*-*n* junction under illumination; the electrons migrate from right to left and the holes from left to right.

energy into more usable forms. The most important of these conversion processes is the generation of electrical energy from light energy using a photovoltaic (PV) device; PV is a word coined from the Greek $\phi\omega\zeta$ (light) and the word *volt*, the unit of electromotive force, named after the Italian physicist, Count Alessandro Volta. The first report of the PV effect was made by Adams and Day in 1877 on selenium, but efficient PV devices were only developed in the 1950s in the famous Bell Telephone Laboratories in the USA. Progress since then has been steady, with prices now well below $5 per peak W and overall efficiencies above 17%. The basic principle of the PV effect is shown in Fig. 1; silicon is a semiconductor, and its electronic structure shows two bands of allowed energy levels. At normal temperatures, the lower of these two bands, the *valence* band, is fully occupied with electrons, and the upper band, the *conduction* band, is empty. Silicon can, however, be made conducting in two ways: the material can be doped with electron donors, which donate freely moving electrons to the conduction band; such a material is termed *n*-type. Conductivity can also be enhanced by doping with electron acceptors, which can remove electrons from the valence band, giving rise to freely moving 'holes'; this material is termed *p*-type. When *n*- and *p*-type devices are combined into a *p–n* junction, their different energy bands combine to form an internal electric field: on illumination with, for example, sunlight, holes, and electrons generated near the junction are separated by this field, giving rise to a voltage and, in the presence of an external circuit, to a current.

Typical thin film PV devices based on amorphous silicon are now commonly available; their basic construction is shown in Fig. 2, and a current–voltage plot for a typical PV cell under standard conditions in normal sunlight is shown in Fig. 3. Modern devices, particularly for use in countries where there is no national grid, are now commonplace, and used for water pumping, refrigeration for vaccines, power for field hospitals and even, in combination with storage devices such as batteries, for street and household lighting. Very large plants have now been built: the Sacramento Municipal Utility District has constructed a 2 MW power plant in California, and there are 500 kW plants in Switzerland and Italy. These plants require little maintenance: there are no moving parts, and the main problem remains slow degradation through dopant migration and other similar effects. However, there remain two major problems: solar energy is both diurnally and seasonally out of phase with our requirements; we need energy in abundance in cold winter's evenings, but the sun supplies most of its energy on hot summer's days. The second problem is that even though solar panels occupy considerable space, and are still very expensive, they do generate power only locally; if we are to continue to drive our cars, we need to store the electrical energy in some way that is easily transportable.

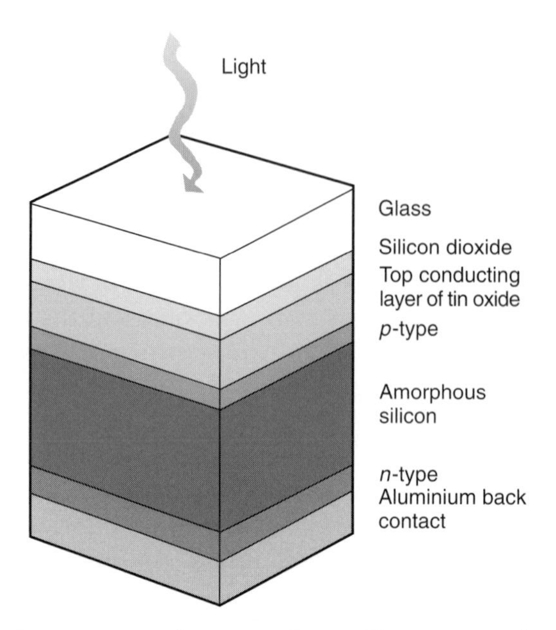


**Fig. 2** Construction of an amorphous silicon photovoltaic cell. The top of the cell is formed from a transparent but electrically conducting layer of tin oxide deposited on the glass. The bottom contact is made of aluminium. In between are layers of *p*-type, intrinsic and *n*-type amorphous silicon.

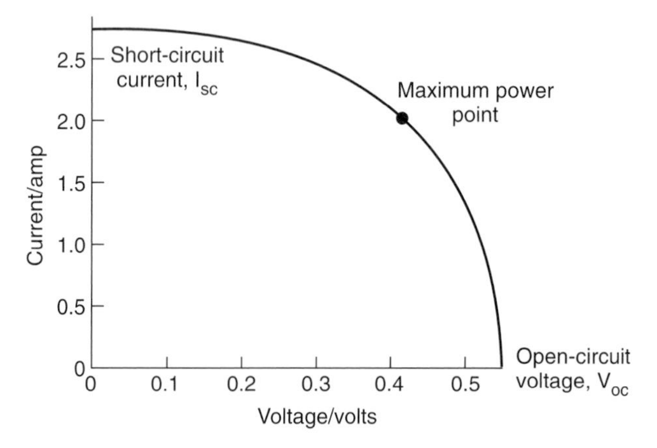


**Fig. 3** Current voltage characteristics of a typical silicon photovoltaic cell under standard test conditions.

This storage can most easily be achieved by conversion of the electrical energy into *chemical* energy, and the most straightforward way of doing this is through the electrolysis of water. In this process, which is shown in Fig. 4, an electrical current is passed through water in which is dissolved a salt (or an acid or base) in order to confer electrical conductivity on the water. At the positive electrode, or *anode*, the water is oxidized to oxygen, and at the negative electrode, or *cathode*, the water is simultaneously reduced to hydrogen. In terms of chemical symbols:

$$\text{Anode} \qquad H_2O \rightarrow 2H^+ + \tfrac{1}{2}O_2 + 2e^-$$

$$\text{Cathode} \qquad 2H^+ + 2e^- \rightarrow H_2$$

and the net reaction is simply $H_2O \rightarrow H_2 + \frac{1}{2}O_2$. The important product is the hydrogen; recombination of hydrogen with the oxygen in the air generates a large amount of energy, and yet hydrogen is extremely stable and can be stored indefinitely and transported relatively easily in the form of compressed gas in cylinders.

Once we have generated the hydrogen, however, we need to find a way of converting it into the form of energy we need: electrical, mechanical, *etc.* In the case of transport, the provision of motive power could be done in principle, in the same way as in an internal combustion engine: the hydrogen could be introduced with air into a piston and sparked. As is well known, hydrogen will react violently with oxygen under these circumstances; the resultant explosion will give rise to a substantial motive force that will drive the piston down as part of the normal internal combustion cycle. While this process will generate primarily water

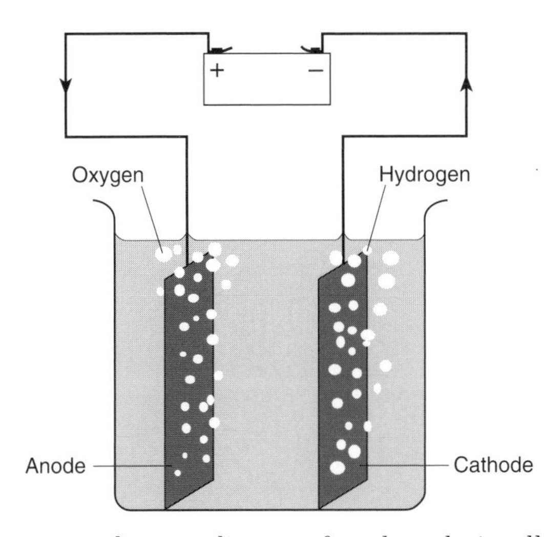


**Fig. 4** Schematic diagram of an electrolysis cell.

as exhaust, it has limited efficiency, and it also has the capacity to generate undesirable by-products such as $NO_x$ species, particularly if the engine is run 'hot'. What is needed is a method of controlled recombination of hydrogen and oxygen which is more efficient than the explosive recombination utilized in internal combustion engines but which will provide motive power. The logical means is to combine the $H_2$ and $O_2$ to form water and simultaneously to draw off the excess energy in the form of electrical power, and the means whereby we can do this is a fuel cell.

The simplest way of visualizing a fuel cell is to imagine the reversal of the electrolysis cell shown in Fig. 4. If, instead of providing a voltage to the cell to break down the water into hydrogen and oxygen, we bubble the two gases over our electrodes and connect them through as resistor, as shown in Fig. 5, then experimentally we find that a voltage is developed between the electrodes and a current will flow through the resistor. Unfortunately, this voltage is lower than that required to break down the water in the first place, and the loss encountered is an irreversible contribution to the overall inefficiency of the solar energy to electrical energy process; it is, in fact, the price we have to pay for the convenience of storing the energy in the first place.

Fuel cells do, however, enjoy some formidable advantages as power sources: in spite of the *caveat* above, they possess relatively high efficiencies, particularly when used with hydrogen as a fuel, excellent part-load characteristics, rapid response times, low pollution emission behaviour, simplicity of mechanical engineering, good power/weight ratio, and their modularity of construction makes them ideally flexible in power provision. Indeed, such advantages should have led to rapid and extensive exploitation of fuel cells in a wide variety of environ-

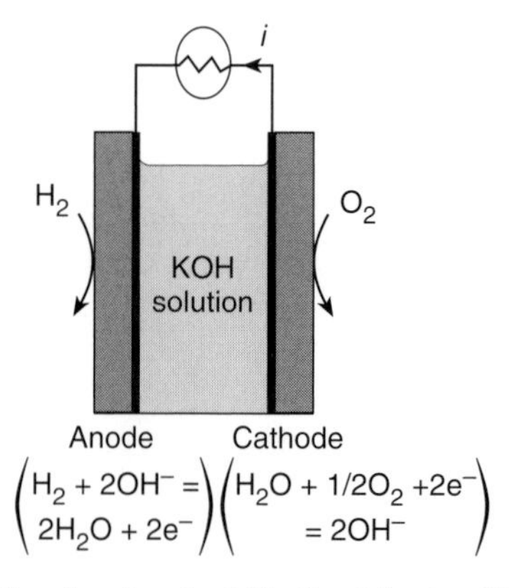


**Fig. 5** Schematic of a simple (alkaline) fuel cell with hydrogen as the fuel and oxygen from air as the combustant.

ments, but historically this has not happened, for reasons that we will explore in more detail below.

The best type of hydrogen-based fuel cell for transport applications is based on the configuration shown in Fig. 6, where hydrogen and air are fed to the back of the two electrodes, which are separated by a very specialized thin polymeric membrane. This membrane is capable of conducting protons, and anode and cathode are, in the simplest designs, formed either directly from metal particles or from catalysed carbon particles bound to the membrane. The current collectors are porous carbon or graphite plates, and the cell reactions are, at the anode

$$H_2 = 2H^+ + 2e^-$$

and at the cathode

$$\tfrac{1}{2}O_2 + 2H^+ + 2e^- = H_2O$$

The low operating temperatures require the use of very active noble metal catalysts, particularly on the cathode side, such as very finely divided platinum alloys. By careful optimization of the construction of the fuel cell and the choice of electrode materials, however, voltages in excess of 0.7 V at current densities in excess of 500 mA/cm$^2$ are attainable in such fuel cells (see below).

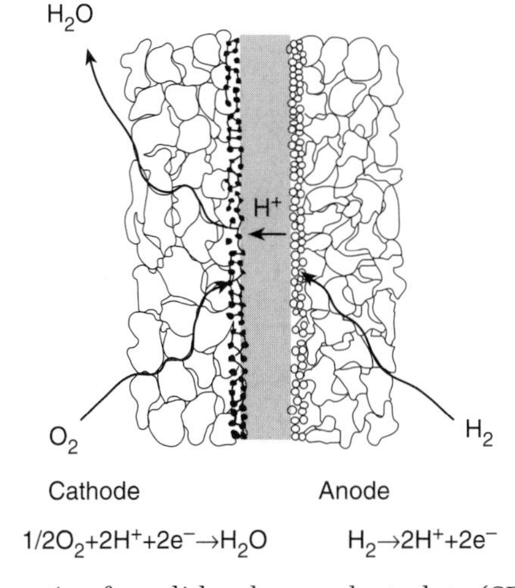


**Fig. 6** Schematic of a solid-polymer electrolyte (SPE) fuel cell comprising a central polymer membrane to which are attached two catalytic layers backed by PTFE-bonded porous carbon layers forming electrical contact.

Critical to the success of this type of solid-polymer electrolyte fuel cell (SPEFC) is the electrolyte, and the revival of interest in this configuration is due to the success of Du Pont and latterly Dow Chemical Co. in developing perfluorinated sulphonic acid polymers. These consist of a PTFE backbone to which are bonded pendant sulphonic acid groups, as shown in Fig. 7(a,b). These materials have the secondary advantage of being dispersible in certain organic solvents, such as ethanol, allowing membrane fabrication at a variety of thicknesses, and also allowing for the use of this sol as a binder for the catalysed carbon particles, thereby greatly increasing the utilization of catalyst in the electrode.

The performance of such cells, particularly under a few bar pressure, is quite remarkable, as shown in Fig. 7(c), where power densities in excess of 1 W/cm$^2$ are seen. This is a substantial magnitude: 10 cells each of area 17 × 17 cm$^2$ and occupying a volume of less than 3 dm$^3$ and would

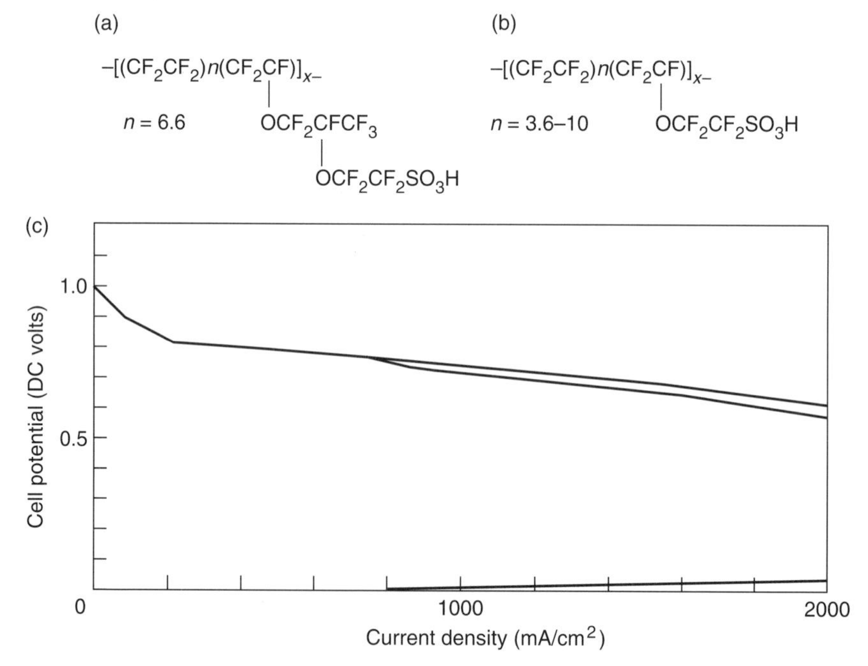


**Fig. 7** (a) Chemical structure of Du Pont Nafion; (b) chemical structure of the Dow membrane; (c) cell voltage (central curve) and half-cell potentials (upper and lower curves, corresponding to cathode and anode respectively) vs. current density for a single cell with a Dow membrane (thickness 125 $\mu$m) and platinum-sputtered Prototech electrodes (Pt-loading 0.45 mg-cm$^2$) operating at 95°C with $H_2/O_2$ at 4-5 bar.

suffice to boil a kettle, and at first sight, therefore, our aim of generating a completely sustainable cycle appears to have been satisfied. Indeed, the high power density of SPEFC's makes them extremely attractive for traction applications, where the size and weight of the fuel cell should be as small as possible. The Ballard Company in Canada have recently designed and constructed a bus that uses compressed hydrogen in cylinders as the fuel, and have been testing this in city driving in Vancouver.

Clearly, fuel cells are related strongly to batteries; both types of device convert chemical energy directly into electrical energy, and the main difference between them is that a battery must carry its fuel with it at all times. Once the fuel is exhausted the battery must be either discarded or recharged, whereas a fuel cell will continue to operate as long as fuel is provided from an appropriate reservoir. The manufacture and recharge of batteries both consume electrical energy, which must originate from a power station, and sustainable operation is really only feasible if this power comes either from a fuel-cell driven system or from hydroelectric sources.

The high performances of the SPE Fuel Cells shown in Fig. 7 should not, however, disguise the fact that if application of the SPEFC to transport is considered, a number of problems arise, the most important of which is the fact that transport of the hydrogen fuel itself carries a substantial penalty. This is because hydrogen is a gas at room temperature, and must be transported either under high pressure in a cylinder or in the form of a metal hydride from which the hydrogen can be released by warming. In both cases, there is a substantial weight penalty, and energy from the fuel cell must be used to transport this essentially parasitic load. It would clearly be far better if the fuel could be transported as a *liquid*, and stored in a conventional fuel tank. Not only could use then be made of the already existing infrastructure in the UK for distribution of liquid fuels, but the safety implications of storage and transport of hydrogen under high pressure would be ameliorated.

Considerable attention has been given to this question, and two approaches have been adopted. The first is to identify a liquid that can be converted easily by chemical means into hydrogen, and this hydrogen then used to fuel a conventional fuel cell of the type described above. A second approach is to identify a liquid that can be directly converted to electrical energy at the anode of a fuel cell specialized to run on that liquid as a fuel. The first of these approaches uses a chemical process termed *reforming*, in which a liquid hydrocarbon, such as gasoline, is converted to hydrogen and $CO_2$ by passage with steam and/or oxygen (from air) over an appropriate catalyst in a reactor called a reformer. The choice of liquid hydrocarbon has itself been the subject of much research, but one of the simplest compounds that contains both substantial amounts of hydrogen and is itself a liquid at room temperature is methanol (*methyl alcohol or wood alcohol*). Reaction with steam in a reformer gives rise predominantly to the reaction

$$CH_3OH + H_2O \rightarrow CO_2 + 3H_2$$

with the formation of small amounts of CO. While this apparently solves the problem, there are some drawbacks: reformers are quite complex and costly, particularly when used on the small scale for transportation. In addition, the reaction above is *endo*thermic (that is, it demands heat), and this heat must either be provided from the fuel cell itself (very difficult for the type of fuel cell shown in Fig. 6 that operates at low temperature), or by partially combusting the fuel, which is wasteful in efficiency terms. Furthermore, the SPEFC of Fig. 6 with conventional catalysts shows very considerable sensitivity to contaminants in the fuel gas; in particular CO must be reduced to well below 10 p.p.m. to avoid deterioration in the anode performance. This type of purity can only be achieved by using multi stage reformers, but considerable effort has also been expended on developing promoted platinum catalysts that are less sensitive to residual CO.

A quite different approach would be to use fuel cell that could *directly* oxidize liquid methanol at the anode, but retained the high power/weight ratio of the SPEFC described above. Such a fuel cell, termed a *direct methanol fuel cell*, is now under active development in Europe and the USA, and the principles are shown in Fig. 8. The anode electrochemical reaction is:

$$CH_3OH + H_2O = CO_2 + 6H^+ + 6e^-$$

and the cathode reaction is:

$$\tfrac{3}{2}O_2 + 6H^+ + 6e^- = 3H_2O$$

The actual construction of a cell of this type is shown in Fig. 9, which can be seen to be similar in form to Fig. 6. However, methanol possesses a number of advantages as a fuel: it is a liquid, and therefore easily transported and stored; it is cheap and plentiful, and the only products of combustion are $CO_2$ and $H_2O$. The advantages of a direct methanol fuel cell are: changes in power demand can be accommodated simply by alteration in supply of the methanol feed; the fuel cell operates at temperatures below *c.* 150°C so there is no production of $NO_x$ and methanol is stable in contact with the acidic membrane.

The basic problems currently faced by the direct methanol fuel cell are: (i) the anode reaction has poor electrode kinetics, particularly at

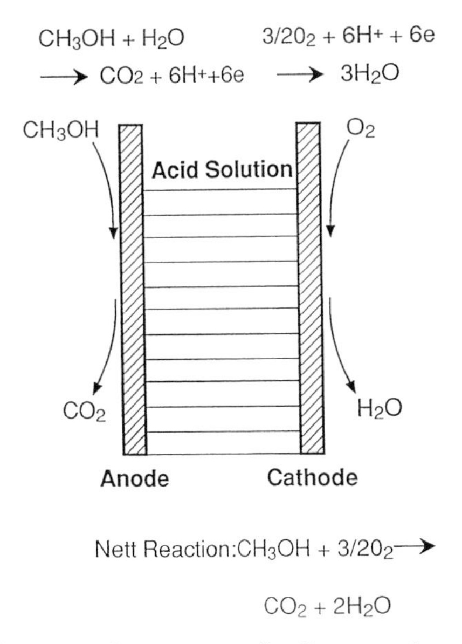


**Fig. 8** Schematic of operation of a direct methanol fuel cell.

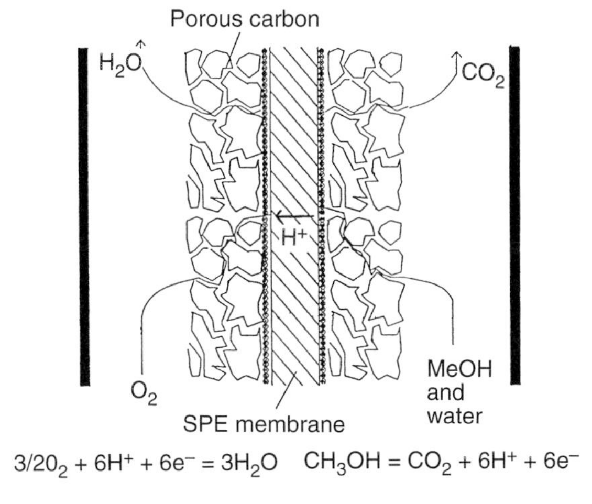


**Fig. 9** Actual construction of a modern DMFC based on the SPE configuration.

lower temperatures, making it highly desirable to identify improved catalysts and to work at as high a temperature as possible; (ii) the cathode reaction, the reduction of oxygen, is also slow, although the problems are not so serious as with aqueous mineral acid electrolytes. Nevertheless, the overall power density of the direct methanol fuel cells is much lower than the 600+ $mW/cm^2$ envisaged for the hydrogen-fuelled SPEFC;

(iii) perhaps of greatest concern at the moment is the permeability of the current perfluorosulphonic acid membranes to methanol, allowing considerable fuel crossover. This leads to degradation of performance, as the methanol can be oxidized at the *cathode*; the result is not only a waste of fuel but a lowering of the potential on the cathode and consequently a lowering of the overall cell voltage. Methanol vapour also appears in the cathode exhaust, from which it will have to be removed.

These problems are now being systematically attacked on an international basis. Perhaps the most important problem is the activity of the anode to the electro-oxidation of methanol; if this can be increased even by a relatively small amount, then the other problems identified above are also reduced in importance.

The oxidation of methanol at the anode does present some very important mechanistic problems. In contrast to the oxidation of hydrogen, in which the strength of the two Pt–H bonds forming at the surface are more than sufficient to break the H–H bond, the initial process in the oxidation of methanol is less clear. The methanol molecule itself consists of a central carbon atom bonded to three hydrogen atoms and an O–H unit. On approach to a flat metal surface, such as platinum, the methanol molecule must displace not one but several water molecules, as the first process that takes place is the successive breaking of C–H bonds by the simultaneous formation of C–Pt and H–Pt bonds, eventually leading to a three-coordinate intermediate. At low potentials, this process is further impeded by the presence of Pt–H bonds on the Pt surfaced arising from water reduction, and these must also be displaced, with the result that until the potential on the platinum surface has risen to rather high positive values, and all hydrogen atoms forming at low potential have been discharged, methanol can only be oxidized very slowly on Pt. This fact leads to low efficiencies for the overall fuel cell; the more positive the potential on the anode the less efficient the cell will be and with bulk platinum alone as the catalyst, cell efficiencies are too low for any sensible operation.

However, a rather different story emerges once bulk platinum is replaced by small platinum particles. The need for particulate catalysts was originally dictated by the expense of platinum; unless the metal could be dispersed in the form of small particles on an appropriate substrate, it would prove far too expensive to use, and this is, for example, the form of platinum found in exhaust catalysts on modern cars. It is also the form of platinum used in fuel cells, where the substrate normally used is very high surface-area carbon. The platinum particles are usually some 2–5 nm in diameter, and such particles will, of course, possess structures very different from normal planar platinum. In particular, as

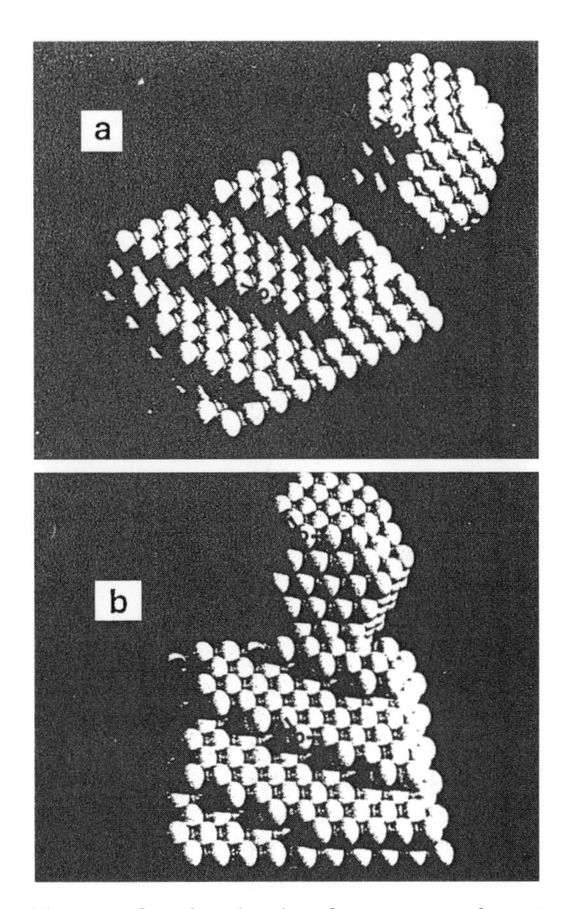


**Fig. 10** (a) Stepped surface (533) with step atoms that mimic edge atoms in the truncated octahedron; (b) step-kinked surface (432) with kink atoms that mimic corner atoms in the truncated octahedron (from Greenler *et al.*, *Surface Science*, 1985, **152**, 338).

shown in Fig. 10, these particles will possess substantial numbers of edge and kink sites, and it is now apparent that these sites are highly active for many electrochemical processes. In the case of methanol, the step sites are known to facilitate the adsorption of methanol: water and hydride species are held less firmly at such sites, and this permits methanol to adsorb even at low potentials in a stepwise process as:

$$CH_3OH \rightarrow Pt{-}CH_2OH + Pt{-}H$$

$$Pt{-}CH_2OH \rightarrow Pt_2{=}CHOH + Pt{-}H$$

The =CHOH species are now doubly bridged along the step edge; further reaction involves the necessity (on geometrical grounds) of forming a third Pt–C bond using a Pt atom not on the edge but on the terrace. This generates a three-coordinate species, which finally loses a fourth hydrogen from the oxygen atom in the $Pt_3{\equiv}C{-}OH$ species with the loss of one or more Pt–C bonds to form a strongly bonded $Pt{-}C{\equiv}O$ species. This species is known to be quite mobile on Pt terraces, and it can move away from the edge to allow further adsorption of methanol.

Our platinum particles then appear to allow quite a facile chemisorption of methanol, but while we have removed the hydrogen atoms, we have not added an oxygen atom, a process we need to effect if the methanol is to be fully oxidized to $CO_2$. Indeed, this second step is rate limiting on platinum particles, and appears to take place either by direct attack of water molecules on Pt–CO species on the terraces, or migration of the CO to edge, step or even kink sites at which again attack by water molecules can take place.

$$Pt{-}CO + H_2O \rightarrow Pt{-}COOH + H^+$$

$$Pt{-}COOH \rightarrow Pt + CO_2 + H^+$$

However, this attack is slow: if we were able to simultaneously chemisorb OH on to the surface of the platinum, a second reaction would become possible:

$$Pt{-}CO + Pt{-}OH \rightarrow 2Pt{-}CO_2 + H^+$$

a reaction that is known to be very much faster. Unfortunately, the surface concentration of Pt–OH is very low at low potentials, and although, therefore, adsorption of methanol is now fast, oxidation of the adsorbed fragments to $CO_2$ remains slow. However, as we and others have shown, by using a second metal such as ruthenium at a promoter, the surface concentration of Pt–OH species may be substantially increased. Provided the ruthenium atoms adsorb on and decorate some of the edge/step sites on the platinum particles, sufficient Pt–OH sites can be generated for rapid oxidation of CO species, giving rise to active catalysts that are likely to form the basis of further development in this field.

It might be objected that methanol itself is no real improvement on gasoline; both are converted to $CO_2$ and both will therefore contribute to the Greenhouse effect. In fact, methanol itself can easily be synthesized from $CO_2$ and $H_2$ in a reaction that is just the inverse of the reformer process described above. By relatively subtle alteration in reaction conditions, methanol can be synthesized in good yield, allowing us, in effect, simply to recycle the $CO_2$ formed. This, coupled with the fact that the DMFC is a low temperature device and therefore does not produce $NO_x$ species, does give it a substantial environmental edge.

The over-riding advantage of the *direct methanol fuel cell*, particularly in the DMFC–SPE configuration is that it is potentially very cheap and therefore competitive with the internal combustion engine, particularly in niche city driving applications, where the low pollution and relatively high efficiency at low load are attractive features. Performances from modern single cells are highly encouraging: an example is shown from our own work in Fig. 11, and it can be seen that in oxygen, power densities of up to 0.35 W/cm$^2$ are possible, in air a power density of 0.2 W/cm$^2$ has been attained with a pressure of 5 bar.

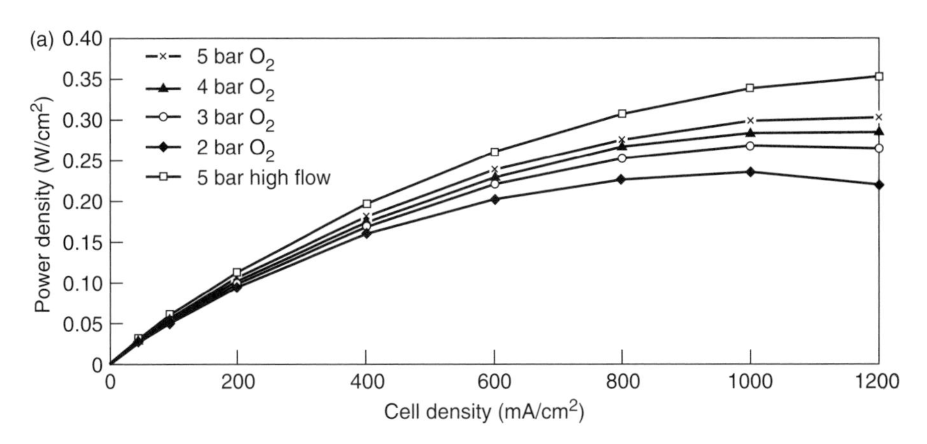


Power density data for MEA type F2. 97°C, 2M MeOH.

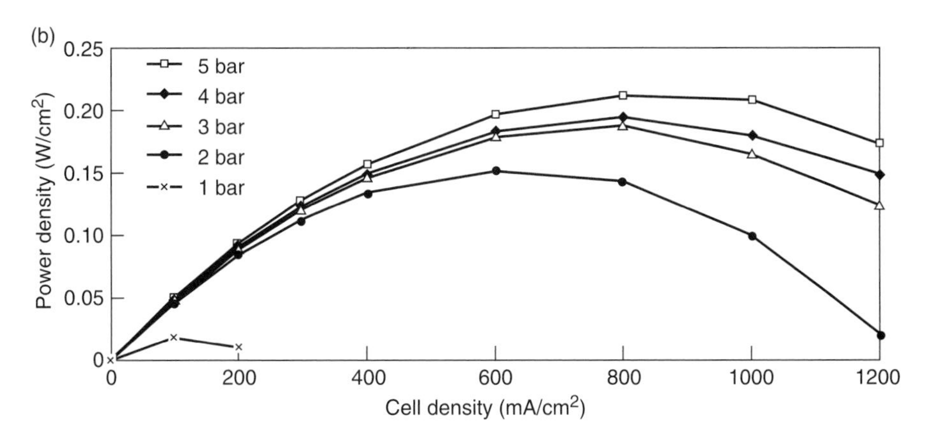


Power density data for type F2 MEA at different air pressures. 97°C, 2M MeOH.

**Fig. 11** Performance of a Newcastle DMFC under (a) oxygen; (b) air. Cells were operated at 97°C and at the pressures of combustant shown; in all cases, the fuel was 10% methanol in water.

## Conclusions

Sustainable energy conversion must be at the heart of our interests in the twenty-first century. We have seen how solar energy can be used to generate hydrogen; in turn this can be reacted with the $CO_2$ in the air to form methanol, and the methanol used directly as a fuel in modern fuel cells. The problems faced by this scenario in the near future are (i) to improve the efficiency of our new devices; (ii) to solve the remaining engineering problems, most of which are associated with scale-up; (iii) to fabricate prototypes of sufficient power to be able to drive somewhat larger vehicles.

Fuel cells remain at the threshold of significant application. Severe reductions in costs are essential if traction applications are to be realized, even given the advantages of environmental beneficence. Nonetheless, the overall picture is a positive one: the extraordinary progress over the last few years has placed fuel cells firmly on the agenda of energy foresight panels the world over, and I have little doubt that we shall start to see significant encroachment, at least in niche markets, in the next decade.

## Further reading

F.C. Treble (ed.). *Generating electricity from the sun*, Pergamon Press, 1991.

A. Johanssen *et al.* (ed.), *Renewable energy sources for fuels and electricity*, Island Press, Washington DC, 1993.

Batteries and fuel cells for stationary and electric vehicle applications (ed. A.R. Landgrebe and Z.-I. Takehara), *Electrochem. Soc. Symposium 93–8, 1993,* Electrochem. Soc., Pennington, NJ, 1993.

*Fuel cell systems* (ed. L.J.M.J. Blomen and M.N. Mugerwa) Plenum Press, NY, 1993.

A.J. Appleby and F.R. Foulkes, *Fuel cell handbook*, Van Nostrand Reinhold, NY, 1989.

K. Kordesch and G. Simader: *Fuel cells and their applications*, VCH, Weinheim, 1996.

ANDREW HAMNETT

Born in 1947, he obtained a DPhil from the University of Oxford. After two years' Postdoctoral study in UBC, Vancouver, he returned to Oxford to work with Professor John Goodenough on a series of novel spectroscopic and electrochemical systems to harness solar energy and to develop direct methanol fuel cells. In 1989, he accepted the established Chair of Physical Chemistry at the University of Newcastle upon Tyne where, in 1993, he was elected Pro-Vice Chancellor. His current interests include physical electrochemistry, developing novel spectro-electrochemical surface probes, and also practical fuel cells and their electrochemical energy conversion devices. He is the co-author of *Techniques and mechanisms in electrochemistry* (Chapman Hall, 1994), and has written a second textbook, '*Electrochemistry*' (Springer Verlag, 1997).

# *To divide or not to divide*

PAUL NURSE

The subject of this discourse is the control of cell division. A cell is the basic unit of life with all living organisms being made up of cells, and the growth and reproduction of living organisms is dependent upon cell reproduction. In order to emphasize the importance of this process I want to begin by introducing two great theories of biology, those of natural selection and the cell theory, and discussing how they have contributed to our understanding of what life is, and how cell reproduction is central to this understanding. I will then consider how the cell division process is controlled, describing the genetical and biochemical approaches to this problem which have recently led to the identification of the cyclin-dependent kinases as universal regulators of cell division. These protein kinases play a central part in all eukaryotic cells controlling whether a cell should divide or not divide.

## What is life?

Perhaps the most interesting question in biology is what distinguishes living from non-living things. There are two ways to approach this question. The usual approach is to describe what a living thing is in terms of its constituents and its properties. This emphasizes the types of constituents, proteins, nucleic acids, and lipids for example, and also specific properties of living organisms, such as the ability to grow, to reproduce, to move, and to be sentient. Listing such constituents and properties describes a living thing, but is rather unsatisfactory because it gives no sense of understanding of what life is.

The second approach is more interesting because it is based on an idea, Darwin's theory of evolution. First articulated by Muller,[1] this idea simply states that living things have properties which allow them to respond to natural selection and thus evolve. There are three properties of living organisms which are central to this idea. First, they must be able to reproduce. Secondly, they must possess an hereditary system whereby the characteristics of the living organism are defined by an information system which is copied and inherited during the process of reproduction. Thirdly, the hereditary information system, although basically stable, must be able to undergo small-scale variation which is inherited during the reproductive process. This provides the variability upon which natural selection can operate, allowing selection for those variants that are advantageous for survival of the organism. As a consequence there is more effective reproduction of those organisms with beneficial variations in the hereditary information system.

Living things clearly exhibit these properties. They basically breed true in the sense that a fly makes flies and a cat makes cats, but variations can occur and are inherited. As a consequence flies with short wings and more legs generate flies with short wings and more legs, while cats with an altered eye colour generate cats with the same altered eye colour. These inherited variants can be selected if they provide the organism with advantages in subsequent reproduction, allowing evolutionary change. This second approach to defining life is more satisfactory because it gives greater insight into what life is.

## Cell reproduction

If life is dependent upon a particular reproductive process which exhibits the three properties described above then it is important to understand the nature of this process. This introduces a second important idea in biology, that is the cell theory. This idea took 200 years for its full development. Cells were first described by Hooke in the mid-seventeenth century when he used a simple microscope to examine plant material.[2] He first looked at cork and observed it was made up of many tiny boxes which he termed 'cells'. These observations were rapidly extended, particularly by the Delft scientist Leewenhoek, who described other cells including single-celled microbes, eggs, and sperm. These observations did not initially lead to any unifying theories but by the late eighteenth century some scientists were beginning to speculate that all living things were made of cells. This was stated explicitly by the German scientist Oken who said that 'all flesh is made of infusorians' where an infusorian is a simple single-celled organism.[3] With the advances in microscopy at the beginning of the nineteenth century it became clear that all plants and animals were made of cells as formulated in the 1830s by Schleiden and Schwann.[3]

Initially, it was not clear how cells arose in the first place and indeed Schleiden and Schwann thought that new cells somehow arose in the body of an original cell. However, by the 1850s Virchow observed cells dividing into two and thus proposed his famous dictum 'omine cellula e cellula' or all cells arise from pre-existing cells.[2] This placed the process of cell division in a central position for understanding the growth of all living organisms. Subsequently, it became clear that cell division underlies all reproductive processes. Although many living organisms are multicellular, they always go through a single-celled phase at some time during their life history. For example, many animals reproduce via eggs and sperm. These are specialized single cells produced by the germ-line, which as Weissman argued in the late nineteenth century, provide the continuity between different generations of animals.[2] An egg fuses with a sperm to produce a fertilized egg, which after repeated rounds of division produces first an embryo and then an adult organism. Therefore, the cell reproductive processes which produce the germ cells and the subsequent embryo and adult are crucial to the whole phenomenon of reproduction.

## The cell cycle

Cells reproduce themselves during the cell cycle, which is defined as the series of events that occur from the birth of a cell to its subsequent division into two new cells. During the cell cycle all components necessary for survival must be doubled in amount and distributed equally to the two newly formed cells at division. This is particularly important for the hereditary material, the deoxyribonucleic acid or DNA. The DNA makes up the genes which are arranged in a linear fashion along the chromosomes. During each cell cycle all the chromosomes must be copied and then equally segregated at cell division.

The core to understanding the copying process is the structure of DNA elucidated in the 1950s by Crick, Watson, and Wilkins. Each gene is made up of a linear sequence of nucleotides. Nucleotides can be viewed as letters which form the basis of the hereditary information system. The type of letter (there are four types of nucleotide, abbreviated to A, T, C, G) and their order encode information which defines what the gene does and how it ultimately contributes to the overall behaviour of the living organism. Therefore, during the cell cycle this linear sequence must be precisely copied, generating two copies which are delivered to the two cells produced at division. The precise copying is possible because of the double helix structure of DNA. A DNA molecule consists of two strands or sequences of nucleotides which are entwined around each

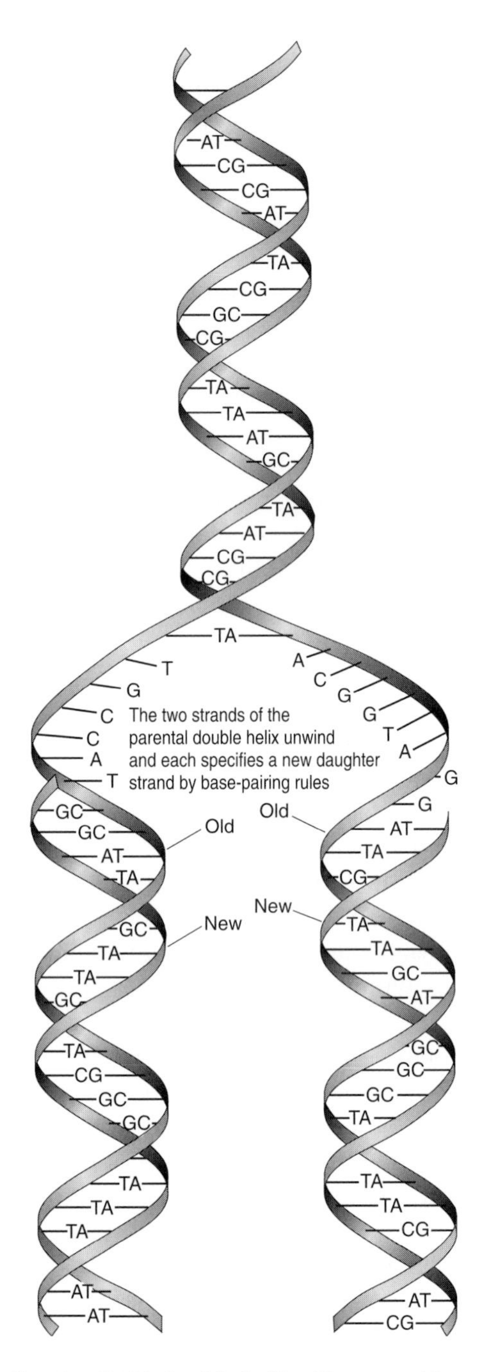


**Fig. 1** Replicating DNA double helix. The two sides or strands of the double helix are made up of the nucleotides A, T, C, G. At replication the strands separate and act as templates for making the new strands.

other. Each strand can be considered as the two sides of a ladder, and the rungs between the two sides are made up of two nucleotides, one associated with each side (Fig. 1). The four types of nucleotides consist of two sets of matching pairs, so that A always pairs with T and C with G. This means that each sequence of nucleotides making up one side of the ladder precisely complements the sequence making up the other side of the ladder. When the DNA molecule is copied, the two sides of the ladder are separated and each serves as a template for making a new DNA strand complementary to itself. This produces two double helices which are precise copies of the original single double helix. Thus the structure of DNA beautifully explains how it can encode information in the nucleotide sequence and how that information can be precisely copied.

This process of copying is called DNA synthesis and occurs during the S-phase which is the first major event of the cell cycle. After the DNA has been replicated there are two sets of chromosomes present in the cell which need to be equally segregated at cell division. The process which achieves this is called mitosis or M-phase, and it is the second major event of the cell cycle (Fig. 2). It is very prominent microscopically and was described by Flemming at the end of the nineteenth century.[2] At the first step of mitosis the chromosomes shorten or condense which allows them to move more easily about the cell. The replicated pairs of chromosomes then line up in the middle of the cell and separate apart with one member of each pair segregating into one of the two new cells formed at division. In this way each cell receives a full set of chromosomes containing all the genes required for the proper functioning of the cell.

Thus the cell cycle is made up of an S-phase and a M-phase. These two events ensure that the two cells receive a full complement of genes and form the hereditary information system. The fidelity of these processes ensures that living organisms basically breed true. However, the processes are not completely faithful. Small changes or mutations in the gene sequences can occur and when this happens the alterations are passed down from one cell to the next. These can arise both because the DNA copying process is not absolutely precise and because the low level DNA damage. The changes are the basis for the inherited variations upon which natural selection works. They give rise to slightly altered germ cells which in turn produce slightly altered adult organisms. Those variants which are most successful have a greater chance of survival and of passing these traits on to the next generation thereby bringing about evolutionary change.

From this discussion, it can be seen that the process of cell reproduction which occurs during the cell cycle is central to understanding the reproduction of all living organisms and to understanding the nature of

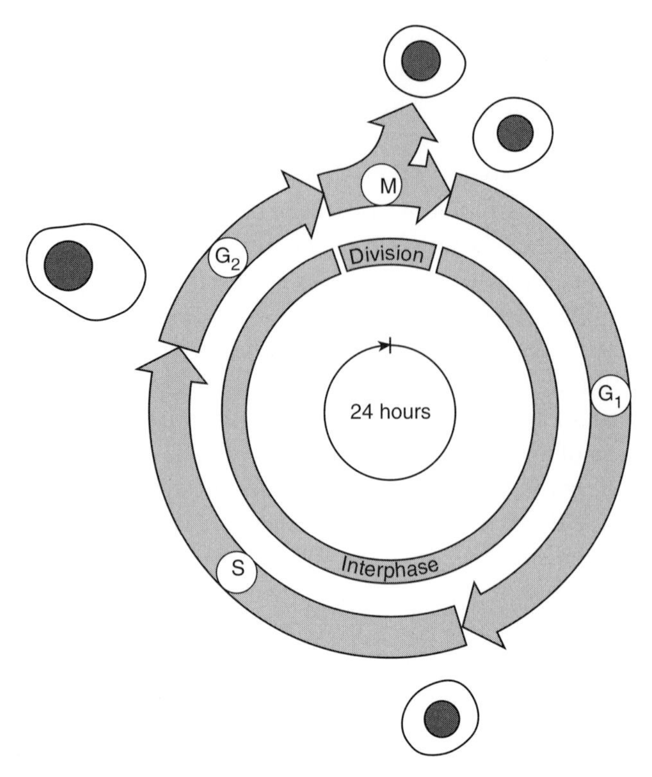


**Fig. 2** Events of the cell cycle. DNA synthesis occurs during S and mitosis during M. Gaps between S and M are called $G_1$ and $G_2$.

life. It is also of medical importance, because when the cell cycle goes out of control cells divide in an unrestrained fashion and can cause cancer. A tumour is made up of a mass of cells formed as a consequence of unrestrained cell division. The presence of such a mass of cells can disturb or even destroy the functions of the tissues or organs in which they occur, making cancer one of the most deadly diseases known to mankind. Therefore, understanding how the cell cycle is controlled is both an important basic biological problem and is also of great significance in medicine. It is this problem of cell cycle control that I will now consider.

## Cell cycle control

An important set of cell cycle controls determines the rate at which cells divide. These controls regulate the onset of S-phase and mitosis and ensure that they are co-ordinated correctly together. It is obviously necessary that the S-phase always precedes mitosis during the cell cycle, otherwise the cell would attempt to segregate unreplicated chromosomes leading to cell death. There should also be only one S-phase per cell cycle otherwise there would be too many gene copies in each cell. The co-ordination of these two events ensures that the hereditary information system is properly inherited.

An important factor in the control of most cell cycles is cell size. During the cell cycle of most normal growing cells the size of the cells doubles. Therefore, after division the newly formed cells are of similar size to that of the mother cell when it was born. As a consequence the onset of S-phase and mitosis is generally correlated with attainment of a critical cell size. This size can vary greatly between different cell types, and in certain specialized cells such as highly enlarged frog eggs, attainment of a critical size is not important and repeated cell divisions can occur without any growth at all. But generally growth to a critical cell size is an important requirement for progression through the major events of the cell cycle, S-phase and mitosis.

## Genetic analysis of cell cycle control

A very powerful approach for investigating cell cycle control has been the use of genetics. The method involves searching for mutants in genes which have an important part to play in the cell cycle. The most useful organisms for this approach have been the yeasts. These are single celled simple living organisms which despite their simplicity are similar to the cells of plants and animals, including humans. All these living organisms are called eukaryotic to distinguish them from viruses and bacteria which are called prokaryotic. Yeast cells with a short generation time grow in simple defined conditions and can be mated easily allowing genetics to be carried out very rapidly. Two types of yeast have been used, the budding yeast and the fission yeast, but because my work has been mainly with the latter I shall concentrate on the fission yeast *Schizosaccharomyces pombe* (Fig. 3). This particular yeast is responsible for the fermentation of the East African beer Pombe, accounting for its name.

So how can mutants altered in genes involved in the cell cycle be discovered? Yeast cells were treated with chemicals which damage DNA and so produce many mutated genes. The mutated yeast cells can than be examined to see if they display alterations in the cell cycle. Two types of cell cycle mutant have been particularly useful. The first are

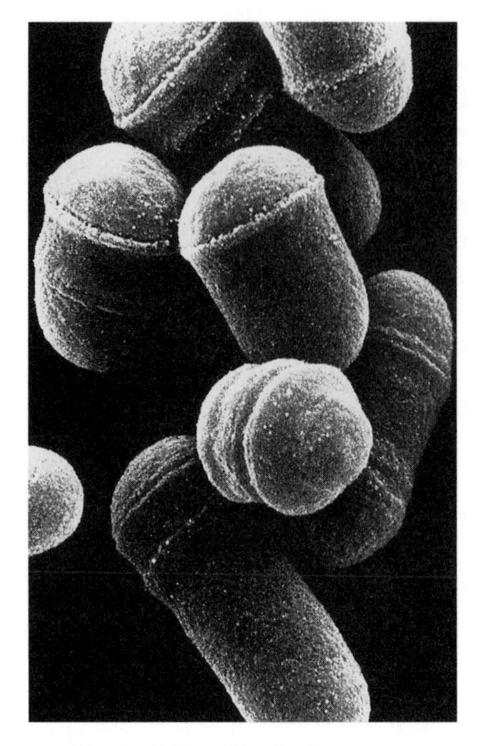

**Fig. 3** Cells of the fission yeast.

called cell division cycle or cdc mutants.[5] These are unable to complete some essential process of the cell cycle such as S-phase or mitosis and as a consequence cannot undergo cell division. Because the basic processes of cell growth are unimpaired, such mutants continue to grow without dividing and become highly enlarged. A fission yeast cell is cylindrical in shape and grows by tip elongation to produce a more extended cylinder. Therefore, inhibition of cell division produce highly elongated cylinders, easily identified under a microscope. Because a failure to divide is lethal, such mutants have to be conditional, meaning that the mutant behaviour is only expressed in a certain set of conditions. The most common condition used has been temperature, so cdc mutants have been sought that grow normally at a low temperature but are unable to divide at high temperature. This is because the mutated gene encodes a product which is temperature sensitive and is unstable at high temperatures.

The second type of mutant has been particularly useful for investigating cell cycle control. These are called wee mutants which are not blocked in cell division but divide at a small size.[6] They were first isolated in Scotland which accounts for their name.[7] These mutants advance cells prematurely through the cell cycle undergoing division before they have grown to the normal size, and as a consequence they divide at a smaller size producing cylinders of reduced length. The crucial characteristic of these wee mutants is that they are advanced into mitosis prematurely. This means that they are altered in some gene that demonstrates the rate of onset of mitosis. Because of the alteration in these particular genes, the cell is now able to undergo mitosis more rapidly than usual, indicating that these genes are rate limiting for progression through the cell cycle and for mitotic onset.

The existence of such cdc and wee mutants is very useful because it allows the alteration of only one particular component, and for the effects of that alteration to be studied on the whole system. The cdc mutants identify genes which when mutated are unable to divide, and therefore are required for some essential event of the cell cycle. In contrast the wee mutants identify genes which, when mutated, allow cell division to occur more rapidly and are therefore important for controlling the rate of cell cycle progression. Studies of these mutants have allowed the regulatory networks controlling the fission yeast cell cycle to be characterized.

## Regulatory gene networks controlling the cell cycle

One gene called *cdc2* was found to be particularly important in cell cycle control because it was defined by both cdc and wee mutants.[8] When the *cdc2* gene function was completely defective as in a cdc mutant, cells could not undergo mitosis and so cell division was blocked. In contrast when the *cdc2* gene function was more active than normal as in a wee mutant, then mitosis occurred more rapidly and so cells underwent mitosis and cell division at a reduced cell size. This identified the *cdc2* gene function as having a rate-limiting role controlling the onset of mitosis.[9] Several other genes were found to interact with *cdc2*. One of these was called *cdc25*, which was required to activate *cdc2*.[10] When *cdc25* was overexpressed *cdc2* became activated early and cells became wee, while in the absence of *cdc25*, the *cdc2* gene function could not become activated and so cells became cdc. In contrast the *wee1* gene inhibited *cdc2*.[11] When *wee1* was defective *cdc2* was activated early and when *wee1* was overproduced *cdc2* was inhibited, delaying mitosis. By experiments of this sort the regulatory network controlling the onset of mitosis was described identifying a crucial role for the *cdc2* gene function (Fig. 4).

As well as this role in mitosis, *cdc2* also has a crucial part controlling S-phase. More detailed analysis of the *cdc2* mutants revealed that as

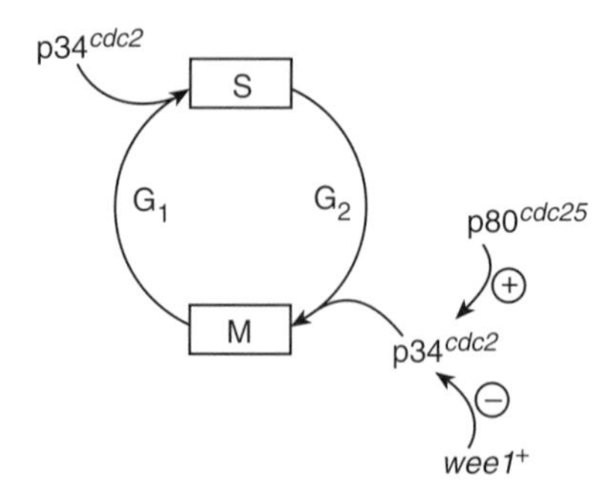


**Fig. 4** Cell cycle regulatory gene networks. The *cdc2* gene makes $p34^{cdc2}$ which acts before S-phase and mitosis. $p34^{cdc2}$ is regulated by cdc25 (making $p80^{cdc25}$) and *wee1* at mitosis.

well as being required for the onset of mitosis, *cdc2* was also required for the onset of S-phase.[12] The *cdc2* gene function is activated at a very early stage in the cell cycle, which has been called Start by Hartwell who discovered this control point in the budding yeast (reviewed in ref. 13). The Start control sets in motion the events which bring about the S-phase and eventually mitosis. An inhibitor of cdc2 acting at this stage of the cell cycle is called *rum1*.[14] When *rum1* is defective the *cdc2* gene function acting at S-phase is activated early and so the onset of the S-phase occurs at a smaller cell size than normal. Therefore, as with the onset of mitosis the *cdc2* gene plays a crucial part in controlling the onset of S-phase.

## Molecular genetic analysis

The genetic and cell cycle analyses described above have produced an abstract description of the controls acting over the cell cycle. The crucial genes are identified but their molecular function is unknown. In order to elucidate these molecular functions it is necessary to clone the genes and work out what they do. Cloning means to isolate chemically the DNA which makes up the gene of interest. This is particularly easy in yeast because genes can be cloned by what is called complementation. The DNA making up the yeast chromosomes is cut up into small pieces and cloned into vector molecules. A vector is usually a DNA plasmid propogatable in bacterial cells into which other pieces of DNA can be inserted. This makes a gene bank which can be introduced into yeast cells. The method used allows one or two different genes to be introduced into each individual yeast cell. The gene bank is then introduced into the mutant of interest, in this case a temperature-sensitive *cdc2* mutant. These mutant cells cannot divide at high temperature. However, if the *cdc2* gene is introduced into one of these cells then it would be able to grow and divide at the restrictive temperature. This is because the *cdc2* gene in the plasmid vector can rescue or complement the *cdc2* mutant in the chromosome. In this way the *cdc2* gene can be identified among the 7000 or so fission yeast genes. A similar procedure can be used to identify all the other genes of interest.

Once the gene has been cloned it can than be sequenced. This involves identifying the type of nucleotide and the order with which they occur in the gene. Having the gene cloned allows large quantities of it to be prepared in bacteria which can then be sequenced after chemical purification. The sequence of the gene sometimes gives insight into its function. This is because the nucleotide sequence determines the amino acid sequence of the protein encoded by the gene. Most genes encode proteins whose functions are determined by the type of amino acid and their order within the protein. Sequences of amino acids from motifs, and particular motifs are found in proteins with similar biochemical activities.

A useful way to investigate a gene function further is to raise antibodies against the protein made by the gene. When the gene is cloned, large quantities of the protein can be produced and used as an antigen for preparing antibodies. Antibodies bind very strongly to the protein, allowing its purification from cell extracts and its biochemical properties to be investigated.

## Molecular basis of cell cycle controls

Using a combination of the methods described above, the molecular function of the genes making up the cell cycle regulatory networks have been elucidated. The *cdc2* gene itself makes a protein cdc2p, which is a protein kinase.[9,15] This is an enzyme which puts a phosphate on to other proteins. A phosphate group has a strong negative charge and its introduction into proteins can alter the way the amino acids making up the protein fold up and function. These changes can have a major effect on the target proteins, activating or inhibiting them. The cdc2p protein kinase phosphorylates a set of proteins required for both the onset of S-phase and mitosis, altering their function, and as a consequence regulates the onset of both processes.

Activation of the cdc2p protein kinase to its highest level occurs at the onset of mitosis, and is inhibited by the *wee1* gene which also makes a protein kinase. wee1p phosphorylates an amino acid in cdc2p reducing overall protein kinase activity.[16] The *cdc25* gene makes a protein phosphatase cdc25p, which removes this phosphate and leads to full activation of the cdc2p protein kinase thus bringing about the onset of mitosis.[17] This phosphorylation control is the ultimate trigger for the onset of mitosis in fission yeast. However, there is another regulatory mechanism provided by the *cdc13* gene, which encodes a protein called cyclin.[18] This class of cell cycle protein was discovered initially by Hunt and Ruderman working with marine invertebrate eggs (reviewed in ref. 15). The cdc13p cyclin must be bound to cdc2p before the cdc2p protein kinase can be activated. The cyclin increases gradually as cells proceed through the cell cycle peaking at mitosis and then falling to a very low level for the start of the next cell cycle. Therefore, the cdc2p kinase activity at mitosis is regulated by both phosphorylation and its ability to complex with the cdc13p cyclin B.

To undergo S-phase, the cdc2p protein kinase is activated to a lower level than that needed for mitosis.[19] Activation is brought about by cdc2p binding to other cyclins, mainly the cig2p cyclin.[20] Before S-phase, cdc2p is inhibited by rum1p, which acts as a direct inhibitor of the cdc2p protein kinase and the rum1p inhibitor must be removed to allow cdc2p activity to rise and bring about S-phase.

Progression through the fission yeast cell cycle therefore seems to be driven by increasing cdc2p protein kinase activity. At the start of the cell cycle cylins are present at a low level and the presence of the rum1p inhibitor keeps cdc2p activity low. Appearance of the cig2p cyclin and disappearance of rum1p leads to low cdc2p protein kinase activity and onset of the S-phase. As cells grow, the cdc13p cyclin accumulates and a complex between cdc2p and cdc13p is formed. Its activity is regulated by the wee1p/cdc25p phosphorylation control and when the complex is maximally activated cells enter mitosis. To exit mitosis and undergo cell division the cdc13p cyclin must be degraded leading to a fall in cdc2p protein kinase activity. The next cell cycle can then begin and the cycle repeated.

## Cell cycle control in humans and other vertebrates

An important question is whether the way the cell cycle is controlled in yeast also applies to other living organisms including ourselves. This was approached by asking whether human beings had a gene that was equivalent to *cdc2*. If similar controls operate in human beings then we should also contain a *cdc2* gene. The usual way to try and identify similar genes in different organisms is to exploit the fact that such genes must have a similar nucleotide sequence. Various methods are available to identify genes related by sequence but there are difficulties using them to identify a gene such as *cdc2* in humans. The first difficulty is the extreme evolutionary divergence between yeast and humans. Their last common ancestor was probably about 1000 million years ago, giving an enormous time span for genes with a similar function to drift apart in sequence. The second difficulty is that cdc2p is a protein kinase, and although it is a unique protein kinase, there are many hundreds of other protein kinases in the cell with a similar structure.

An alternative approach was used to identify the human cdc2 gene. It was cloned by exploiting the fact that it had retained the same function using the same complementation procedure already described for cloning the yeast *cdc2* gene.[21] The only difference was that a human gene library was used instead of a yeast library. Copies of human genes were cloned into the DNA plasmid vector and introduced into a fission yeast temperature-sensitive *cdc2* mutant strain. If a human gene could provide the same function as the defective yeast *cdc2* gene then any yeast cells containing that human gene would be able to grow and divide. In this manner the human gene was cloned. After sequencing the gene it was found to encode a protein over 60% identical in amino acid sequence to the yeast cdc2p. Substitution of the yeast gene by the human version produced a yeast whose cell cycle was controlled by the human *cdc2p* gene, even though the two organisms diverged so long ago. Subsequently, similar work showed that *cdc2* genes exist in many other organisms including flies, worms, marine invertebrates, frogs, and mice.

The above experiments strongly argued that the basic elements of cell cycle control are similar in all eukaryotic organisms. Another demonstration of this came from work with the *Xenopus* frog.[22] This animal produces eggs which develop from oocytes found inside the body of the female frog. They are arrested just before M-phase, and the last stage of oocyte development requires their maturation into eggs by promoting entry into M-phase. Masui injected the contents of a matured egg into an oocyte which they immediately matured into an egg (reviewed in ref. 22). Mature eggs were thought by Masui to contain maturation promoting factor (MPF), which induced entry into the M-phase. Lohka and Maller purified MPF and found it contained two complexed proteins (reviewed in ref. 22). Using antibodies, one was shown to be cdc2p and the other to be a cyclin protein. Therefore MPF, which induced the M-phase in *Xenopus* frogs, is the same as the cdc2p-cdc13p cyclin complex, which induces mitosis in fission yeast.

From this work it was proposed that there are universal regulators controlling progression through the cell cycle in all eukaryotic organisms. These regulators are based on the cdc2p cyclin-dependent kinases which are required for the onset of the S-phase and mitosis and therefore control the overall rate of cell division.

## Complexity of controls and cancer

Work in vertebrate organisms has also revealed that their cell cycle controls are more complex than those found in the yeasts. The major difference is that there are more cyclin-dependent kinases (CDKs) and cyclins. In the yeasts there is only a single CDK while in vertebrates the core cell cycle control machinery requires at least two CDKs. Additionally, other CDKs are important for regulating entry into the cell cycle. In vertebrates many cells are in a resting state and important controls regulate exit from this state and entry into the cell cycle. Yeasts also contain only a few cyclins while there are a wider variety of types in vertebrates. The view which has emerged is that different CDK–cyclin complexes are required at different stages of the cell cycle including entry into the S-phase and mitosis. Activation of these differing complexes as cells proceed through the cell cycle brings about the major events in the correct sequence necessary for an orderly and successful cell division.

The cell cycle of cancer cells is no longer controlled properly and many of the various CDK–cyclin complexes must be inappropriately activated. Study of these cells has shown that changes have occurred in the regulation of certain CDK–cyclin complexes particularly those involved in entry into the cell cycle. Better understanding of these changes should lead to more insight into the way cancer develops, and may provide novel targets for new treatments. Understanding CDK–cyclin complexes and cell cycle controls better may also lead to a greater appreciation of how cancer arises. Cancer is brought about by alterations in a number of genes which influence cell growth and proliferation. These alterations occur because the behaviour of these genes is changed by mutation and by genomic instability. Normally such damage is dealt with by repair processes which block cell cycle progress until the damage is repaired. Increasingly, the CDK–cyclin complexes have been implicated in these cell cycle blocks, and defects within these controls can lead to cell division without damage being repaired. The accumulation of damage by these processes would accelerate the genetic changes which have to occur before cancer develops. Therefore, it is possible that many different cancers may be defective in these controls, providing potential common targets for treating cancer.

## Conclusions

I have argued that the cell cycle leading to cell division is central to the growth and reproduction of all living organisms, and that cell reproduction involves a hereditary information system which allows natural selection to take place. The cyclin-dependent protein kinases play an important part in controlling progression through the cell cycle. In the simple single-celled organism fission yeast, increasing CDK activity through the cell cycle first brings about entry into the S-phase and then entry into mitosis. Although more CDKs are involved in regulating the cell cycle of more complex organisms such as humans, the basic control system is very similar, indicating that the CDKs act as universal cell cycle regulators in all eukaryotic organisms.

It should be noted that the universality of the control mechanisms operating over cell division was proposed at the same time as the development of the cell theory as is clear from the following quotation from Schwann in 1839:[4] 'We have seen that all organisms are composed of essentially like parts, namely of cells: that these cells are formed and grow in accordance with essentially the same laws; hence that these processes must everywhere result from the operation of the same forces'.

However, despite these early speculations it has taken another 150 years to understand the molecular basis of this universal control process.

## Acknowledgements

I thank Jacky Hayles for useful advice on this manuscript.

## References

1. H.J. Muller, *Am. Nat.*, 1966, **100**, 493.
2. A. Pi Suner, *Classics of biology* (translated by C.M. Stern), Philosophical Library, New York, 1955.
3. W. Coleman, *Biology in the nineteenth century*, Cambridge University Press, Cambridge, 1977.
4. E.B. Wilson, *The cell in development and heredity*, Macmillan, New York, 1925.
5. P. Nurse, P. Thuriaux, and K. Nasmyth, *Mol. Gen. Genet.*, 1976, **146**, 167.

6. P. Nurse, *Nature*, 1975, **256**, 547.
7. P. Thuriaux, P. Nurse, and B. Carter, *Mol. Gen. Genet.*, 1978, **161**, 215.
8. P. Nurse and P. Thuriaux, *Genetics*, 1980, **96**, 627.
9. P. Nurse and V. Simanis, *Cell*, 1986, **45**, 261.
10. P. Russell and P. Nurse, *Cell*, 1986, **45**, 145.
11. P. Russell and P. Nurse, *Cell*, 1987, **49**, 559.
12. Y. Bissett and P. Nurse, *Nature*, 1981, **292**, 558.
13. P. Nurse, *The Fungal Nucleus*, Cambridge University Press, Cambridge, 1981.
14. J. Correa-Bordes and P. Nurse, *Cell*, 1995, **83**, 1001.
15. S. Moreno, J. Hayles, and P. Nurse, *Cell*, 1989, **58**, 361.
16. K. Gould and P. Nurse, *Nature*, 1989, **342**, 39.
17. K. Gould, S. Moreno, N. Tonks, and P. Nurse, *Science*, 1990, **250**, 1673.
18. I. Hagan, J. Hayles, and P. Nurse, *J. Cell Sci.*, 1988, **91**, 587.
19. B. Stern and P. Nurse, *Trends Genet.*, 1996, **12**, 345.
20. D.L. Fisher and P. Nurse, *EMBO J.*, 1996, **15**, 850.
21. M.G. Lee and P. Nurse, *Nature*, 1987, **327**, 31.
22. J. Gautier, C. Norbury, M. Lohka, P. Nurse, and J. Maller, *Cell*, 1988, **54**, 433.

## PAUL NURSE

Born 1949 in Norwich, gained a B.Sc. in Biology at the University of Birmingham and a Ph.D. at the University of East Anglia. After carrying out research in the Universities of Edinburgh and Sussex and in London, in 1987 he became Iveagh Professor of Microbiology and then Royal Society Napier Research Professor at the University of Oxford. Since 1993 he has been Director of Laboratory Research at the Imperial Cancer Research Fund in London. He has been awarded the Gairdner, Jeantet, and Pezcoler International Prizes and the Royal Society Wellcome and Royal Medals.

# *The biology of nitric oxide*

SALVADOR MONCADA

The properties of nitric oxide (NO) were studied in 1772 by Joseph Priestley, who called the gas generated by adding nitric acid to various metals 'nitrous air'. By mixing this nitrous air (NO) with common air, in a wide jar over water, then measuring the resultant decrease in gas volume, he devised a new volumetric method for measuring the purity of common air from different sources. The use of this method to test the purity of air for its 'fitness for respiration' was particularly appealing to Priestley as he would no longer need to use live mice to test it. He developed an apparatus, later known as a eudiometer or 'purity measurer' for the accurate measurement of gases and its use extended to many chemical studies, including Priestley's own discovery in 1774 of the generation of 'dephlogisticated air' upon heating of mercuric oxide. The properties of dephlogisticated air were later ascribed to a previously unidentified gas, which was given the name 'oxygen' by Lavoisier.

The naming of 'nitrous air' as nitric oxide is attributed to J.A. Murray in 1806. In the 1840s Walter Crum devised a method for preparing pure NO by shaking together nitric acid, concentrated sulphuric acid and mercury, and in 1908 Fritz Haber described the synthesis of NO in the electric arc.

In more recent years, NO has become known as a constituent of acid rain and photochemical smog. Acid rain is a precipitation containing high concentrations of sulphuric and nitric acids which arise from the emission into the atmosphere of sulphur dioxide and nitrogen oxide from automobile exhausts, some industrial operations and the burning of fossil fuels. These emissions also undergo photochemical reactions in the lower atmosphere, giving rise to smog, which causes reduced visibility and irritant damage to the eyes and lungs.

The identification of NO as a biological mediator came about in the 1980s. At that time I became interested in a substance, then known as endothelium-derived relaxing factor (EDRF), that had been found in the vasculature by Furchgott and Zawadzki in 1980.[1] This novel mediator was highly unstable and accounted for the endothelium-dependent relaxation that occurred in intact vascular tissue, but not in endothelium-denuded tissue, in response to a variety of different chemicals.

Many people, including our group, attempted to identify the chemical nature of EDRF. This elusive substance had been postulated to be many things, including a product of arachidonic acid metabolism, a carbonyl-containing molecule, a product of cytochrome P450 and even ammonia.[2] A number of experiments were carried out at that time in the attempt to transfer biologically active EDRF from one piece of vascular tissue to another. Some of these were sufficiently successful for the half-life of EDRF to be calculated as being between 6 and 50 s. Many other experiments were carried out attempting to identify the chemical structure of EDRF and to inhibit its activity. The difficulties associated with the research on EDRF at that time were related to the indirect nature of the majority of the experiments, as endothelium-dependent relaxation was measured as an indication of EDRF release and furthermore, when direct measurements were attempted, very small amounts of EDRF were generated from vascular strips or rings.

We had a great deal of experience in the technique of superfusion bioassay, which was originally developed by Gaddum in 1959 and later

---

**Fig. 1** Diagram showing a cascade for the bioassay of EDRF. Porcine aortic endothelial cells were grown in culture on microcarrier beads. Between 1–3 ml of beads containing 1–9 × $10^7$ endothelial cells were packed into a modified chromatographic column which was maintained at 37°C and perfused with Krebs' buffer at 5 ml/min (see enlarged detail). The perfusate was allowed to flow over a cascade of four strips of rabbit arterial tissue from which the endothelial cells had been removed. The arterial strips were mounted in heated (37°C) glass chambers and superfused (5 ml/min) with Krebs' buffer for 2–3 h before superfusion with effluent from the column. Changes in length of the tissues were detected by auxotonic levers attached to the transducers, the display of which was recorded. The arterial strips were separated from each other by a delay of 3 s and from the chromatographic column by a delay of 1 s. EDRF was released from the column by 1 min infusions of bradykinin (10–50 nm) through the column (TC). Compounds were also infused directly over the tissues (OT). Using this apparatus, the release of EDRF from endothelial cells can be monitored, its half-life down the column of tissues can be determined, and its biological activity compared with that of known biological agents applied directly over the tissues. The effluent can also be collected and subjected to chemical analysis.

modified by Vane in 1964, whereby tissues isolated from animals are placed in a cascade and a physiological solution or blood flows over the surface of each of them in turn. As each bioassay tissue responds by relaxing or contracting in a characteristic and reproducible way to a given vasoactive agent, the superfusion of a particular combination of tissues allows the recognition and measurement of substances whose biological activity has already been determined. More importantly, the technique permits the discovery of previously unknown substances, to which the bioassay tissues respond in an unfamiliar way. We decided to use this bioassay technique to study the properties of EDRF and to combine it with a means of generating a larger quantity of EDRF than had been used in other studies. To do this we cultured porcine aortic endothelial cells on microcarrier beads and packed them into a modified chromatography column (see Fig. 1). The perfusate through this column was used to superfuse a series of bioassay tissues, which would allow a

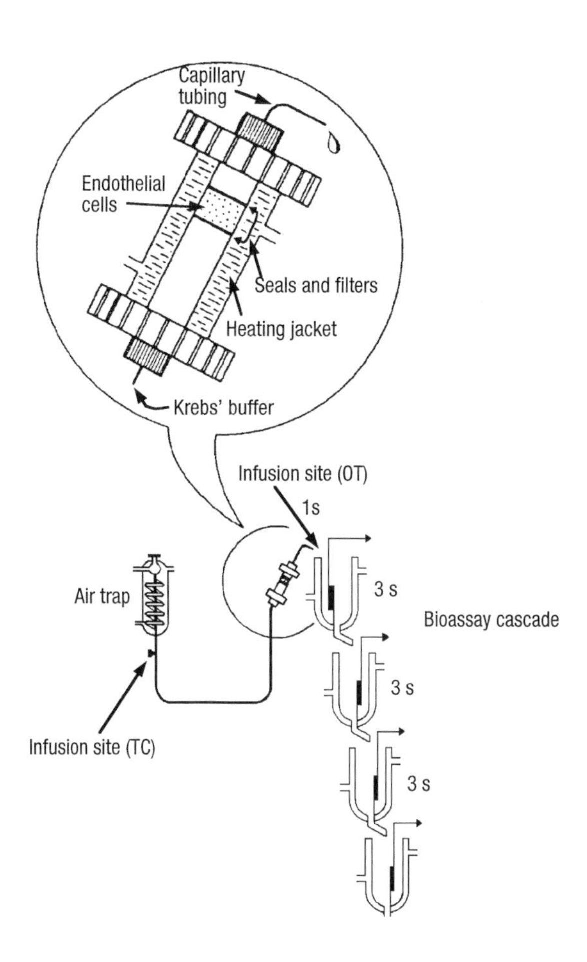

differential bioassay between EDRF and prostacyclin, another vasorelaxant generated in the vasculature which we had discovered some 10 years previously, also using the superfusion bioassay cascade. This system allowed us to study the EDRF generated by large quantities of endothelial cells at once and within a few months of its development we had demonstrated that superoxide anion ($O_2^-$) inactivated EDRF. This led to the elucidation of the mode of action of many inhibitors of EDRF as generators of $O_2^-$. These two papers provided important clues for the later identification of the chemical nature of EDRF (see Fig. 2).

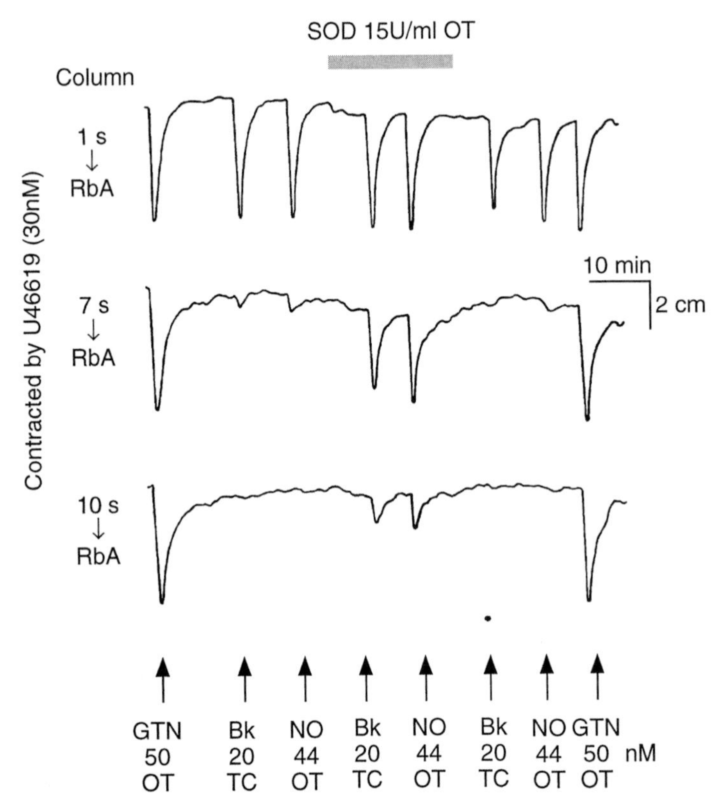


**Fig. 2** Relaxation of rabbit aortae by EDRF and NO. The bioassay tissues were relaxed by calibrating doses of glyceryl trinitrate (GTN), by EDRF released by bradykinin (Bk) and by authentic NO. During an infusion of superoxide dismutase (SOD), which interacts with and inactivates superoxide anion ($O_2^-$), the half-life of both EDRF and NO down the cascade is equally enhanced. This process is reversed after termination of the SOD infusion. This was a key experiment in demonstrating the pharmacological identity between EDRF and NO.

Among the many hypotheses about the nature of EDRF was the suggestion that EDRF might be NO or a related molecule, as both substances are unstable, sensitive to inactivation by $O_2^-$ or haemoglobin and were known to relax vascular smooth muscle via stimulation of the enzyme, soluble guanylate cyclase. This seemed to me to be an extremely interesting and attractive suggestion and therefore, we decided to investigate whether EDRF was in fact NO using two approaches: first by studying the comparative pharmacology of authentic NO gas and EDRF using the superfusion bioassay technique, and second, by trying to develop a method to measure the release of NO by vascular endothelial cells.

After the initial bioassay experiments, in the summer of 1986 I was convinced that EDRF and NO are one and the same substance, as their behaviour was identical in these and other pharmacological studies. These encouraging results led us to try to develop a method of measuring directly the release of NO from vascular endothelial cells. The method we chose was to measure NO by a chemiluminescence technique originally developed to analyse car exhaust and to monitor ambient air. We located a chemiluminescence analyser at the University of Surrey which was being used to detect the release of NO from nitrosamines, a contaminant of food products. Although this machine did not have the sensitivity required to detect NO released from biological tissue, we decided to see whether the experiment could be carried out and found that stimulation of endothelial cells with bradykinin generated a reproducible, but barely detectable, NO signal. Later, by changing the circuitry in the analyser we were able to enhance the sensitivity of this method sufficiently to demonstrate that NO is indeed released from endothelial cells and that it accounts for the biological actions of EDRF (see Fig. 3).[2] These results which we published in 1987 were soon followed by confirmatory findings by other groups. Around that time we received a great deal of critical correspondence including a letter expressing disbelief that a gas 'produced in the upper atmosphere through the energetic intervention of lightning' could have any relevance to biological systems.

We then tried to identify the biochemical source of NO, by feeding endothelial cells with different potential precursors, such as nitrite, nitrate, ammonia, or amino acids, none of which was successful. Formation of nitrite and nitrate from the amino acid L-arginine by activated macrophages had been reported a few months earlier. These papers prompted us to try, again unsuccessfully, to feed our endothelial cells with L-arginine to release NO. The project was abandoned until a few weeks later when we concluded that probably there was already an excess of L-arginine in the culture medium and decided to culture the cells in an L-arginine-free medium for 24 h before the experiment. This solved the

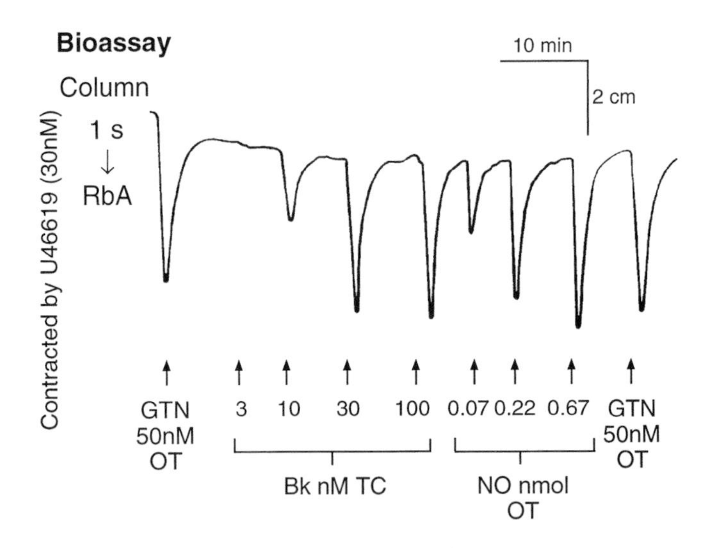


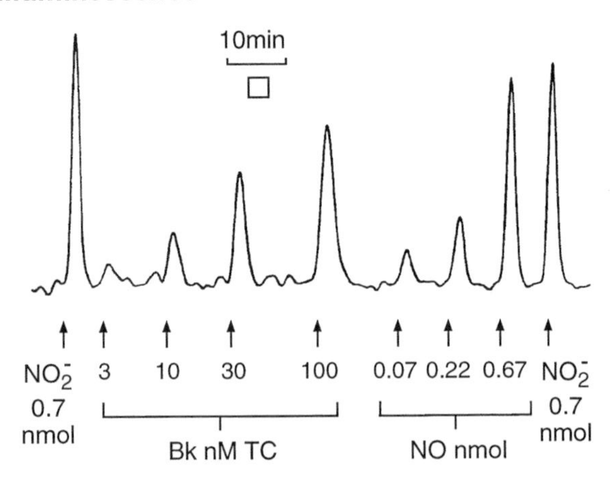


problem, as in this way we could demonstrate that the production of NO by the cells was enhanced by feeding L-arginine. These and other more sophisticated experiments using labelled L-arginine and mass spectrometry demonstrated that NO is synthesized from the guanidino nitrogen atom(s) of L-arginine. This reaction is enantiomer-specific, as the D-enantiomer of arginine is not a substrate (see Fig. 4).

These findings, and our later demonstration of the incorporation of molecular oxygen during the generation of NO, were the main steps in the discovery of what we started to call the L-arginine : NO pathway,[3] which involved an enzyme, NO synthase, converting L-arginine to L-citrulline and NO. Three isoforms of this enzyme have now been purified and cloned–endothelial (eNOS), neuronal (nNOS), and an

**Fig. 3** (A) Bioassay. Relaxation of rabbit aorta by EDRF and NO. The bioassay tissue was relaxed in a concentration-dependent manner by EDRF released from the cells by bradykinin (Bk; 3–100 nM TC) and by NO (0.07–0.67 nmol, OT) dissolved in He-deoxygenated $H_2O$. (B) Chemiluminescence. Release of NO by bradykinin (Bk) from a replicate column of the cells used in the bioassay. The amounts of NO (administered as 1 min infusion into the column effluent) which relaxed the bioassay tissues were also detectable by chemiluminescence. Effluent from the column, or Krebs' buffer into which authentic NO was injected, was passed continuously (5 ml/min) into a reaction vessel containing 75 ml 1.0% sodium iodide in glacial acetic acid under reflux. NO was removed from the refluxing mixture under reduced pressure in a stream of $N_2$, mixed with ozone and the chemiluminescent product measured with a photomultiplier. The amounts of NO detected were quantified after correcting for baseline drift using a polynomial fit and reducing electrical noise, by Fourier transformation and application of a Gaussian function. The areas under the peaks were converted to nmol of NO by reference to a $NO_2^-$ standard curve. Similar results were obtained in two other experiments. □, Area equivalent to 0.22 nmol NO.

---

inducible NO synthase (iNOS); eNOS and nNOS are expressed constitutively in a variety of cells, while iNOS is synthesized by cells following exposure to certain inflammatory cytokines.[4] As predicted, the L-arginine : NO pathway has turned out to be a widespread biochemical pathway in both mammalian and non-mammalian tissues and the synthesis of NO has been shown to underlie an even greater variety of biological functions than we had originally suspected.

Some analogues of L-arginine, such as $N^G$-monomethyl-L-arginine (L-NMMA), have been found to be inhibitors of the synthesis of NO and have proved to be valuable tools in understanding the biological actions of NO.[5] L-NMMA constricts vascular beds, produces a hypertensive response in animals and causes vasoconstriction of the forearm arterial circulation in humans. Its action is entirely endothelium-dependent, and its vasoconstrictor properties result from the inhibition of an endogenous vasodilator mechanism. These findings led us to conclude that there is a physiological, NO-dependent vasodilator tone that is essential for the regulation of blood flow and pressure. This has been confirmed by many different experiments including recent studies in knockout mice. In these studies the gene for eNOS was disrupted so that NO was not produced in endothelial cells of the mice and the resultant mutants have an elavated blood pressure when compared with control mice.

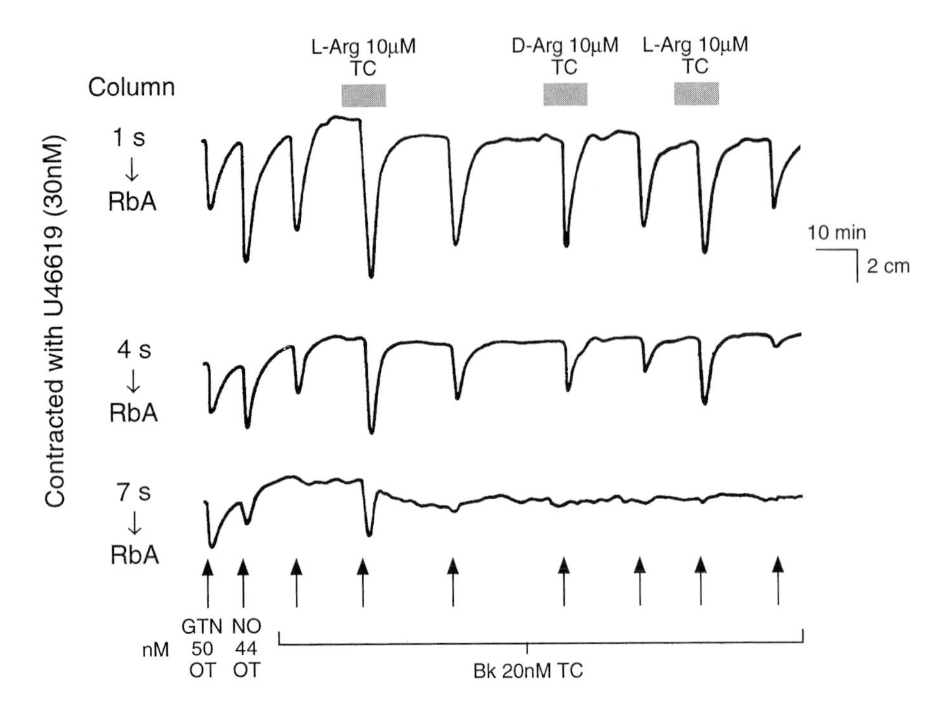


**Fig. 4** The effect of L-arginine and D-arginine on the relaxation of rabbit aortae by NO. A column (1.5 cm in diameter; $1–2 \times 10^7$ cells) packed with endothelial cells cultured on microcarriers for 10–16 days in Dulbecco's modification of Eagle's medium containing 10% fetal calf serum, penicillin (100 U/ml) streptomycin (10 $\mu$g/ml) and gentamycin (5 $\mu$g/ml) and then for 24 h in culture medium without L-arginine, CM(-Arg) cells, was perfused with Krebs' buffer (5 ml/min). The effluent was used to superfuse in a cascade three spiral strips of rabbit aorta denuded of endothelium. The tissues were contracted submaximally by a continuous infusion of 9,11-dideoxy-9$\alpha$, 11$\alpha$-methano epoxy-prostaglandin $F_2\alpha$ (U46619; 30 nM) and were separated from the cells by delays of 1,4, and 7 s, respectively. The amplification of the recorder was adjusted so that the response of each tissue to glyceryl trinitrate (GTN; 50 nM) administered over the tissues was similar. The bioassay tissues were relaxed by NO (44 nM over the tissues); the magnitude of the relaxations declined during passage down the cascade. NO, released by a 1 min infusion of bradykinin (20 nM, administered through the column), caused a relaxation of the bioassay tissues which also declined during passage down the cascade. This release was enhanced by infusion of L-arginine (10 $\mu$m), but not D-arginine (10 $\mu$m) through the column. Similar results were obtained in three other experiments. RbA, rabbit aorta; Bk, bradykinin.

The discovery of this vasdilator tone indicated the existence of an endogenous NO-dependent dilator system, the actions of which are imitated by compounds such as glyceryl trinitrate and sodium nitroprusside. The use of this type of compound for the treatment of angina goes back more than a hundred years.[6] For many years it was believed that the vasodilator action of glyceryl trinitrate was due to its conversion in the circulation to the weak vasodilator nitrite; however, in the 1940s it was demonstrated that a vasodilator dose of glyceryl trinitrate could not yield enough nitrite to account for its pharmacological action. It is now known that these compounds exert their pharmacological actions after their metabolism, by enzymic or non-enzymic processes, into NO. The NO thus liberated reacts with the haem group in the soluble guanylate cyclase of the vascular smooth muscle cell, and the activated enzyme produces more cyclic guanosine monophosphate (GMP), leading to vascular relaxation. These drugs have therefore become known as 'NO donors' as they mimic the actions of the endogenously released NO.

NO inhibits the aggregation and adhesion of platelets, via a cyclic GMP-dependent mechanism. Platelets themselves generate NO and the L-arginine : NO pathway acts as a negative feedback mechanism to regulate platelet aggregation.[5] Interestingly, we have found that NO has a similar role in a species that has existed unchanged for over 500 million years. The haemocytes of the American horseshoe crab (*Limulus polyphemus*) are haemostatic cells that perform a function comparable to that of mammalian platelets. We have shown that these haemocytes also produce NO, suggesting that the L-arginine:NO pathway originated very early in evolution. There is now evidence for the existence of this pathway in species ranging from humans and other mammals, through birds and insects to protozoa and slime moulds. Recent evidence suggests that plants produce NO, a finding which, together with the earlier identification of soluble guanylate cyclase in plants, suggests that the L-arginine : NO pathway may not be confined only to the animal kingdom. As yet no role for NO has been identified in plants.

With the growing knowledge of the biological actions of NO have come new uses for NO donors. Some of these new uses will greatly depend on the possibility of developing NO donors with selective actions away from the vascular tissue. Like NO, some NO donors are potent inhibitors of platelet aggregation and adhesion *in vivo*. Studies using *S*-nitroso-glutathione have shown that this compound potently inhibits platelet aggregation and adhesion at doses that cause only minimal vasodilatation. *S*-nitroso-glutathione inhibits the platelet activation that occurs, despite concurrent treatment with aspirin, glyceryl trinitrate, and heparin, in patients undergoing balloon angioplasty for coronary artery blockade. Moreover, it seems to be effective in the treatment of HELLP syndrome (haemolysis, elevated liver enzymes, and low platelet count), a rare and severe form of pre-eclampsia in which platelet clumping in the vasculature may be a major component of the condition. Selective targeting to platelets without accompanying hypotension will allow the use of this type of compound in thrombotic disorders, where the NO donor may be used alone or in combination with other antithrombotic agents.[7]

As NO plays a number of parts in the physiological functioning of the cardiovascular system, it is hardly surprising that reduced generation of NO has been linked to a number of clinical disorders, including hypertension and atherosclerosis.[8] Interestingly, administration of L-arginine, the substrate for formation of NO, appears to enhance the generation of NO. Thus, L-arginine causes a rapid reduction in systolic and diastolic pressures when infused into healthy humans and patients with various forms of hypertension. Infusion of L-arginine also improves endothelium-dependent vasodilatation in cholesterol-fed rabbits and in hypercholesterolaemic patients. Furthermore, dietary supplementation with L-arginine reduces platelet reactivity in hypercholesterolaemic rabbits. Administration of L-arginine may also be beneficial in preventing restenosis after balloon angioplasty, as it reduces intimal hyperplasia in rabbits.

In 1988 it was demonstrated that rat cerebellar cells stimulated with N-methyl-D-aspartate (which activates certain receptors for the neurotransmitter glutamate) release an EDRF-like material and have elevated levels of cyclic GMP. At about this time we were looking for the L-arginine : NO pathway in the brain, as we had learned from the literature that L-arginine can activate the soluble guanylate cyclase in brain cells and tissue. We found that addition of L-arginine to cytosol from rat brain synaptosomes does indeed result in the formation of NO; this process was inhibited by haemoglobin and L-NMMA, showing that the brain possesses the NO synthase. Now it is known that nNOS is widely distributed in the central nervous system where it is stimulated by the action of glutamate on a specific N-methyl-D-aspartate receptor. The NO thus produced mediates a variety of functions including synaptic plasticity, regulation of cerebral circulation and cerebrospinal fluid production, induction and regulation of the circadian rhythm, the induction of hyperalgesia, and the development of tolerance to and withdrawal from morphine. NO also appears to play a part in the development of the nervous system, either via a trophic action on developing neurons or by eliciting programmed cell death (apoptosis).[9,10]

nNOS has been found in similar locations in primitive and higher species. For example, immunohistochemical studies in the locust

(*Schistocerca gregaria*) brain have shown that the enzyme is particularly abundant in the olfactory processing centres, the antennal lobes. NOS-containing local interneurones have also been identified in the mammalian olfactory bulb, suggesting that NO performs analagous functions in locust and mammalian olfactory systems. It is now clear that NO released in the olfactory bulb plays a part in the formation of olfactory memory, as sheep treated with inhibitors of NO synthase or of soluble guanylate cyclase fail to recognize their own offspring. It is likely that in the locust NO also plays a part in the formation of olfactory memory.

In the peripheral nervous system, NO is now known to be the mediator released by a widespread system of nerves, previously recognized as non-adrenergic and non-cholinergic, and now known as nitrergic. These nerves mediate some forms of neurogenic vasodilatation and regulate certain gastrointestinal, respiratory, and genitourinary functions.[11–13] Thus in the gastrointestinal tract, NO is responsible for some forms of relaxation such as occurs during peristalsis, sphincter relaxation, and the dilatation of the stomach in response to increased intragastric pressure. NO contributes to relaxation of tracheal muscle and of the urinary bladder.[8] In addition, the L-arginine : NO pathway is responsible for the relaxation of the corpus cavernosum and thus penile erection in animals and humans. Electrically evoked relaxation of the corpus cavernosum *in vitro* is prevented by inhibitors of NO synthase and mimicked by NO donors. Immunohistochemical evidence of nerves containing NO has been found in penile tissue from different species, including humans (Fig. 5). Furthermore, small doses of an inhibitor of NO synthase reduce electrically induced penile erections in rats. Thus, NO is the final common mediator of penile erection. These physiological actions of NO are mediated by activation of the soluble guanylate cyclase and

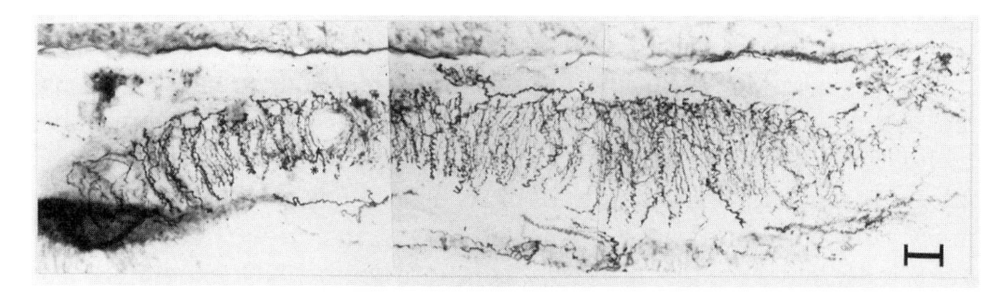

**Fig. 5** Immunostaining with nNOS antibody of an arteriole in the human corpus cavernosum. The arteriole is horizontal; note undulated nNOS-positive fibres encircling the vessel. Scale bar is 800 $\mu$m.

consequent increase in concentration of cyclic GMP in target cells. These nitrergic nerves are proving to be as important as adrenergic, cholinergic and peptidergic nerves, and their dysfunction may lead to a variety of disorders, including hypertrophic pyloric stenosis in infants, achalasia, hyperactivity of the urinary bladder, and male impotence.[8]

NO synthase is induced in activated white cells and the NO produced by this iNOS accounts for the L-arginine-dependent cytostatic and cytotoxic actions of these cells against tumour cells, bacteria, fungal cells, protozoan parasites, and viruses.[14,15] The importance of NO in the killing of pathogens can clearly be shown using knockout mice for iNOS. Activated peritoneal macrophages taken from normal (wild-type) mice are very efficient in killing the intracellular protozoan parasite *Leishmania major*. In contrast, activated macrophages from iNOS-deficient macrophages engulf the leishmania but are unable to kill them (F.Y. Liew, personal communication). Thus the L-arginine:NO pathway acts as a primary defence mechanism against intracellular microorganisms; in addition NO released by the white cells can kill pathogens that are too large to be engulfed by these cells.[16]

iNOS is induced in endothelial and vascular smooth muscle cells, as well as other cells and tissues in a number of pathological situations such as septic shock. NO generated by this enzyme accounts for the profound vasodilatation, resistance to vasoconstrictors, and vascular leak syndrome associated with this condition. In patients with septic shock, low doses of L-NMMA added to standard therapy have been shown to restore blood pressure. It remains to be demonstrated whether this increase in blood pressure will result in reduced mortality in patients, who at present have no more than approximately a 50% chance of survival. Selective inhibitors of iNOS are being developed that are likely to provide improved therapy as, unlike L-NMMA, they will allow NO to continue to be produced by the constitutive NO synthases.[7]

Increasing evidence indicates that NO may also play a part in acute and chronic inflammation. Enhanced production of NO, citrulline, or nitrite has been demonstrated in patients with asthma, ulcerative colitis, and arthritis. The presence of iNOS has been demonstrated immunohistochemically in macrophages in the inflamed tissue around loosened joint replacement implants. iNOS or its mRNA has also been detected in microglial cells and brain tissue after viral infection, treatment with cytokines, or in experimental encephalitis; the NO generated may contribute to the neuronal damage associated with such conditions. Treatment with inhibitors of NO synthase reduces the degree of inflammation in rats with acute inflammation or adjuvant arthritis, whereas L-arginine enhances it. Immune complex-induced vascular injury in rat lungs and dermal vasculature can be attenuated by inhibitors of NO synthase. Furthermore, inhibitors of NO synthase ameliorate experimentally induced chronic ileitis. All these data suggest that selective inhibitors of iNOS may be beneficial in the treatment of acute and chronic inflammation and this has led to the search for such compounds by the pharmaceutical industry worldwide. In support of this, iNOS knockout mice are resistant to endotoxin and display a reduced inflammatory response to phlogogenic agents when compared with their wild-type controls. Interestingly, these animals also have a reduced capacity to fight invading micro-organisms. As NO is a general defence mechanism, the degree to which selective iNOS inhibition will impair this vital function remains a worrying prospect to be studied when the compounds become available.[7]

In summary, NO has turned out to be a very significant biologically active molecule. Its actions range from regulation and control of many physiological systems to defence mechanisms and pathophysiology. What was totally unexpected was the fact that so many different functions could be mediated by such a simple molecule. Whether NO is unique or is part of a family of gases with biological actions is not clear at this stage. What is clear is that whichever functions are mediated by NO, they have been highly conserved throughout evolution which indicates the efficiency of this unique mechanism.

## Acknowledgement

I am very grateful to Annie Higgs for assisting in the preparation of this manuscript.

## References

1. R.F. Furchgott, *Annu. Rev. Pharmacol. Toxicol.*, 1984, **24**, 175.
2. S. Moncada, R.M.J. Palmer, and E.A. Higgs, *Thrombosis and haemostasis*, Leuven University Press, 1987, p. 597.
3. S. Moncada, R.M.J. Palmer, and E.A. Higgs, *Biochem, Pharmacol.*, 1989, **38**, 1709.
4. R.G. Knowles and S. Moncada, *Biochem. J.*, 1994, **298**, 249.
5. S. Moncada, *Acta Physiol. Scand.*, 1992, **145**, 201.
6. S. Moncada, R.M.J. Palmer, and E.A. Higgs, *Hypertension*, 1988, **12**, 365.
7. S. Moncada and E.A. Higgs, *FASEB J.*, 1995, **9**, 1319.
8. S. Moncada and E.A. Higgs, *N. Engl. J. Med.*, 1993, **329**, 2002.
9. J. Garthwaite, *Trends Neurosci.*, 1991, **14**, 60.

10. S.H. Snyder and D.S. Bredt, *Sci. Am.*, 1992, **266**, 68.
11. J.S. Gillespie, X. Liu and W. Martin, *Nitric oxide from L-arginine: a bioregulatory system*, Elsevier, Amsterdam, 1990, p. 147.
12. M.J. Rand, *Clin. Exp. Pharmacol. Physiol.*, 1992, **19**, 147.
13. N. Toda, *Nitric oxide in the nervous system*, Academic Press, Orlando, 1995, p. 207.
14. C.J. Nathan and J.B. Hibbs, Jr, *Curr. Opin. Immunol.*, 1991, **3**, 65.
15. A.K. Nussler and T.R. Billiar, *J. Leukoc. Biol.*, 1993, **54**, 171.
16. J. MacMicking, Q-W. Xie, and C. Nathan, *Annu. Rev. Immunol.*, 1997, **15**, 323.

## SALVADOR MONCADA

Born 1944 in Honduras, he received a medical degree in El Salvador and came to London in 1971 to do a PhD on the mode of action of aspirin-like drugs at The Royal College of Surgeons. After a brief return to Honduras he joined the Wellcome Research Laboratories in 1975, where he worked on the unstable metabolites of arachidonic acid. He initiated the work that led to the discovery of the enzyme thromboxane synthase and of prostacyclin. His studies contributed to the understanding of how small doses of aspirin prevent cardiovascular episodes, i.e. myocardial infarction and stroke. He obtained a DSc in 1983 and in 1986 was appointed Director of Research at Wellcome. In 1988 he was made a Fellow of the Royal Society and in 1994 he became a Fellow of the Royal College of Physicians as well as a foreign member of the National Academy of Sciences of the USA. Since 1986 his major research interest has been the biology of nitric oxide. In 1995 he became the Director of the Wolfson Institute for Biomedical Research, formerly known as the Cruciform Project at University College London. He has published over 700 papers and edited a number of books on nitric oxide.

# *An imaging renaissance—new opportunities for medicine*

IAN ISHERWOOD

Medical images in mediaeval times were hand produced, finely illuminated manuscripts—in most circumstances the only means of visual communication about medical situations and usually the only visual aid for practical surgical procedures (Plate 1). The publication in 1543 by Vesalius of *De Humani Corporis Fabrica—On the structure of the human body*—heralded a new era of communication technology in medicine. Vesalius perceived that the printed textbook could be used to provide reproducible anatomical teaching material without risk of corruption (Fig. 1). The advent of the microchip in recent years and the concept of reproducible images from digital data have had much the same revolutionary impact on traditional pictorial radiology. It is no longer necessary, or even desirable in some instances, for *in vivo* biological events to be recorded pictorially. The 'image' may be entirely numerical, graphic, or spectral. The radiologist of the 1990s might paraphrase the thoughts of Stephen Daedalus as he walked the shore in James Joyce's *Ulysses*—'Signatures of all things I am hear to read'.

The work of Vesalius, by its very reproducibility, was a challenge to orthodoxy which shifted the paradigm of medical and scientific communication in contemporary Europe. It is often easy to dismiss or ridicule radical developments and what to some appear to be unorthodox deviations. It is sometimes less easy to detect their value and their catalytic potential. I would like to suggest to you tonight that the evolution and renaissance of medical imaging over the last 100 years has not only been the product of remarkable technological advances but has also involved challenges to orthodoxy which in themselves have led to further and dramatic paradigm shifts.

**Fig. 1** Illustration from *De Humani Corporis Fabrica*, Vesalius 1543.

At the turn of the nineteenth century there occurred four major and influential scientific events all of which had a profound effect on twentieth century medical imaging. While the discovery of X-rays by Röntgen in November 1895 overshadowed the others in the public mind, the discoveries of radioactivity by Becquerel, of the electron by J.J. Thomson, and the splitting of the spectral lines of light by Zeeman and Lorentz were all of seminal importance. Radioactivity led to new fields of both diagnosis and treatment. An understanding of the electron exposed the entire panoply of twentieth century science to medicine. The Zeeman effect—predicted by Faraday—proved to be a powerful tool in the unravelling of atomic structure and decisive for the later discovery of electron spin.

The scene was set for the scientific explosion by the conjunction of three nineteenth century developments—the vacuum, electricity, and photography. The account of Röntgen's discovery, almost exactly 100 years ago on 8 November 1895, of the penetrating powers of a new ray emanating from a Lenard cathode ray tube by the chance observation of fluorescing crystals nearby, is perhaps too familiar to recount in detail. Suffice it to say that all three nineteenth century developments were involved in the observation and its recording. It is perhaps less well known that others had made similar observations but failed to recognize their significance. In 1784 William Morgan, a Welsh mathematician, in the course of boiling mercury to create a vacuum had observed the emitted light change from green to violet to blue and then to 'invisible light'—the first documented account of X-ray production. Sir William Crookes complained during his experiments with a vacuum tube in the 1890s that photographic plates nearby were being fogged. Pasteur said 'chance favours the prepared mind' but, as Röntgen later admitted, it was not so much the science as the image of Frau Röntgen's hand (Plate 2) which took the world by storm!

Only two people in Great Britain are known to have received Röntgen's startling paper directly—Arthur Schuster, Professor of Physics in Manchester and Lord Kelvin, Professor of Physics in Glasgow (Fig. 2). Schuster was so intrigued that he kept his wife and daughter waiting in a cab outside the physics building on a cold winter evening for over an hour while he read it. Kelvin was sceptical and, it is said, made three predictions—that radio would never work, heavier than air machines would never fly, and X-rays would prove to be a hoax! He was soon to retract, however, and generously acknowledged Röntgen's work. The news was reported in the *Daily Chronicle* on the 6th of January 1896. On the 7th, C.H. Lees, assistant to Schuster, spoke to the Manchester Literary and Philosophical Society while Schuster himself wrote to the *Manchester Guardian* on the 8th. In Glasgow Lord Kelvin was ill and gave Röntgen's paper to the physicist Bottomley who in turn gave it to John Macintyre. Macintyre both ENT surgeon and hospital electrician—a combination which might now have qualified him for performance-related pay!—was a true visionary. He immediately recognized the potential for both diagnosis and treatment and went on to establish the first hospital X-ray department in the world. Much of the clinical work in other centres in the early years was carried out in physics laboratories or

**Fig. 2** Professor Arthur Schuster reclining in his study in Manchester. Note the portrait of Lord Kelvin on the bookcase in the background. Reproduced by courtesy of the Wellcome Institute Library, London (Schuster Collection).

chemist's shops (Fig. 3). The former complained of interruption to scientific research while the latter found it necessary to contract out their services to local hospitals.

The first X-rays in Britain were obtained in London on the evening of the 7th of January 1896 by A.A. Campbell Swinton, an electrical engineer who having read the newspaper accounts, constructed his own apparatus. The *Lancet* referred to 'the new photography' while the *British Medical Journal* considered the discovery to be 'a feat sensational enough and likely to stimulate even the uneducated imagination'. The *Electrical World* advertised 'X-ray proof underclothing especially for the sensitive woman'. Punch was more concerned with British resentment at the Kaiser's intervention in colonial affairs (Fig. 4) and such satirical comments as

**Fig. 3** Isenthal's laboratory in London 1898. Isenthal was an early supplier of X-ray equipment and a founder member of the Röntgen Society, now The British Institute of Radiology, the oldest radiological society in the world.

We do not want, like Dr Swift
To take our flesh off and to pose in
Our bones, or show each little rift
And joint for you to poke your nose in

Over a 1000 scientific papers were published on the subject in 1896 but the effect on the public imagination was unprecedented. Apparatus was easy to construct and, without immediate knowledge of any harmful effects, the results easy to demonstrate.

The key influences on the rapid acceptance and development of diagnostic X-rays in the medical field were undoubtedly related to the novel facility offered for the demonstration of fractures and the detection of foreign bodies—notably bullets. The construction of the Manchester Ship Canal in 1894 had already necessitated the development of a service for the organized care of the civilian injured. A number of military campaigns including the war in the Sudan and subsequently the Boer War provided the opportunity to put this organization and the new technology to the test. X-ray tubes and glass photographic plates were

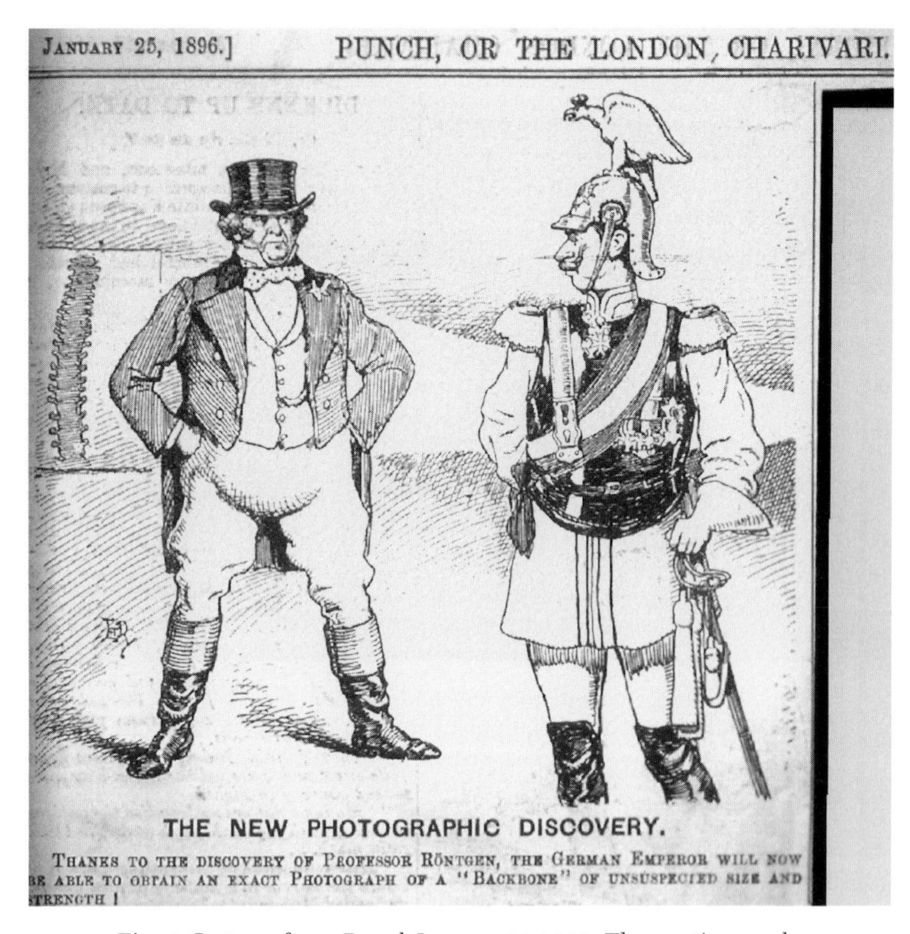


**Fig. 4** Cartoon from *Punch* January 25 1896. The caption reads "Thanks to the discovery of Professor Röntgen, the German Emperor will now be able to obtain an exact photograph of a 'backbone' of unsuspected size and strength!" Reproduced by courtesy of *Punch* (25 Jan 1896).

both heavy and extremely fragile while the means of generating electricity and the facilities for photographic processing were primitive. Climatic conditions were frequently hostile. Nevertheless with considerable effort and not a little ingenuity (Fig. 5) the difficulties were overcome. From the point of view of both surgeon and patient in the field the benefits were obvious yet the War Office was initially sceptical of this unorthodox addition to military surgery.

The years before the First World War were years of rapid progress with developments in both apparatus and technique; the bismuth meal

**Fig. 5** Major Battersby and his orderly X-raying a casualty at the battle of Omdurman in the Sudan in 1899. The primitive equipment was powered by a tandem bicycle. Battersby recorded 'Having carefully adjusted the current, my warrant officer took his seat on the bicycle and commenced pedalling. As the resistance became greater, a sensation of riding uphill was experienced and the services of an additional orderly were requisitioned for the front seat. The practice was carried out in the shade of a temperature of 110°F. At the end of half an hour we unanimously agreed that some other form of scientific amusement was desirable!' Reproduced by courtesy of the British Institute of Radiology (Battersby J *Archives of the Röntgen Ray* 1899, 3, 74–8).

examination of the stomach, for example, exploiting the use of the radio opacity of high atomic number elements was introduced in the first decade of the twentieth century. The Coolidge X-ray tube was available in 1913. The potential hazards of overexposure to X-rays were, regrettably, appreciated only slowly and the unfortunate victims—doctors and technicians—now commemorated as the 'X-ray martyrs', suffered both burns and local malignant tumours.

The years between the two world wars saw many new clinical applications. Film replaced glass plates, intensifying screens improved, and the Potter Bucky diaphragm made its appearance. Patient couches and screening stands became more elaborate. Exposures were measured in seconds and fractions of seconds rather than minutes. The use of air as a negative contrast medium in the brain, developed in 1919, and iodine as a positive contrast in the urinary and vascular systems added new dimensions to diagnosis. Egas Moniz in Portugal attempted to opacify the brain by arterial injection of contrast medium in 1927. He failed in this unorthodox endeavour but as a result invented cerebral angiography. Tomography, the ability to image slices of tissue, although patented by Bocage in 1922, was only developed for practical clinical application in the 1930s. Other developments in the 1930s were directed towards greater precision and safety.

In the 1950s a great influence on the development of diagnostic radiology was the urge to investigate the cardiovascular system. Until that time most developments had been concerned with the links between the radiation source and the film. The relationship between detector and observer was not exploited until image intensification became a practical possibility, making dark adaptation with red goggles unnecessary and permitting more normal visual acuity and contrast discrimination. The introduction of percutaneous catheterization of blood vessels by Seldinger in 1953, the advent of rapid filming devices and above all the development of safer iodinated contrast media enabled rapid advances to be made (Fig. 6). The addition of closed circuit television finally brought such radiological procedures into open daylight providing the opportunity for even more sophisticated interventional activities.

On 6 February 1896, only 2 months after Röntgen's announcement Thomas Edison had received a cable from the *New York Journal* signed by Randolph Hearst. 'Will you as an especial favour to the Journal undertake to make a cathodograph of the human brain. Kindly telegraph at our expense'. Edison, the versatile inventor, already experimenting with fluoroscopy accepted the challenge. He was, of course, unsuccessful due, as the rival *New York Herald Tribune* correctly observed, to the then insuperable barrier imposed by the skull. Some years earlier in 1877 Edison had invented the phonograph. In 1898 by an inversion of the name the Gramophone Company was formed employing the well known bemused dog listening to an Edison machine as a trademark. The subsequent merger of the Gramophone Company with its rivals Marconiphone and British Columbia led, in 1929, to a new firm—Electrical Musical Industries (EMI) who, ironically, were later to play such a major part in what Edison had failed to do. The first 75 years of radiology had been a classic period of development influenced by the technical advances of the twentieth century and galvanized intermittently by the introduction of new clinical applications. Despite the

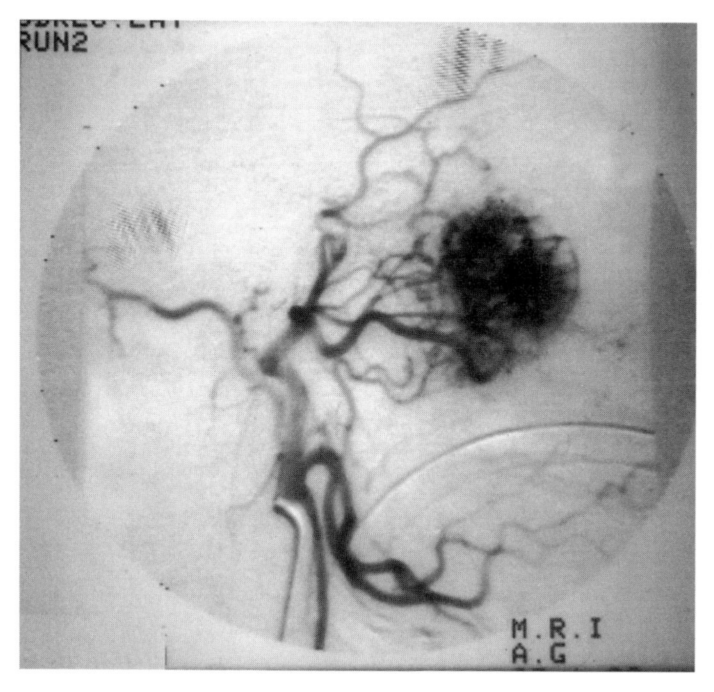


**Fig. 6** An external carotid angiogram of a vascular tumour in the nose obtained by selective Seldinger catheterization of the carotid artery from the femoral artery. The image has been obtained by video subtraction of the background bony structures.

many advances, however, the X-ray image remained essentially the same: a two-dimensional shadowgram of the three-dimensional patient, little different in concept from the X-ray of Frau Röntgen's hand in 1895. In addition, the phenomena capable of providing quantitative data lay entirely outside the patient—in the laboratory.

In April 1972 Hounsfield and Ambrose announced the first clinical results of Hounsfield's invention—computed tomography (CT) or, as it became popularly known, the EMIscanner (Fig. 7)—displaying, in cross-section, the living brain with intracranial pathology (Fig. 8). This stunning challenge to radiological orthodoxy which provided the ability to express, to manipulate and to quantify the content of a cross-sectional image *in vivo* led to a shift of the radiological paradigm comparable only with Röntgen's discovery in 1895. The effect on diagnostic medicine and surgery was dramatic. The technique liberated, as it were, the brain of both patient and investigator from the constraints of traditional imagery. While initially confined to investigation of the brain the method was extended in 1975 to the rest of the body (Plate 3), The clinical success of

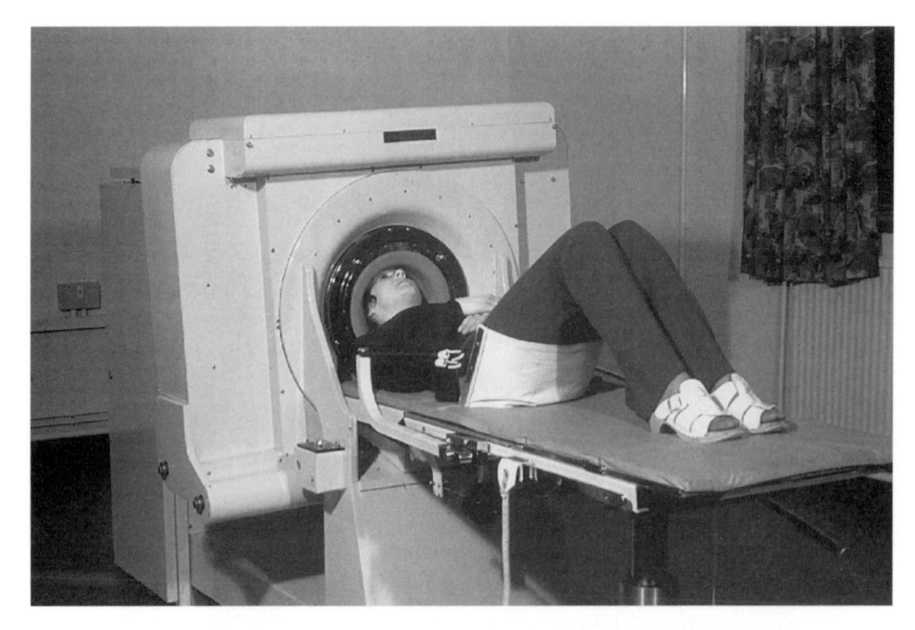

**Fig. 7** The first commercial computed tomographic system (EMIscanner) in the world installed in Manchester 1973. The prototype was developed by Hounsfield and Ambrose in the Atkinson Morley's Hospital London. Note the restriction of the 'rotate-translate' scanning planes to the head at this time by a water bath required for calibration purposes.

Hounsfield's invention, which gained for him the Nobel Prize in 1979, was made possible by the huge advances in computing and microtechnology which followed. Images, 1 cm in thickness on an $80 \times 80$ matrix which required many minutes of acquisition and processing time, quickly became possible with sections less than 1 mm in thickness and a $512 \times 512$ matrix in seconds. Fast spiral scanning employing slip ring technology to permit continuous data collection, together with three-dimensional reconstructions of the contiguous cross-sectional data now provide new opportunities for both structural and quantitative functional studies and for the production of virtual endoscopic images (Plate 4).

CT is an X-ray procedure. Ultrasound, however, does not use ionizing radiation. It is relatively inexpensive and highly mobile but operator dependent. It relies for its imaging ability on the reflected echoes produced by high frequency vibrations of a piezoelectrical crystal. Pierre Curie described this phenomenon 15 years before Röntgen's discovery of X-rays and the technique was applied by Langevin in the early 1900s to detect submarines. An obstetrician, Donald of Glasgow, however, was

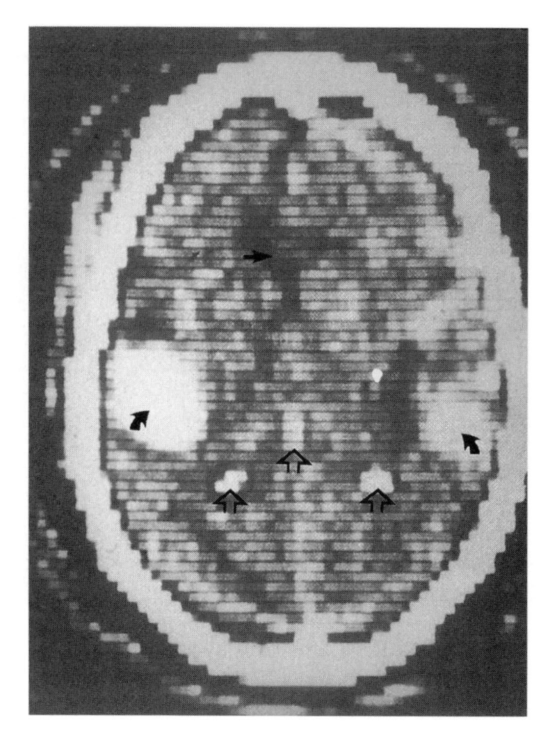

**Fig. 8** One of the first brain scans carried out in Manchester in 1973 of a patient with a head injury. The image even with an 80 × 80 matrix clearly demonstrated bilateral intracerebral haemorrhages (curved arrows), calcification in the normal pineal gland and choroid plexuses (open arrows) and cerebrospinal fluid in the cerebral ventricles (straight arrow).

the first to exploit the technique for the demonstration of the developing fetus and the procedure is now a standard component of obstetric practice (Fig. 9). Real-time ultrasound of most parts of the body is now widely applied in diagnostic medicine. The Doppler effect, i.e. the familiar change in the sound of a passing fire engine produced by a shift in frequency can, because it is both directional and quantitative, be used to display and to measure blood flow (Plate 5). Applications of the Doppler effect are currently being employed for the better detection and demonstration of vascular tumours (Plate 6). Micro-bubbles of air or even biodegradeable polymers can provide useful contrast agents for ultrasound procedures.

The fact that certain atomic nuclei might behave like small bar magnets by virtue of their spin and associated electrical charge had been postulated by Pauli in 1924 and the possibility of a resonance method

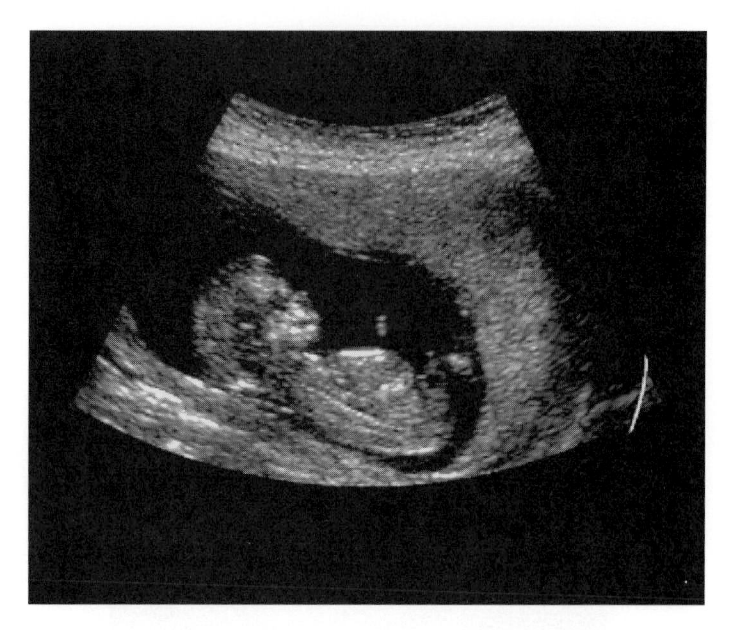

**Fig. 9** Ultrasound image of a normal fetus *in utero*.

for detecting such magnetic moments discussed by several workers in the 1930s. It required the stimulus of the Second World War, however, to provide the appropriate electronic techniques to enable the first demonstration of nuclear magnetic resonance (NMR) in matter. The results were announced simultaneously in 1946 by Bloch in Stamford and Purcell in Harvard. NMR became a fundamental tool in physics and chemistry laboratories and found practical applications in many fields of science. The concept of magnetic resonance imaging (MRI) was first mooted by Lauterbur in a letter to *Nature* in 1973. He argued that if a gradient field were to be applied to a structured object each nucleus would respond with its own resonance frequency determined by its position and that reconstruction of the object could then be accomplished by the type of algorithm being proposed for CT (Fig. 10). There were many who were sceptical of this unorthodox approach but the first live human proton image—a cross-section of a finger—was reported by Mansfield in 1976. Demonstration of other parts of the anatomy quickly followed leading to the elegant demonstrations of brain and body to which we are now accustomed (Plate 7). These images which are digitally derived can be obtained in any plane without any ionizing radiation. Selection of appropriate radio-frequency pulses and suitable postprocessing permits three-dimensional imaging and, for example, the demonstration of blood vessels without the use of contrast agents

**Plate 1** Six scenes from an early fourteenth century manuscript showing stages in the surgical treatment of a fractured skull. Reproduced by courtesy of the British Library in association with The Wellcome Institute for the History of Medicine ('Medieval Medical Miniatures' by Peter Murray Jones, 1984, Plate IX).

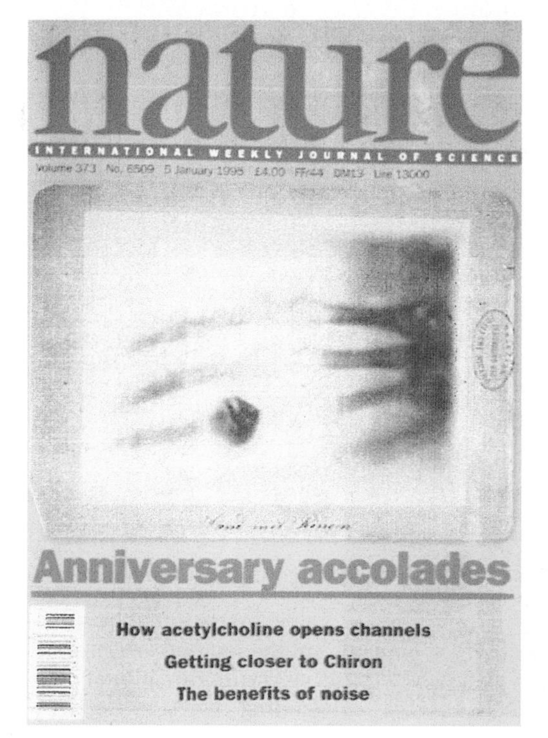


**Plate 2** The radiographic image of Frau Röntgen's hand produced by W.C. Röntgen in November 1895 and featured on the front cover of *Nature* in the centenary year 1995. Reproduced by courtesy of *Nature* (5 Jan 1995).

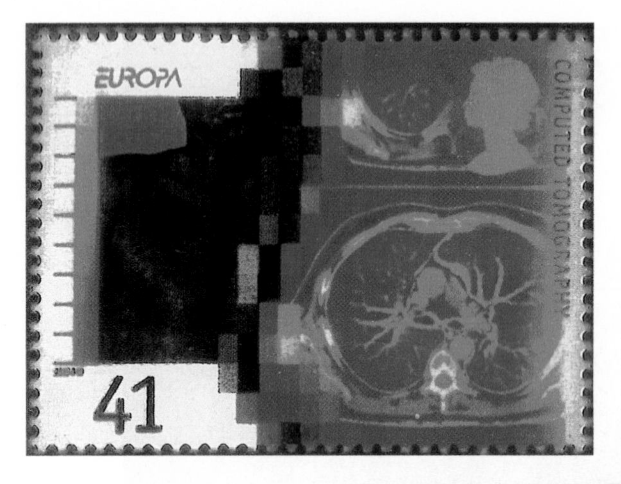


**Plate 3** A 41p postage stamp issued in 1994 illustrating a CT scan of the thorax. Reproduced by courtesy of the Post Office.

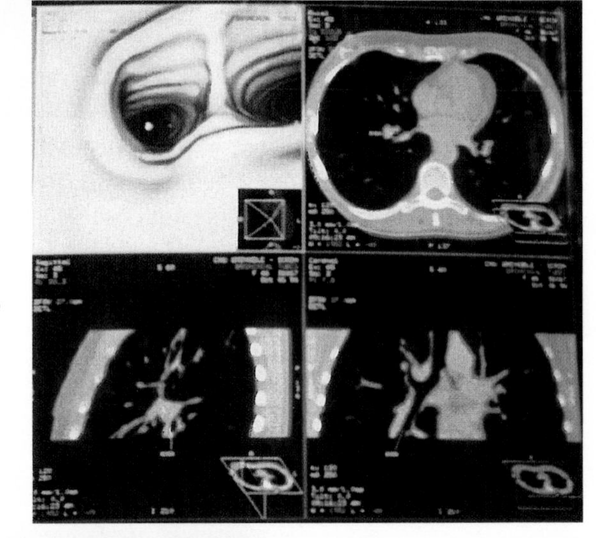

**Plate 4** In clockwise order: (a) virtual endoscopy of the bronchial tree obtained from a three-dimensional high resolution block of CT attenuation values; (b) transverse; (c) sagittal; and (d) coronal sections from the same data.

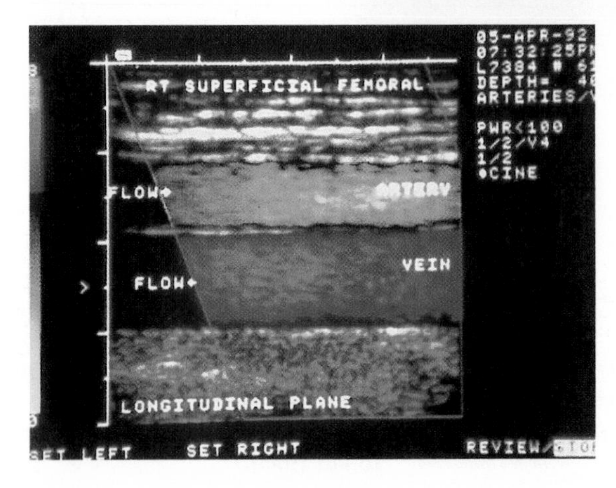


**Plate 5** Doppler ultrasound demonstration of blood flow in the femoral artery and vein with directional colour coding.

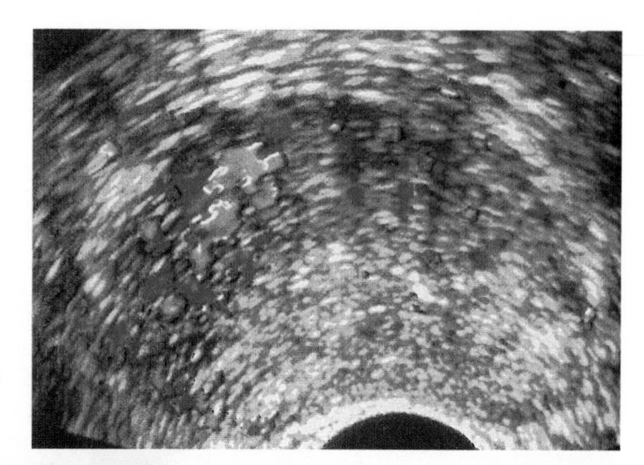

**Plate 6** Transrectal Doppler ultrasound image of a vascular tumour in the prostate gland.

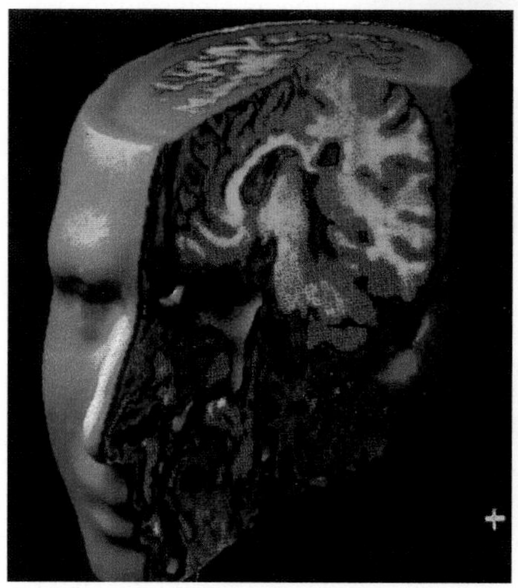

**Plate 7** Three-dimensional MR 'cut away' image of the head and brain. Reproduced by courtesy of *Diagnostic Imaging* (Feb 1995).

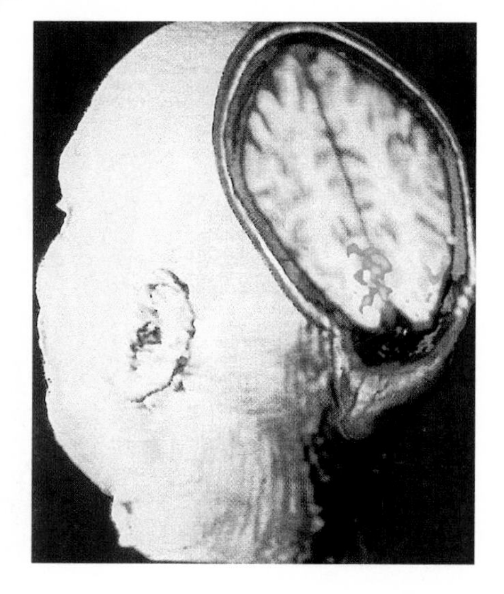

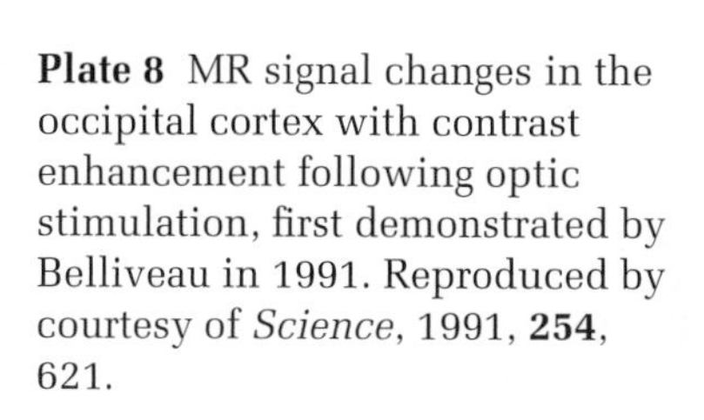

**Plate 8** MR signal changes in the occipital cortex with contrast enhancement following optic stimulation, first demonstrated by Belliveau in 1991. Reproduced by courtesy of *Science*, 1991, **254**, 621.

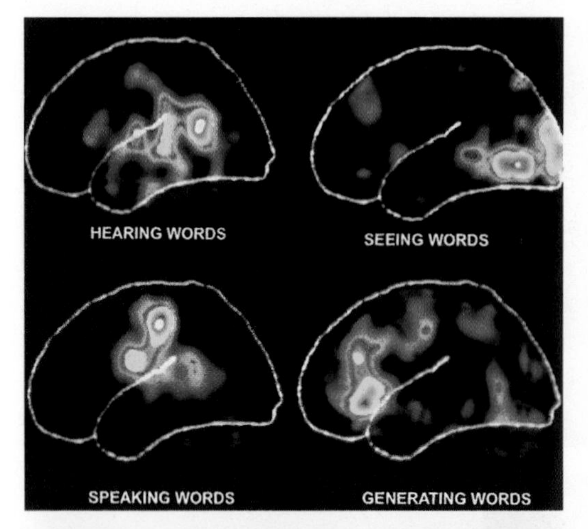


**Plate 9** PET scan of the left hemisphere of a human subject performing a series of intellectual tasks related to words. Blood flow shifts to different locations depending on the task. Reproduced by courtesy of Marcus E. Reichle, Washington University School of Medicine.

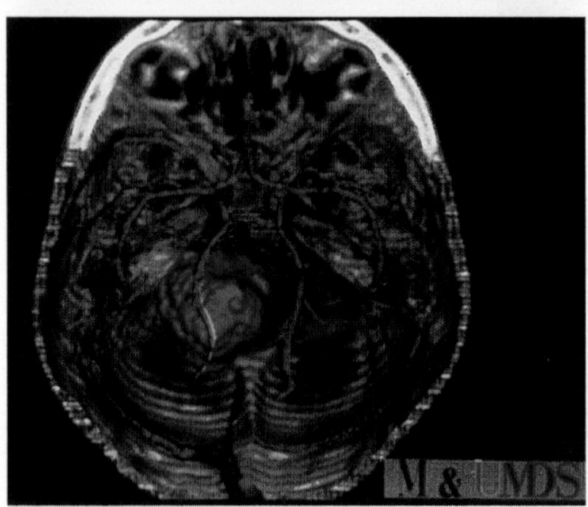

**Plate 10** Three-dimensional colour rendered view of a patient with a large acoustic neuroma. The tumour from MR imaging is green, the blood vessels from MR angiography are red and the bone from CT is grey. Reproduced by courtesy *Radiology* (Hill, D.L.G. *et al.*, 1994, **191**, 447–54).

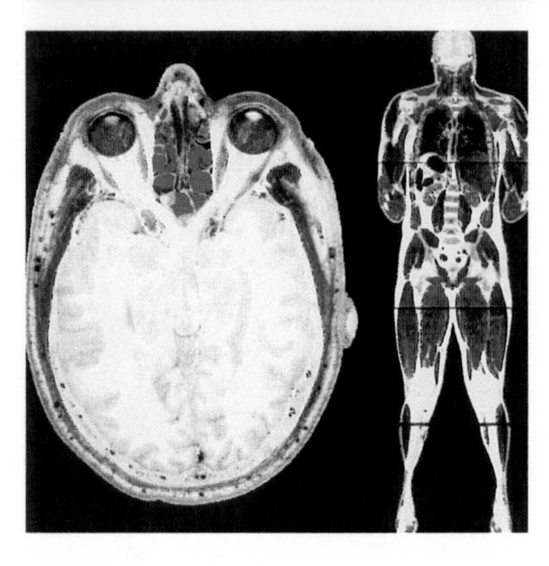

**Plate 11** Images from the Visual Human Dataset pioneered by the University of Colorado and now available on the Internet.

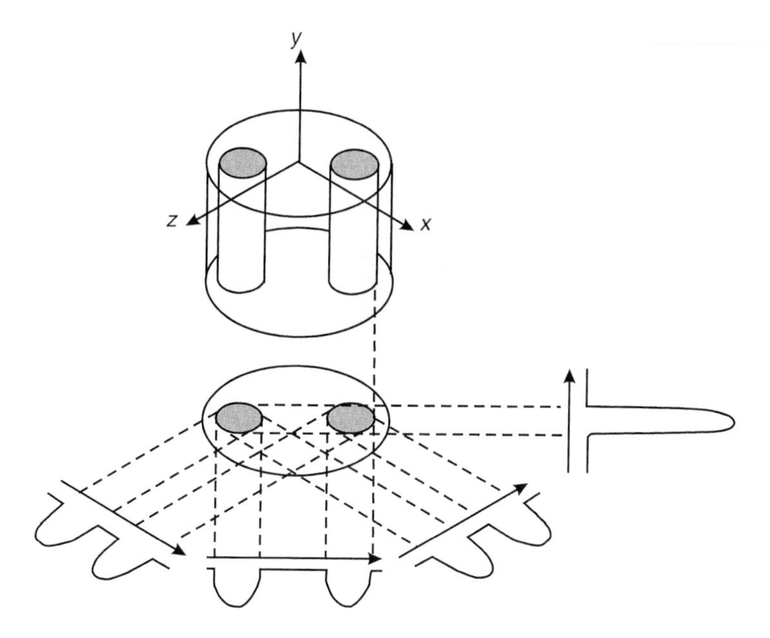


**Fig. 10** First published NMR image—two glass tubes of water in a field of magnetic gradients—by Lauterbur in 1973. Reproduced by courtesy of *Nature* (Lauterbur, P.C., 1973, **242**, 190–1).

(Fig. 11). How are we to convey such a rich vein of quantitative information contained within our radiological landscape to other clinical colleagues and to the patient. In the evocation of time and place the power of human language must sometimes transcend the image. Or to put it another way

> I'm grateful Lord for MRI
>   For X-rays and CT
> But may a method ne'er be found
>   Which might dispense with me

More detailed characterization of tissue and the assessment of its functional capacity can be approached in a variety of ways—the source image, the quantification of image parameters or derived properties, the use of paramagnetic agents and biological markers and microscopy—or at least the highest attainable resolution. The multiplicity of measurable parameters including proton density, relaxation times, chemical shift, flow, susceptibility, diffusion and perfusion provide unprecedented opportunities to explore the morphology, pathology, physiology, and biochemistry of living tissue. The paramagnetic gadolinium chelates, the

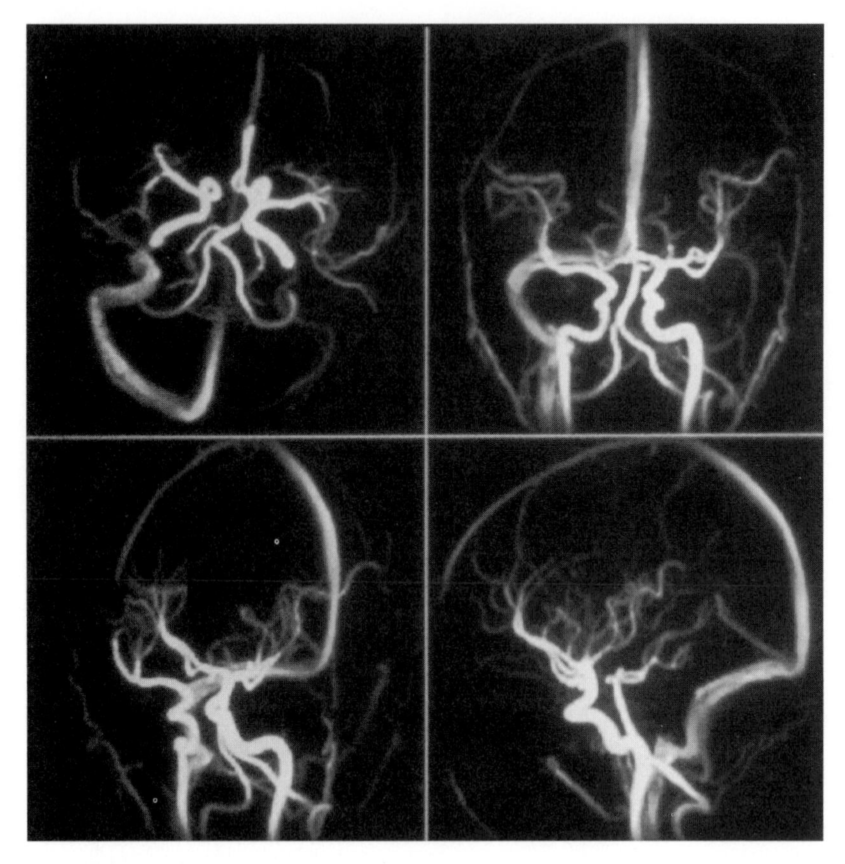

**Fig. 11** Magnet resonance angiography (MRA) in multiple planes without contrast medium injection. The selective demonstration of blood vessels is achieved by the use of appropriate radio-frequency pulse sequences.

most commonly used intravascular contrast agents for MRI, are taken up by vascular tissues or in certain brain tumours, where the blood–brain barrier has been destroyed (Fig. 12).

High-resolution MRI means optimizing spatial and chemical resolution in a clinical setting but very detailed and valuable anatomical and clinical information can be achieved (Fig. 13). True MR microscopy of course, requires high field gradients and high sensitivity radio-frequency coils. Nevertheless, Cho and co-workers operating at 7 Tesla—more than three times that used in conventional clinical practice—have been able to follow cell lineages, using gadolinium DTPA dextran micro-injected as a biological marker in a single cell of a blastomere at the eight-cell stage, with a resolution of 12 $\mu$m.

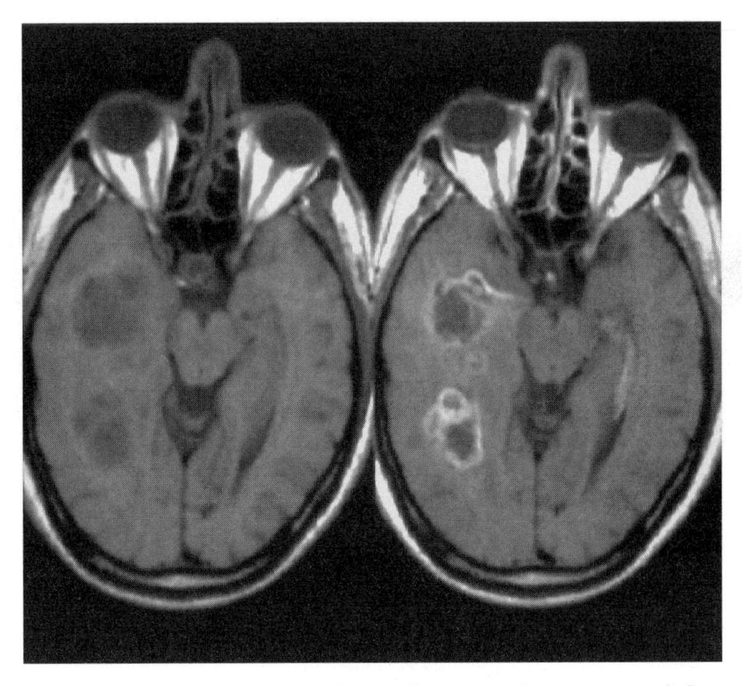

**Fig. 12** MR transverse sections of metastatic tumours of the brain imaged before and after the intravenous injection of the paramagnetic agent gadolinium DTPA.

MRI has significantly changed attitudes towards the management of disease. The less desirable effect has been that the high cost of equipment has limited its availability to the industrialised countries. As T.S. Eliot said:

> Between the idea and the reality
> Between the motion and the act
> Falls the shadow

Two-thirds of the world's population have no imaging at all. In the medically disadvantaged countries, including parts of eastern Europe 30–60% of the equipment available is not functioning and of that that is, most is located in the major cities. Can we address this moral dilemma?

What are the key issues for clinical radiology in the next century? Patients certainly come first and patients' expectations in the developed world are now exceedingly high. Indeed in some cases, patients are as knowledgeable about medical technology as the average doctor. In this country the Patient's Charter quite rightly encourages a demand for more and better services. Clinical radiology is a provider and in the current health service environment, driven by the purchasers. In the eighteenth

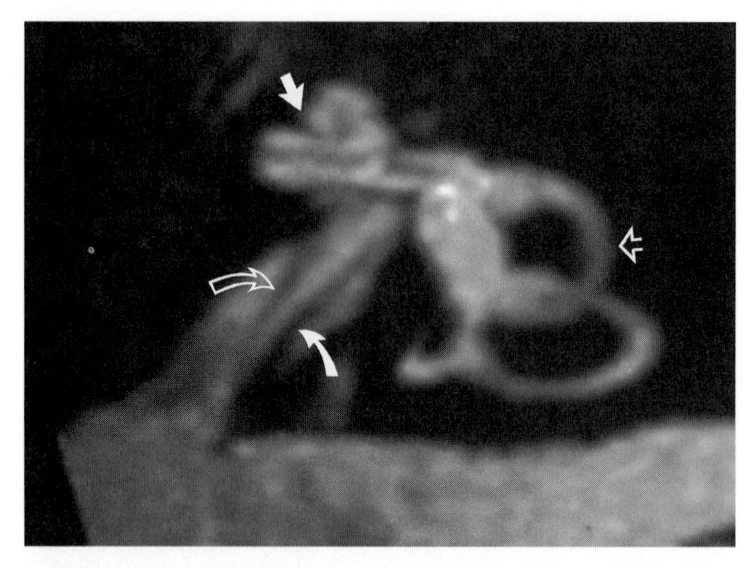

**Fig. 13** High resolution MR image of the inner ear demonstrating the cochlea (straight arrow), semicircular canals (open straight arrow), auditory nerve (curved arrow) and facial nerve (open curved arrow).

century Voltaire said that 'Medicine is a means of assessing the patient whilst nature cures the disease'. There is a danger that in the twenty-first century nature may have to be recalled to the front line as our diagnostic capabilities exceed our financial ability to treat. Nevertheless, it is reasonable to predict that the three most significant issues for radiology in the next century will be function, intervention, and digital networks.

Functional imaging implies the detection, demonstration, and quantification of one or more functional parameters. These might include blood flow, diffusion, perfusion, cell metabolism, or the use of targetting agents, although the differentiation of form and function can sometimes be difficult. Garrison Keeler, in his poem 'Mrs Sullivan', expressed it well in terms of architecture:

> 'Function follows form'
> Said Louis Sullivan one warm
> Evening in Chicago drinking beer.
> His wife said 'Dear,
> I'm sure that what you meant
> Is that form should represent
> Function, its function that should be followed'.
> Sullivan swallowed

And looked dimly far away
And said, 'OK,
Form follows function then'.
He said it again
'Form follows function then'.
A three word spark
Of modern arch-
Itectural brilliance
That would dazzle millions.
'Think I should write it down?'
He asked with a frown.
'Oh yes' she said, 'and here's a pencil'
He did and soon was influential.

Functional MR of the brain, in some studies, exploits the relationship between blood flow and neuronal activity—a relationship observed by Sherrington over 100 years ago—to produce signal changes in response to stimuli with or without the use of pharmaceutical agents (Plate 8). The absence of ionizing radiation is clearly attractive, although the precise origin of the unenhanced signals is as yet unclear. They are presently considered to be due to a decrease in tissue susceptibility reflecting an associated decrease in deoxyhaemoglobin.

Nuclear medical techniques employing gamma emitting radioactive isotopes have played a significant part in the study of function for many years and constitute true chemical imaging. More recently positron emission tomography (PET) has become a well established, although expensive, sensitive targetting technique using positron emission from very short lived isotopes not only to detect and identify particular pathological processes but also in some cases to demonstrate normal behaviour, for example, in the study of human cognition (Plate 9).

Contrast media to highlight both structural and functional change in various body systems have been explored since 1896. The market now is huge. Six million doses of iodinated and 3 million doses of paramagnetic agents are administered every year. By the year 2000 it is estimated that some 100 million patients per year will be subjected to vascular administration of contrast media. In the meanwhile industry and medicinal chemists are producing new molecules of pharmaceutical interest. The opportunity to design new and safer contrast agents for radiology by computer generation will be of particular significance. The human genome project is identifying with increasing frequency, genes that place individuals at risk from conditions that range from cancer to osteoporosis. The molecular diagnosis of gene mutations and the targetting of enzymes for cancer therapy will have profound effects on medicine. Almost any biological mediator or receptor can now be labelled and or targetted by nuclear medicine techniques—truly molecular medicine. An important step in the design of antigen-targetted receptor agents for magnetic resonance has been the development of intravenously administered ultrasmall particulate iron oxide (USPIO) superparamagnetic tracers capable of passing through capilliary endothelial junctions into the extravascular space. USPIOs have been specifically directed to asialoglycoprotein receptors on hepatocytes in the liver for example, enabling tumours to be demonstrated by their inability to express these receptors. Further surface modifications should make a variety of receptor antibody-specific agents possible.

Interventional radiology is a term used to describe a wide range of image guided techniques in diagnosis and treatment. There are four broad categories of description—diagnostic, palliative, alternatives to surgery, and unique procedures where conventional surgery is not possible. Techniques range from simple biopsy to the most sophisticated procedures. Stenosed ducts and blood vessels can be opened up by stents or angioplasty for example, while ruptures of blood vessels can be closed off by balloons, coils, or glue. Minimally invasive therapy (MIT) is a term used currently by both radiologists and surgeons to encompass both radiological interventions and a variety of endoscopic surgical procedures. It is estimated that in 10 years time 70% of surgical procedures will be conducted by MIT, the advantages of which, it is generally accepted, are reduction in pain, morbidity, and hospital stay with a more speedy return to normal activity for the patient.

Imaging has come to play a major part in the planning, guidance, monitoring, and control of treatment in many situations. Real-time image display by non-ionizing MR and the full perception of three-dimensional space are the real keys to the future of MIT. MR guided therapy employs novel magnet designs and instruments with position sensing devices to align the three-dimensional image with the surgical field in any plane. Furthermore, the operator's hands can remain in the imaging field. Such systems can track the device in the operator's hand and automatically move the scan plane relative to that device providing, in effect, an operating 'M*a*croscope' for a limited surgical field of view which can of course, be endoscopic. The operator can point to a target area on the image and then be enabled to see through the direct vision operating field to features beneath. This type of guidance capability provides frameless stereotaxis and on-line surgery with direct vision from any angle. Thus, real-time imaging and three-dimensional modelling with multimedia systems link the reality of the patient with the virtual reality of the image from the operator's perspective (Plate 10). Many of these concepts are in routine military use. Head up displays augmenting the real world with additional information on flight paths, enemy identification, and targetting are readily available facilities for fighter pilots. Miniature displays, light and flexible, for mounting in the eyepieces of pilot's helmets are currently being developed with direct laser scanning of images on to the retina. The technology for virtual reality and telepresence exists. As an industrial colleague recently remarked—'Under the great threat of peace we are compelled to diversify into medicine'.

Virtual reality is now more than a Nintendo game. It has become a computer science in which not only imaginary artificial worlds but also truly natural three-dimensional environments can be created. Once a simulated world of, for example, the cranial cavity, has been created, it can be entered and, with appropriate tactile feedback systems, the operator can be immersed in a virtual three-dimensional world and interact with structures as if they existed, which of course they do, in the real world. More than one person can enter such a virtual world enabling collaborative procedures to be undertaken. The opportunities for training and practice are considerable. Simulated organs derived from mathematical computations and provided with superimposed tissue maps can indicate real-time geometric deformations at the site of a virtual instrument grasp.

Telepresence means that the operator can be at a surgical workstation with sufficient virtual reality information to feel physically present at the remote site. As the virtual world is a digital one, the availability of satellite networking means that distance is no object. Teleconferencing, telemedicine, and the transmission of digital images are routinely practised. Radiology in the Hammersmith Hospital here in London, through an effective programming, archiving, and communication system (PACS) is now entirely digital with no requirement for film-based images.

The Visible Human Project, a multiple gigabyte three-dimensional anatomical atlas incorporating colour photographs, X-rays, CT, and MR in microthin slices in any plane providing the facility to dissect and reassemble is available to all on Internet (Plate 11). The technology for both teleimaging and telesurgery in distant locations, static or mobile, is presently available providing perhaps some hope for the solution of our moral dilemma. Future technological developments are likely to come from improved visual systems and instrumentation, robotics, and microengineering. Artificial hip replacements and prostatectomines have already been carried out by telerobotics.

The notion of our domination by machine depicted by Chaplin in *Modern Times* (Fig. 14) is being replaced by nanotechnology—the application of techniques which enable devices to be made with dimensions

**Fig. 14** Charlie Chaplin in *Modern Times*.

of nanometres (Fig. 15). Such devices which have already entered the fields of both robotic and integrated circuits undoubtedly carry the potential to provide more effective drug delivery systems.

Radiological development in the 1980s was mainly concerned with hardware. In the 1990s, and for the foreseeable future, it is, and will be, more to do with software and computing speed. The storage requirements for three-dimensional imaging, for example, will be of the order of a trillion bits per $cm^3$. The number of functions per chip is increasing and by the year 2000 is likely to have increased by a factor of over a 1000. Computing speeds are—from gigaglops to terraflops per second—enabling significant advances to be made in diagnostic imaging. Applications will include reasoning with uncertainty or 'fuzzy logic' allowing a computer to proceed with incomplete information, object-orientated modelling, and perhaps most importantly, with the development of neural networks—image understanding by machine. The impact on clinical radiology, in the next century will be profound. Team organization will be essential, although team members may change to accommodate changing circumstances. Technology will force fundamental revision of of both radiological and surgical training programmes. Hospital design of the future will have to take into account patient handling, shorter hospital stays, and the need, for both financial and practical reasons, to locate imaging and treatment facilities centrally.

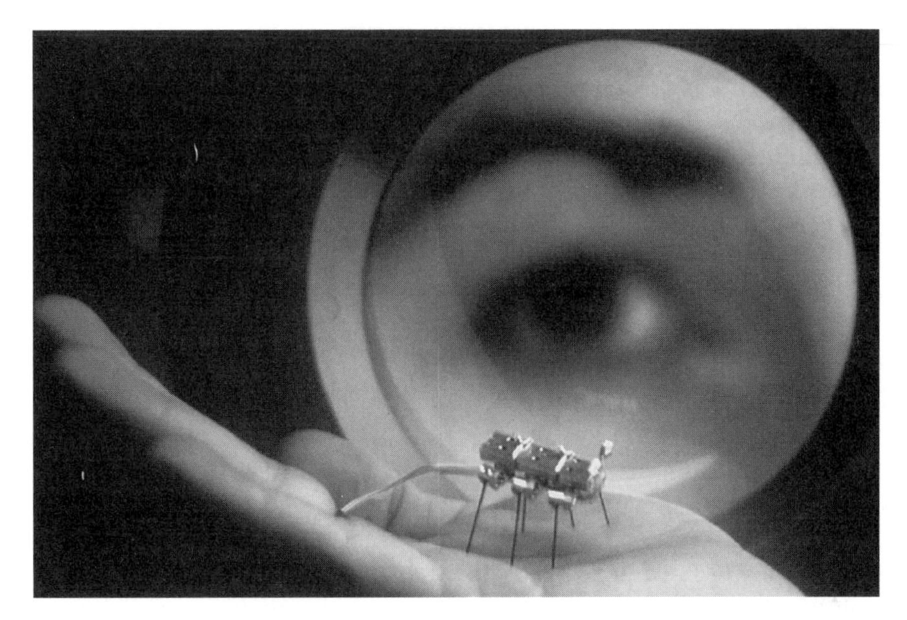

**Fig. 15** The 'colon crawler', a miniature mobile device able to carry scanning facilities within the intestinal lumen. Reproduced by courtesy of Mr J. E. Wickham

Fortunately, the world's most advanced computer—the brain—does not require a single chip! It has organic molecules and a sophisticated neural network. It can calculate, perceive, manipulate, self-repair, think, and feel. Technology has no imagination and no intuition—yet! Nevertheless, image understanding by machine might represent the ultimate challenge to radiological orthodoxy!

Only twice in history—in Ancient Greece and the Renaissance—have systematic attempts been made to approximate the image to reality. Perhaps others will come to look back on radiology of the twentieth century as a third such period.

In the words of Keats, the bicentenary of whose birth we also celebrated in 1995: 'The rise, the progress, the setting of imagery, should—like the sun—come natural'.

## Acknowledgements

'Mrs Sullivan' reproduced courtesy of Faber & Faber. (In *We are still married*, Garrison Keeler, 1989.)

## IAN ISHERWOOD

Born 1931 he is Emeritus Professor of Diagnostic Radiology in the University of Manchester, Dean of the European College of Radiological Education, President of the British Society of Neuroradiologists, and Chairman of the Radiological History and Heritage Charitable Trust. Recently President of the Röntgen Centenary Congress celebrating 100 years of radiology, and past President of the European Association of Radiology (EAR), the British Institute of Radiology, and the Radiology Section of the Royal Society of Medicine (RSM). Awards include an Honorary Doctorate of Medicine from the University of Zaragoza, the Gold Medal of the Royal College of Radiologists, the Jephcolt medal of the RSM and the Boris Rajewsky Medal of the EAR. He is an Honorary member of 10 foreign national societies of radiology and an Academician of the Russian Academy of Medical Sciences. He was involved in the early development and clinical application of Computed Tomography (CT scanning) and Magnetic Resonance Imaging (MRI), and is author of over 250 scientific papers in the fields of diagnostic imaging, the history of medicine, and Egyptology.

# *Electricity, magnetism, and the body*

ANTHONY T. BARKER

## Introduction

Most of the major functions of the body are controlled by electricity. If I hold my hand in the air and wiggle my fingers it is electricity that makes them move. Electrical impulses start in the motor cortex area of my brain, pass down my spinal cord and the motor nerves in my arm. They cross from nerve to muscle fibres at electrochemical junctions called synapses and, as the impulses then travel along the muscle fibres, they cause them to contract resulting in the desired movement.

Sensation is electrical. If I hit my thumb with a hammer, electrical impulses travel up sensory nerves to my brain, and I feel it! The reason you can see me is electrical. Light is reflected from my body and focused by the lenses of your eyes on to the cells of the retina. These convert the light into electrical signals which travel along the optic nerve to the visual centres of the brain, and you see me.

The reason you can hear me is due to electricity. My voice creates sound pressure waves in the air. These vibrate structures in the ear which are designed to convert them to electrical impulses. These impulses pass along the auditory nerve to the auditory cortex, and you hear me. Even your very thoughts consist of millions of electrical impulses whizzing around inside your brain.

Given all this natural, on-going, electrical activity it is not surprising that the body will sense and react to externally applied electric and magnetic fields in some circumstances. Much of the electrical activity of the body can also be detected, recorded, and used to assess function. We shall look at selected examples of effects, both proven and speculative, of electric and magnetic fields on the body, and at how these can sometimes be harnessed to measurement techniques to help in the diagnosis of disease.

## Getting electricity into the body

There are three ways in which electric current can be made to flow in the human body. Direct electrical connection can be made to it, such that it forms part of a circuit. Alternatively, the body can be exposed to a time-varying electric field or a time-varying magnetic field. At high frequencies electric and magnetic fields are inextricably linked to form an 'electromagnetic' field, but at the frequencies of interest here (primarily a few kilohertz or less) they can be considered to be independent.

### *Direct electrical connection*

The most familiar examples of direct electrical connection to the body are accidental, and result in electric shocks. These can come from a variety of natural and man-made sources, such as lightning (where the connection is made via a conducting plasma created in the air), static electric charges, and the domestic and industrial electricity supply system. Lightning strikes cause about three deaths per year in the UK but some 100 deaths per year in the USA, with about 200 additional cases of serious injury. While lightning can disrupt the beating of the heart, the greatest damage it usually causes is due to very rapid heating of tissue by the large currents which pass through the body (so-called $I^2R$ heating), and the resultant internal burns.

Shocks from static charges, although a nuisance and sometimes unpleasant, do not present a real hazard in themselves (although the resultant 'startle response' may make you spill a cup of hot coffee over yourself, or jerk backwards and stumble over a cliff edge!).

Injuries due to high voltage man-made sources (> 1 kV) usually occur from accidental contact with power distribution cables and fatalities are normally due to the resultant burns or falling.

The major hazard of contact with voltages below 1 kV is that of causing fibrillation of the heart. Such incidents normally occur domestically by accidental contact with mains voltages via either exposed wiring or faulty appliances. The effect of a power frequency (50/60 Hz) electric shock depends on the *current* which passes through the body, and the path it takes, rather than on the contact voltage directly. The internal resistance of the body between two limbs at power frequencies is a few hundred ohms. The resistance of small areas of intact skin is much higher, and is determined primarily by the thin surface layer of dead cells known as the stratum corneum. The resistance of the stratum corneum can vary widely. If the skin is completely dry it can be as high as several hundred thousand ohms when measured through an area of 1 $cm^2$, and as low as a few thousand ohms through the same area if it is saturated with sweat or water. Hence, when the skin is wet, even quite small areas of contact will result in a low resistance connection being made to the body (which is why electric shocks in the bathroom are particularly hazardous), and even low-voltage sources can be dangerous.

The physiological effects of a 50 Hz current applied to the body for a few seconds or more depend on its amplitude and the path it takes. The most hazardous situation commonly encountered is the limb-to-limb shock (Fig. 1), in which the current path passes through the chest. This usually occurs between the arms, or between an arm and a leg, although leg to leg shocks are occasionally reported.

At low current levels ($< \sim 6$ mA) the only effect of a limb-to-limb shock is to cause a slight tingling sensation due to stimulation of the sensory fibres in the relatively high current density areas at the contact points. Exposure for several minutes is not harmful but such relatively low currents rarely remain stable. The source of most electric shocks is usually a constant voltage, and the skin resistance varies with time and physiological factors such as sweating. Indeed, any discomfort experienced is likely to increase sweating, thus decreasing skin resistance and result in greater current through the body.

As the current increases into the range 7–24 mA the 'let-go' value is reached for the individual concerned (average 16 mA for adult males, 99% fall between 10 and 22 mA). This is so-called because, if a live conductor is touched with the palm of the open hand such that a current

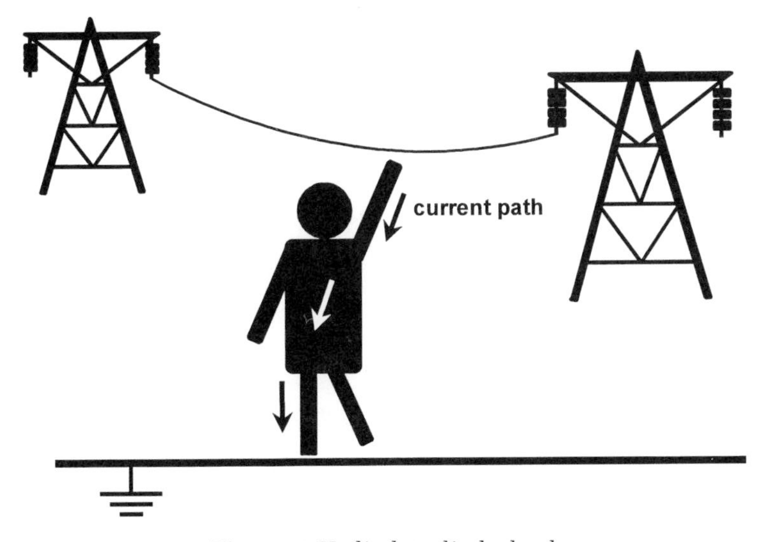


**Fig. 1** 50 Hz limb-to-limb shock.

of this value or above passes into the body, a continuous (or tetanic) contraction of the muscles will result, and the hand will involuntarily grip the conductor. Positive feedback then occurs—as the hand closes it makes better contact with the live conductor, current increases and the muscular contraction becomes stronger. To release the hand it is necessary either to prise it open or break the circuit to stop the flow of current. Currents in this range, although not normally lethal, do have some unpleasant side-effects. They can be painful, cause difficulty in breathing by interfering with the respiratory muscles, and they are psychologically distressing owing to the inability of the subject to release themselves. However, unless the current flows for an extended period of time, long-term effects do not normally occur.

Limb-to-limb currents above approximately 25 mA and which last for a few seconds or more are very dangerous, and are likely to cause the heart to go into ventricular fibrillation (Fig. 2). Then the normally co-ordinated contractions of the heart cease, and are replaced by irregular, desynchronized movements, resulting in the loss of pumping action. The heart is not normally able to resynchronize itself once fibrillation has commenced, even if the electric shock ceases. Loss of the pumping action of the heart results in irreparable brain damage in about 2–4 min. Fibrillation of the heart can also be caused by events other than electric shock, such as cardiac muscle damage due to blocked coronary arteries,

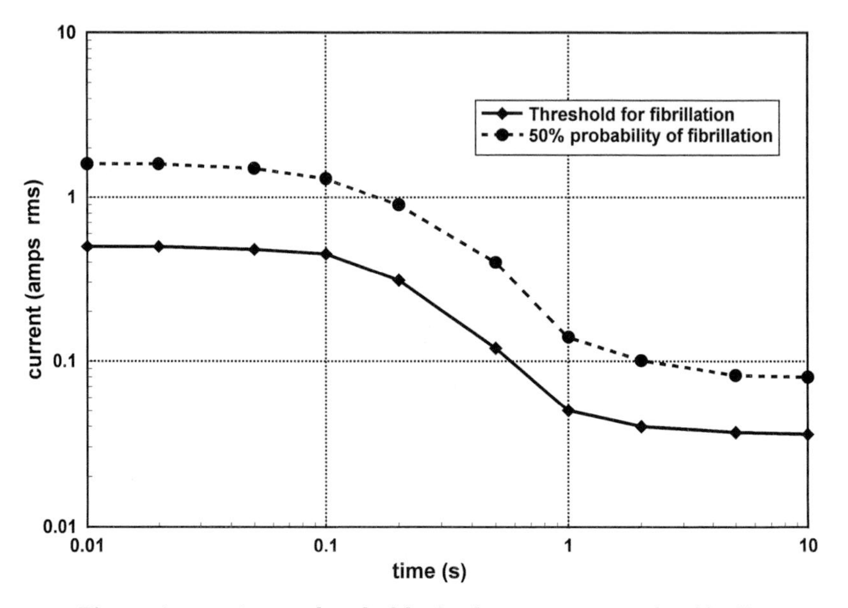


**Fig. 2** Approximate thresholds for human ventricular fibrillation versus duration of 50 Hz current (from IEC 479-1, 1984).

drug or anaesthetic overdoses and electrolyte imbalances of the body fluids. Ventricular fibrillation can be interrupted and the heart re-synchronized by the use of a defibrillator—a device which discharges up to 400 J of electrical energy (at voltages of up to 3 kV and peak currents of up to 50 A) into the chest—but that's another story.

Direct electrical connection can also be used to deliver electrical energy intentionally to the body at much lower levels than those used by the defibrillator. This has many applications in medicine such as stimulating nerves and muscles.

## *Time-varying electric fields*

The second way in which electricity can be made to flow in the body, or any other conducting object, is to expose it to a time-varying electric field. Examples of man-made electric field sources include overhead power transmission lines, wiring inside houses and some electrical appliances. The current depends on the geometry of the object, whether it is earthed, and the frequency and amplitude of the electric field. For a typical adult, in contact with electrical earth and standing in a field of 1 kV/m, the current flow to earth is about 14 $\mu$A. (Fig. 3).

## *Time-varying magnetic fields*

Electromagnetic induction using time-varying magnetic fields is the final method for getting electricity into the body. The history of electromagnetic induction is intimately linked to the Royal Institution—indeed

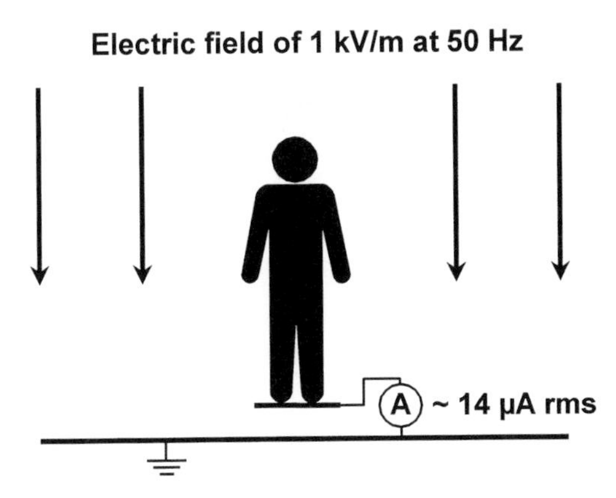


**Fig. 3** Body current due to 50 Hz electric fields.

it was in this very building that Michael Faraday first discovered the phenomenon. His work laid the foundation stone of modern electrical engineering and, along with his other electrical discoveries such as the dynamo and the transformer, changed all our lives.

On 29 August 1831 Faraday described in his diary an experiment on the production of electricity from magnetism:

> Have had an iron ring made (soft iron), iron round and seven-eighths of an inch thick and ring six inches in external diameter. Wound many coils of copper wire round one-half, the coils being separated by twine and calico—there were 3 lengths of wire, each one about 24 feet long and they could be connected as one length or used as separate lengths. By trial with a trough each was insulated from the other. Will call this side of the ring A. On the other side but separated by an interval was wound wire in two pieces together amounting to about sixty feet in length, the direction being as with the former coils; this side call B.
>
> Charged a battery of 10 pairs of plates 4 inches square. Made the coil on B side one coil, and connected its extremities by a copper wire passing to a distance and just over a magnetic needle (3 feet from iron ring). Then connected the ends of one of the pieces on A side with battery; immediately a sensible effect on needle. It oscillated and settled at last in original position. On *breaking* connection of A side with Battery again a disturbance of the needle.

The actual iron ring with which this experiment was performed, still wrapped with its original coils, is on permanent display in the museum of the Royal Institution and is, for me, one of the great artefacts of science.

It is not necessary to use iron-cored coils in order to achieve electromagnetic induction, as Faraday demonstrated on 1 October 1831 using interleaved coils wound on a wooden block. Voltages and currents can also be induced in all conductors, not just in wires. As already discussed the human body conducts electricity, albeit with a higher resistance than a wire. The possible health effects of *small* voltages and currents induced in the body by time-varying magnetic fields are highly controversial, but recent developments have led to very promising medical applications of *large* induced current pulses.

Having discussed the methods by which an electric current can be caused to flow in the body let us now look at the effects, both proven and speculative, of electricity and magnetism on body function and health. There are at present two main areas of controversy, does long-term exposure to electromagnetic fields cause chronic health problems, and can applied fields be used for therapeutic purposes?

## Is long-term exposure to low level, low frequency, electromagnetic fields bad for human health?

Man-made electric and magnetic fields are ubiquitous in developed countries and we are all, to varying degrees, exposed to them in our everyday lives. They come from a variety of sources. At low frequencies the most common source is the electricity supply network (which has a frequency of 50 Hz in the UK and 60 Hz in the USA), and appliances directly powered from it. At higher frequencies there are radio and radar transmitters, microwave communication links, terrestrial and satellite television transmitters, and mobile phones with their attendant base stations. While there has recently been increased interest in the possible biological effects of radio frequencies, largely because of the rapid growth in the use of mobile phones, it is the possible effects of 50/60 Hz fields that have so far stimulated the most research and public interest.

Up until about 18 years ago the possibility that weak, low frequency electromagnetic fields could affect human health had not been seriously considered in the West. In the 1960s a few studies from the Soviet Union reported non-specific complaints, such as headache and fatigue, in staff working in electrical switchyards where line voltages of up to 500 kV could occur, but these were largely dismissed in the West because of poor study design and interpretation.

Attitudes changed when, in 1979, Wertheimer and Leeper, working in Denver, Colorado, reported an epidemiological study which showed an association (or link) between the type and proximity of electricity supply cables outside the home and the incidence of childhood cancer. This study had a major impact both on the general public and on the scientific community, and stimulated much subsequent research. Controversy has continued to grow, partly fuelled by sensationalist coverage in the media. Today some members of the public regard the high voltage pylon not as an eyesore, nor as supplying energy to their homes and workplace, but as a hazard to their children's health.

### *To what field levels are we routinely exposed?*

No discussion of this topic would be complete without an ominous picture of pylons looming over houses—so here is one (Fig. 4). Both time-varying electric and magnetic fields will cause currents to flow in the body, and overhead lines produce both. Maximum values of electric and magnetic field from a 400 kV line (of the most common 'transposed' type), measured 1 m above ground level are typically 11 kV/m (dropping

**Fig. 4** An extreme example of overhead power cables near houses.

to less than 1 kV/m beyond 25 m from the centre of the line) and 40 $\mu$T (dropping to less than 4 $\mu$T beyond 40 m from the centre of the line), respectively. If the same power is transmitted via an underground cable the electric field is almost entirely abolished by a combination of the cable being armoured, the conductors being closer together, and the screening effect of the ground. The magnetic field falls of rapidly with distance from an underground cable but, because it can usually be approached more closely than an overhead cable (unless it is buried very deeply), its maximum magnetic field can be several times higher. Values vary considerably from installation to installation and are of order 100 $\mu$T measure 1 m above ground level immediately over the cable, dropping to typically 10 $\mu$T at a distance of 8 m from the cable line. Houses tend to act like Faraday cages and their interiors are partially screened from the external electric field, reducing the value inside the house (depending on its construction) to approximately 1% of the undisturbed external field. The magnetic field, by contrast, is not affected by normal building materials.

Although high voltage pylons are the most visible aspect of the electricity distribution network only about 0.1% of homes in the UK are within 50 m of one. Most people receive the bulk of their exposure to 50 Hz fields from other sources. The background magnetic field in the majority of houses comes from the 415 V distribution system, which is usually underground. Background fields are relatively uniform across a house and are usually below 100 nT. A mean value of about 40 nT has been found in a sample of houses across the country. These fields are caused primarily by the 'net current'—that current which diverts out of the neutral conductor and returns to the substation through water pipes, gas pipes, sewers, or even the ground itself. These background fields vary mainly because of the different earthing arrangements that are in use.

We are also exposed to fields from electrical appliances. These can be quite high close to the appliance, but decrease rapidly with distance from it. Figure 5 shows the 50 Hz magnetic field amplitude recorded at three distances from a selection of common household appliances.

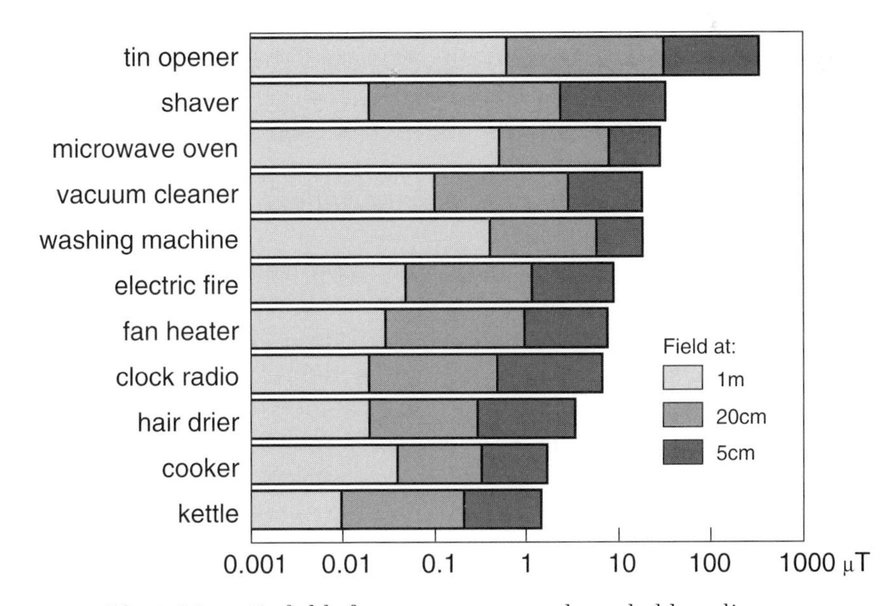


**Fig. 5** Magnetic fields from some common household appliances.

While we do not usually spend much time close to appliances, *if* the relevant measure of exposure 'dose' is the time-weighted average (and no one knows what that measure should actually be), then the dose to the surface tissue of parts of the body very close to the appliance, such as the hands or head when using a razor or hairdryer, can be higher than that due to background fields. An example is my wife's hairdryer. A magnetic field meter, measures a maximum field at the surface of the drier of approximately 155 $\mu$T. If it is used for 5 min per day the time-weighted average exposure is 0.54 $\mu$T on its surface, more than 10 times higher than typical domestic background levels.

Little information exists concerning domestic electric field levels from appliances, which are very variable and can be both distorted and screened by nearby metal or other conducting objects. Electric fields in houses immediately underneath 400 kV lines will be about 100 V/m, although this figure is very dependent on the construction of the house. Office workers (who work in an electrical environment not grossly dissimilar to a domestic one) have been monitored with body-worn data loggers and found to be exposed to electric fields in the approximate range of 2–60 V/m.

Some occupations result in higher exposures than would be encountered domestically. Examples are those which involve working with high voltages or currents, the extreme values encountered being about 10 kV/m for linesmen working with high voltage cables and 1 mT for generator workers. Workers in the vicinity of induction furnaces have been reported to have maximum magnetic field exposures as high as 10 mT. For present purposes I shall, however, confine my comments primarily to the implications of domestic exposure levels.

## *The thermal noise problem*

One difficulty that many scientists have, when considering the plausibility of reported effects of low level, low frequency electromagnetic fields, is the 'thermal noise problem'. All electrical conductors, including the human body, have thermal noise fields due to the random thermal motion of electrical charges (often referred to as 'kT' noise). It is thought by many scientists to be unlikely that applied fields much below kT noise levels can affect biological systems. For example, the noise field across the membrane of a typical cell (assumed to be 10 $\mu$m in diameter, with membrane thickness $5 \times 10^{-9}$ m and membrane resistivity $10^6$ $\Omega$m) in a 100 Hz bandwidth is approximately 500 V/m. External applied electric fields are attenuated by a factor of about $5 \times 10^7$

inside the body but, because the cell membrane is much less conducting than its surroundings, this figure is reduced by a factor of about 3000 to an attenuation of some $1.7 \times 10^4$. Thus a very high external electric field of 10 kV/m would result in a membrane field of about 0.6 V/m, some 800 times lower than thermal noise levels. If typical indoor domestic fields are assumed to be less than 100 V/m then they will result in levels some 80 000 times lower than the noise fields across the cell membrane.

A similar analysis can be carried out for applied 50 Hz magnetic fields. The electric field E induced in a loop of radius *r* by a magnetic field of amplitude B Tesla is given by:

$$E = \frac{dB}{dt} \times \frac{r}{2}$$

Assuming the human body has a radius of 0.2 m then at 50 Hz:

$$E \approx 30 \times B$$

Again the field across a cell membrane will be amplified by the ratio of 1.5 × membrane thickness/cell diameter ≈ × 3000. Thus the membrane electric field is approximately $10^5$ B V/m and a high whole-body exposure of 50 $\mu$T would give a value of 5 V/m, some 100 times lower than thermal noise levels.

If cells are able to average a sinusoidal signal they can improve its signal-to-noise ratio by the square root of the number of cycles averaged. To achieve a signal-to-noise ratio of 1 : 1 for an external field of 10 kV/m would require $800^2$ cycles at 50 Hz to be averaged, taking over 3 h, and for a 100 V/m field averaging would need to take place for over 4 years. To achieve such long averaging times the frequency acceptance of the cell must be very narrow, some 0.0001 Hz or less, and precisely centred on 50 Hz in the UK or 60 Hz in the USA. No such extremely narrow resonances are known to exist in biological systems. A similar calculation for a 50 $\mu$T field shows a required averaging time of 200 s and a frequency acceptance bandwidth of 0.005 Hz.

For those unfamiliar with the concept of signal to noise a simple demonstration can be made using audio signals. A 0–2 kHz noise signal sounds like the sea breaking continually on a shingle beach. A 1 kHz tone of the same root mean square amplitude as the noise signal is analogous to the 50 Hz fields we are concerned with. If the tone and the noise signal are combined a 1 : 1 signal to noise ratio results, and the former can just be heard over the latter. However, if the amplitude of the tone is decreased by a factor of 10 to give a 1 : 10 signal to noise ratio, it becomes completely inaudible. The inability of the human ear to detect a 1:10 signal to noise ratio, despite its intrinsic frequency discrimination and sophisticated postprocessor (the brain) is a graphic illustration of the signal to noise problem.

## *The epidemiological evidence for the effects of low level fields*

The 1979 study of Wertheimer and Leeper reported an association between 'wire codes' and the risk of childhood cancer (wire-codes are a method of estimating the magnetic field in the home from the type of distribution wiring near to it). Since then there have been a number of other studies looking at both adult and child cancers as well as other disorders. Several extensive reviews of these studies have been published and it is beyond the scope of this Discourse to cover the entire literature. Instead I will try to give a flavour of the subject by looking only at 'case–control' studies of childhood leukaemia.

Case–control studies are those which compare cases (subjects who have the disease being investigated) with controls (who do not but are similar to them in every other respect). The cases and the controls are classified as 'exposed' to (in this case) magnetic fields or 'non-exposed'. The results of the study are expressed as an odds ratio (OR) defined as:

$$\mathrm{OR} = \frac{\text{exposed cases}}{\text{non-exposed cases}} \times \frac{\text{non-exposed controls}}{\text{exposed controls}}$$

If the odds ratio is unity the incidence of the disease is not associated with the exposure, if it is greater than unity then there is a positive association. As with all statistical analyses, confidence limits can be placed on the odds ratio. The 95% confidence limits (for example) delimit the range of the odds ratio within which it is 95% certain that its true value lies. A common way of interpreting the confidence limits is to see if they include unity. If they do not, it is 95% certain that the elevated (or depressed) odds ratio is not due to chance, and the study is regarded as having found a statistically significant association between the disease and the exposure.

A very abbreviated summary of the findings of 15 case–control studies which have looked at childhood leukaemia is presented in Fig. 6. In general they show odds ratios somewhat elevated above unity (the mean values are all less than 4), and few of the studies are statistically significant.

There are considerable difficulties in carrying out such studies. A large study size should narrow the width of the confidence limits and

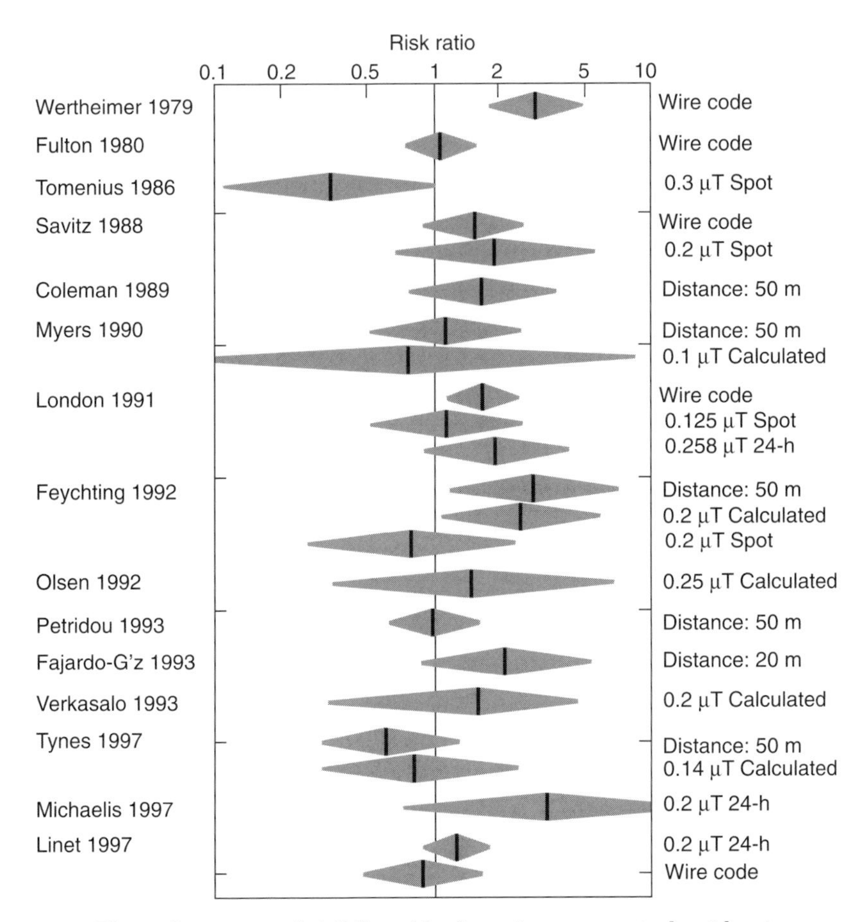


**Fig. 6** Summary of childhood leukaemia case–control epidemiological studies.

enable smaller effects to be detected. However, childhood leukaemia is a rare disease and it is necessary to study very large populations in order to detect small increases in odds ratios with statistical confidence.

It is not known what exposure parameters are relevant. For example, are both magnetic and electric fields relevant, and is the 'dose' the average or the peak exposure? Body worn data-loggers are now available, but obviously cannot give retrospective information about the exposures which were occurring at the time the disease was contracted. Most studies have used surrogate estimates of exposure such as wire-codes, calculated fields based on power line load currents, or single measurements of fields in homes.

It is also important that the controls are closely matched to the cases in all aspects, except in their exposure to the fields being studied. For example, if the controls come from a different socio-economic group than the cases, their diets, parental occupations and the proximity of their housing to busy roads (where airborne carcinogens are known to be higher) are also likely to be different. Attempts are made in most studies to adjust for these 'confounding factors' because they can lead to spurious associations, but as the causes of childhood leukaemia are not well understood, it is hard to be confident that they have all been eliminated.

The findings of the last study summarized in Fig. 6 by Linet and co-workers from the US Childrens' Carer Group have been published recently. This is the first of what might be called the third generation studies. The earliest studies were characterized by being relatively small in size and have been criticized for their methods of exposure assessment, selection of control subjects, and their handling of possible confounding factors. The 'second generation' studies were larger, with better control selection, and with exposures calculated from the currents flowing through the power lines outside the home. This latest study looked at the population of nine USA states (over 65 million people in total), and with a total of over 600 cases is four times larger than previous comparable studies. As well as calculating 'wire-code' scores for the homes being studied, 24-h magnetic field measurements were made in the child's bedroom along with shorter measurements elsewhere around the house. The findings of the study can best be summarized by quoting the conclusions from the abstract.

> *Our results provide little evidence that living in homes characterised by high measured time-weighted average magnetic-field levels or by the highest wire-code category increases the risk of acute lymphoblastic leukaemia in children.*

Other scientists have since looked at the published data of Linet and co-workers. and have drawn the opposite conclusions from it, namely that it *does* show an association between exposure and leukaemia. On balance I think that the conclusions of the original authors, who have no vested interest in the outcome, who have been intimately involved in the design and running of the study from its outset, and who have seen and analysed all the data, are likely to be more reliable. However, the varying interpretations placed on the same data highlights the very real problem of detecting weak associations, even in studies which encompass several million people.

## In vivo *and* in vitro *studies*

Many animal and cellular studies have looked for effects of low-level electromagnetic fields. The range of models and exposure conditions that have been used is vast, ranging from high electric field exposures of primates down to cell cultures exposed to combinations of weak alternating and static magnetic fields. The majority of these studies claim to show effects of the applied fields. Unfortunately, many of them have not had adequate control of the variables involved, most of the findings are contradictory and few attempts have been made to explore the claimed effects over a range of exposure amplitudes and frequencies. In view of the absence of established mechanisms of interaction, and the inconsistent nature of the existing findings, I believe that the key to progress with such models lies in the replication of already published studies.

Only a few attempts have been made to replicate existing studies—replication is not perceived as a glamorous scientific activity, funding is hard to obtain, and the publishing of findings which contradict those of the original authors can prove difficult and cause ill-feeling. Additionally, it is general tenet of science that findings, published as a scientific paper, must be correct.

Where attempts at replication have been made, they are usually unsuccessful. For example, along with colleagues in Sheffield I have been involved in replication studies as diverse as neurotransmitter release from cultured nerve cells, bone growth in chick embryos, calcium efflux from human lymphocytes and rate-limited studies of myosin phosphorylation. In every case we have failed not only to show the same effect as the original authors, but also to show *any* effect of the electromagnetic fields under test, despite having selected the studies because of their apparently robust initial findings.

In order to try and clarify this complex and confused literature I would suggest that a moratorium should be placed on the publication of new studies until they have been independently replicated. As well as leading to a large decrease in the number of artefactual studies which are published, scientists could then concentrate on models which showed real promise. It would also save a lot of trees!

## *Literature reviews issued by scientific panels or government bodies*

The large and increasing literature on this controversial subject has been studied in depth by over 80 bodies, including a working party of the Institution of Electrical Engineers of which I am the chairman. None of them has concluded that exposure to power-frequency electromagnetic fields causes cancer or any other disease. The latest major review was carried out by the US National Academy of Science which concluded in 1996:

> Based on a comprehensive evaluation of published studies relating to the effects of power-frequency electric and magnetic fields on cells, tissues and organisms (including humans), the conclusion of the committee is that the current body of evidence does not show that exposure to these fields present a health hazard. Specifically, no conclusive and consistent evidence shows that exposures to residential electric and magnetic fields produce cancer, adverse neurobiological effects, or reproductive and development effects.

## Conclusions

It is always unwise for scientists to 'stick their head above the parapet' and make pronouncements or prophecies based on (inevitably) incomplete data. My favourite example of this is when Lord Rayleigh wrote in 1882: 'Yesterday I had an opportunity of seeing the telephone which every one is talking about .... it is certainly a wonderful instrument, though I suppose not likely to come to much practical use.' But I'm going to do it anyway! In my opinion, the lack of consistent and robust scientific evidence of harmful biological effects due to low level, low frequency electromagnetic fields from both epidemiological and laboratory studies, despite an extensive research effort over nearly two decades and an expenditure of many tens of millions of pounds, coupled with the lack of any plausible mechanism of interaction, means it is now time to move on. If these fields produce deleterious effects, and in my judgement they do not, those effects must be small, otherwise they would by now have been unequivocally observed. In the last 40 years there has been approximately a three-fold increase in domestic electricity consumption and the average domestic magnetic field exposure has increased by a similar amount. Despite this there has been very little increase in the incidence of childhood leukaemia, the disease that has been most frequently linked to electromagnetic field exposure (Fig. 7). The slight increase that is shown is most likely to be due to improved accuracy in the reporting of the statistics.

There are many real risks in society. In England and Wales there are, for example, approximately 27 000 deaths each year due to pneumonia and influenza, 4000 due to traffic accidents, 4000 suicides, 3000 due to accidental falls, and some 300 murders. *If* the fears of those who believe in the hazards of low level electromagnetic fields are correct, extrapolation of the epidemiological studies suggests that living near overhead

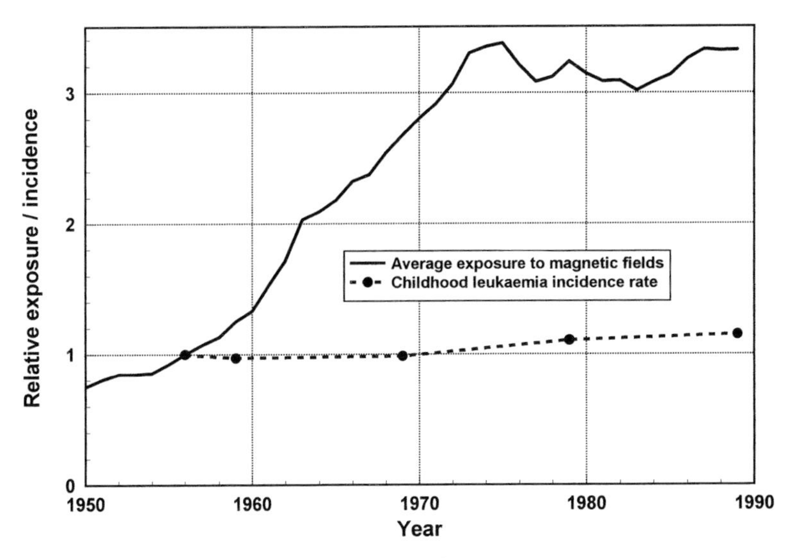


**Fig. 7** Leukaemia incidence rate and average exposure to 50 Hz magnetic fields since 1956.

high-voltage cables might result in one extra death due to childhood leukaemia per year in the UK. One often suggested approach is to place existing overhead cables underground. To put just one mile of high voltage overhead cable underground (of a total of about 4000 miles in England and Wales) would cost about £10 million. I estimate that every schoolchild in the UK could be equipped with reflective armbands for the cost of just 2 miles of underground cable. About 40 miles would buy reflective jackets and bicycle crash helmets for them all, or fund four major teaching hospitals for a year. Any of these alternative uses of resources would save many lives. Making minor decreases in the electromagnetic fields to which we are exposed will not. Our resources can be better spent than on an endless quest to detect and avoid very small, or non-existent, risks. Constant and unjustified public speculation, by both scientists and the media, about hazards from electromagnetic fields merely fuels paranoia in the general public and diverts attention from real issues of public safety. We all have more important things to worry about.

## Can electric or magnetic fields be used for therapy?

The idea that electric and magnetic fields are therapeutic has existed since long before they were understood or could be generated in a controlled fashion. One of the earliest reports, in AD 25, speaks of gout being cured by stepping on an electric eel. Whether electric or magnetic fields really do have beneficial effects depends on the type and strength of fields being considered. Fields which cause direct stimulation of nerves or muscles, or produce significant thermal effects, have genuine and demonstrable biological effects and are used successfully in a range of therapies. Effects due to fields whose frequencies or energy levels rule out such mechanisms are much harder to demonstrate, and their therapeutic value remains controversial.

## *Direct neuromuscular stimulation*

There are a number of therapeutic applications in which nerve or muscle tissue is directly stimulated by electric current, injected into the body via either surface or needle electrodes. The most well known examples of these include cardiac pacing, functional electrical stimulation (FES), cochlear implants, and bladder stimulation.

Cardiac pacing uses an implanted, battery powered stimulator to augment or replace the natural electrical pacemaker, which controls the rate at which the heart beats, in circumstances where it is malfunctioning.

FES is used to stimulate muscles whose nerve supply has been damaged, and thus restore function. Perhaps the most well known example of its use is to aid paraplegics (patients who legs are paralysed, usually due to accidental damage of the spinal cord) in standing and walking. The technique, which is still largely in a development phase rather than in routine use, usually employs an external, multichannel stimulator connected to surface electrodes placed over the muscles to be activated, although implantable systems have also been developed.

Cochlear implants are multichannel electrical stimulators with electrodes positioned in the inner ear such that they can directly stimulate fibres of the auditory nerve. They are helpful to very deaf patients who gain little or no benefit from conventional amplifying hearing aids. The incoming audio signal is processed in an external unit which can be programmed to suit the needs of the individual patient and which transmits both power and data to the implant via radio frequency coils.

Bladder stimulators are usually implanted in patients with spinal cord injuries, to aid bladder emptying and diminish incontinence by stimulating spinal nerve roots. They have an external controller, which is activated by the patient when required, to transmit energy to the implant and generate the appropriate stimuli.

### *Electric or magnetic therapies based on thermal effects*

Radio-frequency (r.f.) generators have been used for therapeutic purposes in medicine for over a century. In 1892 D'Arsonval described an experiment in which he passed a high-frequency current of three amps through his own body and experienced heating effects. In 1893 he invented a non-contact method of applying high frequency currents to humans, which he called 'autoconduction'. The subject was positioned in a large solenoid, which was driven with a radio frequency signal (Fig. 8). The resultant magnetic field induced current in the body and caused heating. Known as 'short-wave diathermy' the technique is today widely used by physiotherapists for the treatment of soft tissue injuries. Power outputs of up to 100 W at 27 MHz are applied locally to the area requiring treatment, either by inductive coupling using a tuned coil placed on the skin or by a

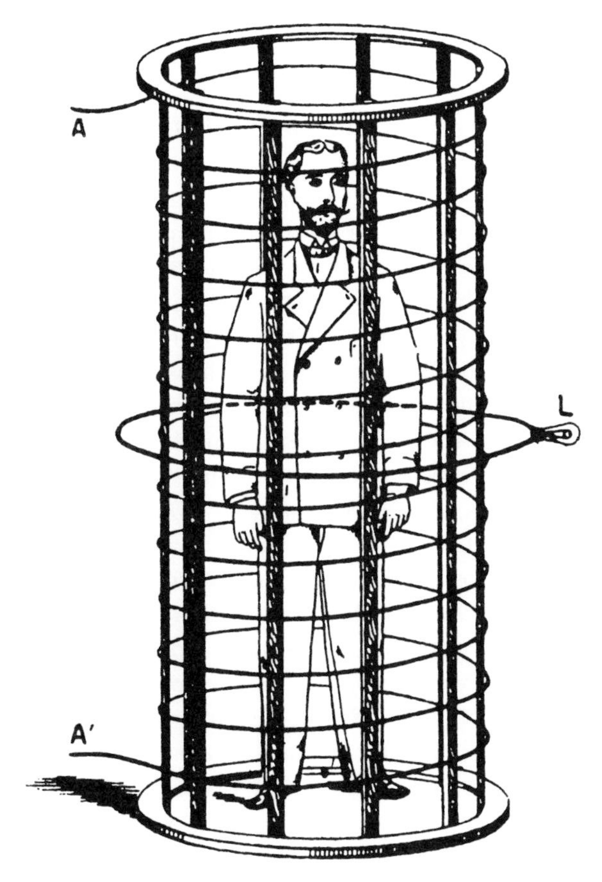


**Fig. 8** Whole-body radiofrequency heating using D'Arsonval's 'Autoconduction' apparatus (1892).

pair of capacitatively coupled electrodes. Demonstrable heating effects occur and lead to an increase in blood flow and hence tissue oxygenation. This thermal effect can often be beneficial—it is somewhat similar to lying in a warm bath but has greater depth of penetration and is easier to apply to localized regions of the body.

Unfortunately, this well understood and useful modality has been mysticized in several widely marketed products, which generate high power pulsed (as opposed to continuous) r.f. outputs and are claimed to have beneficial effects that are non-thermal. Such claims are not well substantiated. Peak powers of up to 1 kW and average powers of up to 30 W are applied locally and can clearly be demonstrated to cause vasodilation and increased blood flow. That these powers produce significant thermal effects in the body is not surprising. The heat output of the human body at rest (the basal metabolic rate) is typically 70 W. Hence 30 W applied locally is greater than that produced by normal metabolism, and represents a significant thermal input to the body.

## *Other electric or magnetic therapies*

The popularity of electric and magnetic therapies that do not cause heating or direct neuromuscular stimulation appears to be cyclical and, if judged by the plethora of devices which are commercially available today, is again on the increase. Its last resurgence was in Victorian times when devices such as 'galvanic spectacles', the 'electric corset', and the 'electromagnetic brush' (Fig. 9) were in fashion. In general there are no scientifically established mechanisms which can explain their claimed beneficial effects, and the thermal noise limits previously discussed in the context of 50 Hz fields are again relevant. Modern devices fall into three broad categories, r.f. generators, low frequency pulsed magnetic field generators, and static magnetic field devices.

Typical r.f. generators produce just a few milliwatts (or often much less) of pulsed power. The number of disorders which they claim to cure seems to be inversely proportional to their power output. At least one device, targeted at physiotherapists, claims to be effective for over 40 disorders, including such diverse complaints as the menopause, gout, and laryngitis, as well as soft-tissue injuries. Several hand-held r.f. devices are available for sale direct to the general public and are claimed to be beneficial in the treatment of complaints such as lumbago, sciatica, arthritis, migraine, and chronic pain.

The most widespread clinical use of low frequency pulsed magnetic fields is in the treatment of bone 'non-unions'—broken limbs which

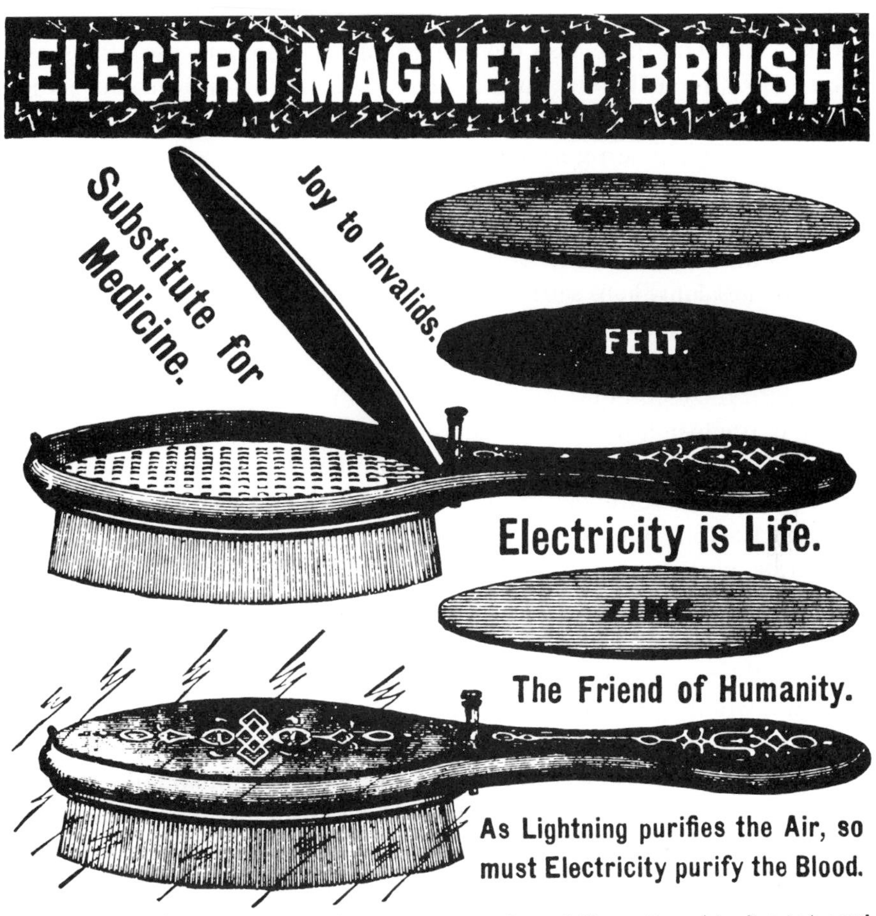


**For** Imbuing the Human Organism **with New Life, Health, and Strength,** and for Developing and Invigorating the **Brain.**

**HUMAN LIFE PROLONGED; BY ITS USE.**

This Brush contains over five hundred flexible **magnetized** steel teeth, with a complete **galvanic battery** in the back of the brush, as shown in the engraving. Electricity and Magnetism can be applied to the head or affected parts of the body more evenly and effectually than by any other process ever discovered. The teeth are so arranged that they transmit a gentle current of Electricity and Magnetism, producing **friction** without irritation.

By its daily use there will be no **sleepless nights,** no **Softening of the Brain,** no **Paralysis,** no **sudden death,** caused by over-taxed mental powers, or exhaustion of the **nervous system.** The life of a Business Man is one of energy, activity, and ambition. In the interest of his profession he often forgets the mental taxation he is enduring, until tired Nature gives way under the constant and severe strain of business, and either **Softening of the Brain,** or sudden death, is the result.

The daily use of the Electro-Magnetic Hair and Flesh Brush will prevent either of the above terrible calamities. The Electricity and Magnetism given to the brain by the application of our Brush is its **food,** its **strength,** its **vigor, and its rest.** No element soothes or stimulates the nervous system to healthy activity like Electricity or Magnetism. We guarantee the use of this Brush to be a **sure cure** for **Rheumatism, Partial Paralysis, Skin Diseases, Neuralgia, Functional Diseases of the Heart, Mental Despondency, Loss of Muscular or Nerve Power, Nervous Headache, Kidney Disease, Dyspepsia, Catarrh, Nervousness,** and **Sleeplessness.**

These Brushes should be used **daily** by the Aged, and by Persons with feeble constitutions; and they are particularly useful to Professional Men, Students, and Children attending school.

**We manufacture two kinds of Electro-Magnetic Brushes.** The Electro-Magnetic **Hair** Brush, which gives five hundred and forty currenls of electricity, to be used for the head only. The action of electricity on the roots of the hair **often** brings out a fine growth of hair on **bald** heads. Price, **$2.50.**

The "Electro-Magnetic Flesh Brush" gives seven hundred and four currents of electricity, to be used on any part of the body, **except the head.** Price, **$3.00.** We will send both brushes to one address for **$5.00.** These brushes will be seut by mail or express. If ordered by mail, **eighteen cents extra** must be sent for postage. Address,

**ELECTRO-MAGNETIC BRUSH CO.,**

**192 West Fifth Street, CINCINNATI, O.**

**Fig. 9** One of many Victorian electromagnetic 'therapies'.

have failed to unite after being put in a plaster cast for a number of weeks. The technique was developed and patented in the 1970s by a group of orthopaedic surgeons in the USA. Patients are treated using a pair of near-Helmholtz coils placed on either side of the limb. These are driven to produce a highly specific magnetic field waveform consisting of a 5 ms burst of pulses repeated every 65 ms. Within each burst are pulses having a rise time of 200 $\mu$s and fall time of 20 $\mu$s. All waveform parameters, along with the value of peak magnetic field in the centre of the coils (~1.5 mT) are said to be critical for the therapeutic effects to be obtained. How such a unique combination of parameters was discovered is not clear. Tens of thousands of patients have been treated worldwide with the technique, and its success rate (over 80%) is not in doubt. What is unclear however, is whether the magnetic field to which the patients are exposed has added to the undoubtedly beneficial effect of the superior immobilization of the fracture that these patients receive, and the 12–16 h of enforced daily rest they get for a period of typically 6 months while treating themselves at home.

Recently, a number of hand-held low frequency pulsed magnetic field devices have been advertised in the national press. Their literature usually gives no indication of the type of output they produce, or even that they are magnetic field generators at all. Typical peak output fields measured at the surface of such devices are usually less than a few tens of $\mu$T and, because they are generated by small coils or solenoids within the units, the fields decrease very rapidly over a distance of just a few millimetres. Incredibly, one device is claimed to replace 'missing frequencies' from the electroencephalogram (the EEG—a recording of the electrical activity of the brain), the absence of which are said to be the cause of complaints such as migraine and myalgic encephalomyelitis (ME). It is programmed for each individual user after a recording is made which purports to show these missing frequencies, but which bears no resemblance to a human EEG. Another device, whose virtues are extolled in its advertising literature by a well known radio presenter, has a set of four switches which can be set by the user to 'sleep mode, communications mode, memory mode, stress mode and optimism mode'. If only I thought it worked I would certainly buy one!

Static magnetic field devices such as magnetic necklaces, bracelets, and shoe inserts are numerous. They produce magnetic fields at their surface of order 1–100 mT. The usual therapeutic claims are for improved circulation, relief of stiffness, and the healing of muscle sprains; some devices are recommended for migraine sufferers, the stimulation of wound healing, and the cosmetic improvement of scars. Many are available by mail order, and some even from major High Street chemists. Not long ago I received through the post a few sample of a 'small, strong, adhesive magnetic disk' for 'the treatment of muscle stiffness and tension'. The device consists of a circular plaster with a small ferrite magnet at its centre. Its manufacturers claim to sell over 20 million boxes of these disks annually throughout the world.

## *Scientific evaluation of electric and magnetic therapies*

How should the claims for these therapies be scientifically evaluated? It is important to bear in mind that the absence of a plausible theoretical mechanism of interaction between the output of these devices and the body must never override reliable experimental or clinical observations of a real effect. The claims for most electric and magnetic therapies can be readily confirmed or refuted using a standard scientific test, the double-blind clinical trial, which is routinely used to assess (for example) new drugs.

The basic principle of such a trial is to select a population of volunteer patients, to whom the purpose and design of the study has been fully explained. These patients are then divided at random into two groups, a test group which receives the treatment, and a control group which does not, but is treated identically in all other respects. The study design is called 'double-blind' because neither the patient themselves, nor the personnel administering the treatment or evaluating its outcome, know which group the patient is in. At the end of the study the patients are grouped using the (until then) secret treatment allocation code, and the outcome of the two groups is compared statistically.

The double-blind trial is well established in medicine for the objective evaluation of new therapies. It enables a true effect of the treatment to be distinguished from those of observer bias, unconscious selective encouragement of the patient by their therapist, concurrent improved general medical care, the natural healing process, and placebo effects (which can often be surprisingly large). This type of trial is particularly suited to the evaluation of low level electric and magnetic therapies because their outputs cannot be detected by our senses. Hence the blinding process is easy to achieve, and patients are unable to tell if they are being treated with an active device or one whose output has been disabled. In the case of an electrically powered device used by a therapist, it can usually be modified such that, when the patient's trial number is keyed in, the device becomes active or disabled as appropriate, while all the externally visible functions such as indicator lights remain unchanged. For therapies which are used at home, the patient can be supplied with either a working or disabled unit or, in the case of permanent magnet devices, one which is either magnetized or has been demagnetized.

None of the therapies described above have been demonstrated to have consistent and significant non-thermal therapeutic effects when evaluated by independent double-blind trials. A limited number of such trials have been carried out on selected devices, including some which have been run by colleagues and myself, but they have proved negative.

Many people believe in the effectiveness of electric or magnetic therapies, but there is no reliable evidence of efficacy to support those beliefs. The public are often misled by unscrupulous advertising which makes claims that are unproved by any scientific measure. These claims are usually 'supported' by statements such as 'over 50 000 patients have received relief in the last 5 years'—but these patients are never reported in the scientific literature. Uncontrolled assessment of efficacy is, at best, of very limited value because these therapies are usually used to treat disorders which are self-healing or have natural periods of remission, thus allowing ample opportunity for wishful thinking on the part of both the patient and the therapist.

Doctors, health-care professionals, and members of the public at present appear to have little protection against such dubious but persuasively presented claims, which are often wrapped in scientific mumbo-jumbo to lend them an air of credibility. Electric and magnetic field therapies come under the general category of 'medical devices', and as such are exempt from the Medicines Act which regulates the claims made for drugs. Some Trading Standards officer make valiant efforts to tackle the problem but do not have the specialist expertise or resources required to evaluate the claims made for these electrotherapies.

This subject has been highlighted by scientists and engineers before. A biography of Silvanus P. Thompson, an illustrious past president of the Institution of Electrical Engineers, records that, in 1882 he 'forcefully condemned' the 'quacks and rogues who deal in so-called magnetic appliances and disgrace alike the sciences of electricity and medicine while knowing nothing of either'. Surprisingly little has changed over one century later.

If society wishes this branch of medicine to develop objectively and honestly, some form of licensing is needed so that products are only allowed to be advertised after their claims for efficacy have been independently scrutinized. Help may, however, be at hand. From June 1998 most medical devices, including electrotherapies, must meet the requirements of the Medical Devices Directive (MDD) before they can be CE marked and sold within the European Union. One of these requirements is 'that devices must achieve their intended purpose as claimed by the manufacturer'. If the 'Notified Bodies' (who are responsible for checking that individual products meet the requirements of the MDD before they can be CE marked) are rigorous in applying this requirement, then society and individuals may, at last, be protected from the 'quacks and rogues'.

## Electricity, magnetism, and diagnosis

There are many uses of electricity and magnetism for diagnostic purposes. Well known examples include the recording of the electrical activity of the heart (the electrocardiogram) and imaging of the body using a combination of static and pulsed magnetic fields (magnetic resonance imaging). I have chosen to describe (and demonstrate) a slightly less familiar example, which combines both the injection of electricity into the body and the recording of the resultant response from nerves and muscles. In particular, the recently developed technique of magnetic nerve stimulation is another example of the application of Faraday's principle of electromagnetic induction to medicine.

### *Electrical properties of the nervous system*

The nervous system forms a digital signalling network which transmits small electrical impulses, of about 1 ms duration, around the body. The conduction velocity of individual nerve fibres is proportional to their diameter (approximately 1.7 m/s per $\mu$m of diameter). This is increased by an additional factor of about 3.5 if the fibre is surrounded by a layer of fatty insulation called the myelin sheath. The range of conduction velocities in the body, from the smallest unmyelinated fibre to the largest myelinated fibre, is about 0.2–100 m/s. The body has evolved to use large, fast fibres only where speedy communication is important. Slow conducting fibres are quite adequate for the control of, for example, sweat glands or the movement of food through the digestive system, but a 10 s delay between the brain issuing a command and a leg moving would be rather inconvenient, particularly if being chased by a sabre-toothed tiger! The majority of sensory and motor fibres have conduction velocities in the range 35–80 m/s. Intensity of information, be it the strength of either a sensation, or of the muscle contraction required, is transmitted by 'frequency modulation' of the electrical impulses—for example a strong muscle contraction is produced by firing the appropriate nerve fibres more times per second than for a weak contraction.

Diseases of the nervous system, mechanical damage, or entrapment of nerves all lead to a decrease in conduction velocity and hence measurement of this and other electrical parameters of nerves and muscles can be an invaluable aid in diagnosis.

## *Recording electrical activity from nerves and muscles*

To record directly electrical activity occurring within the body it is necessary to convert the flow of ions, that carry the current in the tissue, into a flow of electrons which can pass down external wires to the recording equipment. This is done using either needle of surface electrodes, the former being inserted through the skin such that their tips are close to the source of signal, the latter being placed on the surface of the intact skin over the source. Needle electrodes are usually made of stainless steel, surface electrodes of silver, coated with silver chloride to minimize electrical noise. Signals recorded from nerves using surface electrodes can be up to 50 $\mu$V and from muscles up to 20 mV. Needle electrodes can give signals several times larger than those from surface electrodes.

It is also possible to record the external magnetic fields caused by electrical activity within the body. These fields are, however, very small, specialized equipment such as SQUIDs (Superconducting Quantum Interference Devices) are needed to detect them, and this approach is not widely used at present.

One can demonstrate the recording of electrical signals from muscles by fastening a pair of silver/silver chloride electrodes over a superficial hand muscle (such as abductor digiti minimi) and connecting them, along with a reference electrode relatively remote from the muscle, to a high input impedance amplifier. The resultant output can be displayed on an oscilloscope-type screen. On contracting the muscle a complex signal is recorded, of amplitude a few millivolts, formed by the superimposition of the signals from many 'motor units' (groups of muscle fibres fed by a signal nerve fibre) firing repetitively but at different times.

Because muscle signals have a bandwidth of a few hundred Hertz they can be played through a loudspeaker and heard as a kind of rumbling sound. The *eye* is not very good at analysing such signals; for example, if a microphone was used to display my voice on the same screen you would not be able to tell what I was saying by looking at the waveform. The *ear* however, is very good at analysing complex patterns, hopefully you have no difficulty in understanding me, even when I speak quickly! Clinical neurophysiologists are trained to be able to detect some muscle abnormalities by 'listening' to the signals picked up by needle electrodes.

## *Electrical stimulation of nerves*

In order to measure nerve conduction velocities it is necessary to start the impulses on the nerve at a known time, normally using electrical stimulation. The abductor digiti minimi muscle from which recording can be made is connected to the ulnar nerve, one of the three main nerves of the arm. Placing a pair of stimulating electrodes about 1 cm apart over the ulnar nerve at the wrist, and injecting a current pulse of 100 $\mu$s duration into them once per second, causes the hand to twitch and the electrical activity of the muscle is seen on the screen. The movement is completely involuntary and cannot be stopped by the subject. A current of about 20 mA is needed to stimulate all the individual nerve fibres simultaneously and obtain a 'supramaximal response'. The response is detected about 3 ms after the stimulus is applied, the delay being due to a combination of transit time down the nerve, synaptic delay at the neuromuscular junction, and conduction down the muscle fibres to the recording electrodes. If, for example, the ulnar nerve was trapped, compressed, or diseased between the stimulating and recording sites this would be revealed by an abnormally long conduction time.

## *Magnetic stimulation of nerves*

Electrical stimulation of nerves is widely used for diagnostic purposes, but has three main limitations: it can be painful, it is difficult to stimulate deep nerves without using needle electrodes, and structures with an overlaying layer of bone, such as the brain, are virtually inaccessible.

The technique of magnetic nerve stimulation, developed at the Royal Hallamshire Hospital and the University of Sheffield, overcomes some of these limitations. It is based on Faraday's principle of electromagnetic induction. We saw earlier, with a large air-cored coil, that voltages can be induced in the human body using a time-varying magnetic field. Passing a large pulse of current through a smaller coil from a purpose designed magnetic nerve stimulator, the resultant magnetic field pulse will also induce voltage (and hence current) in my body. The basic principle of magnetic nerve stimulation is shown schematically in Fig. 10 and the first clinical stimulator in Fig. 11. We can show that a magnetic field pulse is being generated with the aid of an ordinary aluminium disk. If it is placed on the stimulating coil and the stimulator fired, the resultant induced currents will have their own magnetic field associated with them which will oppose the driving field, causing the disk to be repelled and fired across the room. And if I place the coil on my arm and

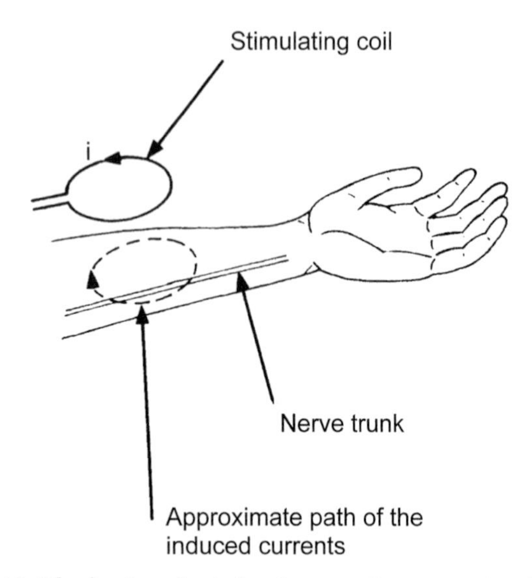


**Fig. 10** The basic principle of magnetic nerve stimulation.

fire the stimulator, I twitch! It is interesting to note that, unlike electrical stimulation, the magnetic field passes through clothing. Indeed the coil need not even touch the body as it will also pass through air without significant attenuation. 'Magnetic stimulation', although a well established and convenient name for the technique, is a slight misnomer. Stimulation is, of course, caused by the induced voltages and currents rather than by the magnetic field directly.

Magnetic stimulation has a number of disadvantages when compared with electrical stimulation. First, the equipment costs more and is bulky, one can build an electrical stimulator in a cigarette packet with ease, a typical magnetic stimulator weighs about 20 kg. This is because magnetic stimulators need to switch some 3000 V across the coil, to give a current pulse of some 8000 A and achieve the required magnetic field pulse of approximately 2 T with a risetime of 100 $\mu$s. Secondly, fast repetition rates are much harder to achieve than with electrical stimulation because of the large energies involved and of problems with coil heating. However, these difficulties are gradually being overcome by a combination of innovative circuit design and good engineering, and commercial machines are now available that can stimulate at a few tens of Hertz for short periods. Thirdly, the site of stimulation is not well defined because of the relatively large current loops that are induced in the tissue, although this can be somewhat improved by using 'figure-of-eight' shaped stimulating coils.

Magnetic stimulation has a number of advantages when compared with electrical stimulation. No electrical contact is needed with the

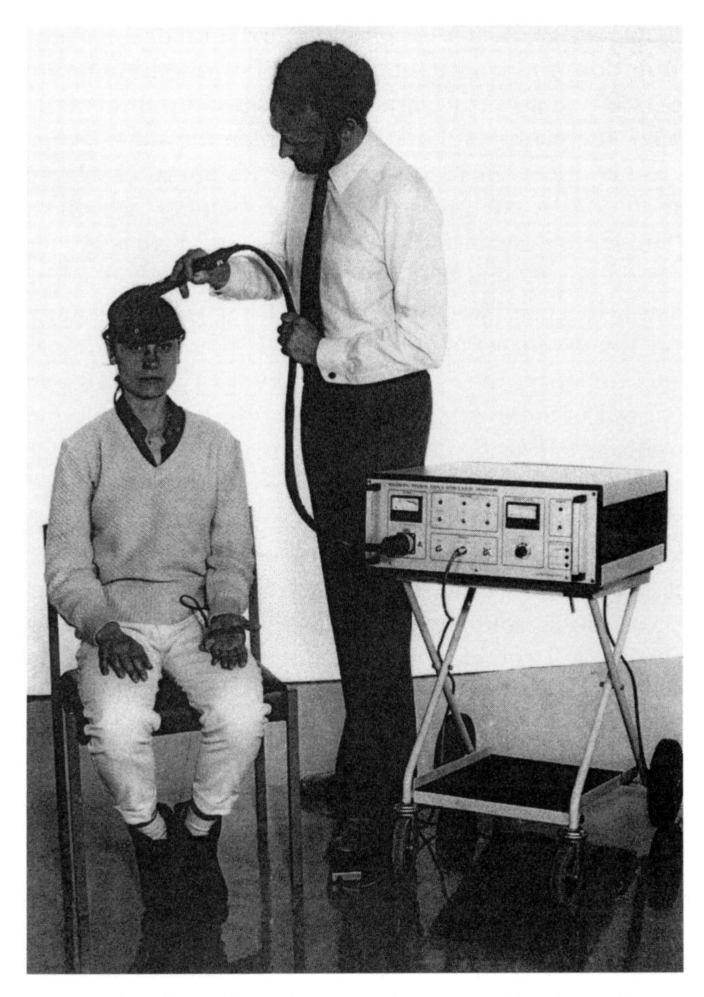

**Fig. 11** The first clinical magnetic nerve stimulator in use, Sheffield, November 1985.

subject and the stimulus goes through clothing. It is virtually pain free, can stimulate deep nerves and is also able to stimulate the human brain with ease. Let us return to the electrical activity of abductor digiti minimi in the hand. The ulnar nerve, which is connected to it, leaves the spine in the brachial plexus region of the shoulder, but is fairly deep in the body at that point, which makes it difficult to stimulate with surface electrodes. Placing the stimulating coil between the shoulder and the neck and firing the stimulator successfully activates the ulnar nerve, the arm moves and a response is recorded from the muscle. On moving the coil up to the top of the head and firing the stimulator the motor cortex is stimulated, the hand moves again and another response is recorded.

Figure 12 shows two traces recorded from a normal subject using magnetic stimulation and one using electrical stimulation. As would be expected, it can be seen that the time between stimulus and response (the 'latency') increases with distance between the sites. If required, the length of the conduction pathway in the periphery can be measured, and the time converted into a conduction velocity for that pathway. Figure 12 indicates how magnetic stimulation can be used to aid in diagnosis. If the latency from shoulder to hand is subtracted from that of head to hand, the resultant value (about 9 ms in this example) is the conduction time from head to shoulder. This path is almost entirely within the brain and spine—the central nervous system. Diseases which affect the speed of conduction of signals in the central nervous system, such as multiple sclerosis, result in an increase in this conduction time, which can be several times longer in severe cases.

The ease with which the brain can be stimulated is noteworthy. Prior to magnetic stimulation the only method of non-invasively stimulating the motor cortex was to put a 1 kV pulse between two electrodes on the scalp, the very high voltage being needed to punch sufficient current through the skull. I can report, from personal experience, that it was quite unpleasant! The lack of pain, ease of use, and ability to study previously inaccessible pathways in both normal volunteers and patients

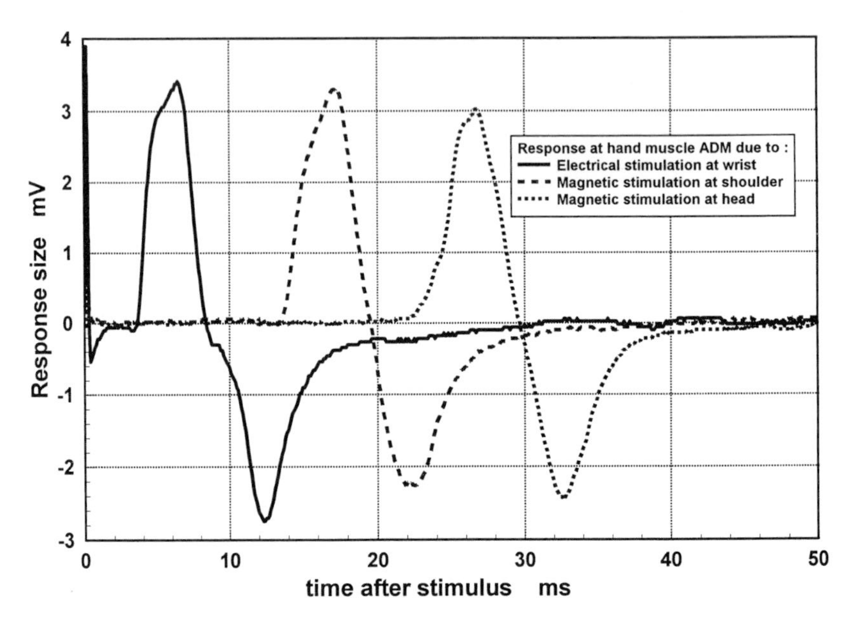


**Fig. 12** Recordings from abductor digiti minimi in response to single electrical and magnetic stimuli.

has led to a rapid growth in the use of magnetic nerve stimulation, both for routine clinical use and in research. Today, magnetic stimulators are available from several manufacturers and over 3000 stimulators are in use world-wide.

## *Magnetic stimulation and depression*

A few years ago reports started to appear suggesting that patients examined with magnetic stimulation of the brain experienced an unexpected improvement in their mood. The simultaneous development of rapid rate magnetic stimulators provided the necessary tool to investigate this phenomenon. There are now a number of groups around the world whose results suggest that it may become possible to treat severe depression with rapid rate magnetic stimulation (typically using repeated 5 s duration bursts of 10 Hz stimuli) applied in the region of the left prefrontal cortex. At present such patients are often treated with electroconvulsive therapy (ECT) which, while effective, can have unpleasant side-effects. The generation of convulsions does not appear to be necessary for effective treatment when using magnetic stimulation. Hence general anaesthesia is not required and side-effects, such as memory loss, have not been reported.

If these enthusiastic early reports of efficacy are confirmed by larger and more detailed controlled studies, magnetic nerve stimulation may have an important part to play in the treatment of depression. Much work still remains to be done to optimize such parameters as site of stimulation, stimulating coil geometry, stimulus frequency, and amplitude, and to identify the mechanism of action. However, if this can be achieved, the technique may also prove useful in less severe cases of depression, as well as those for whom ECT would be recommended today.

## Conclusions

This account has only scratched the surface of the many effects of electricity and magnetism on the human body and their applications in medicine. The examples I have chosen can best be summarized by looking at some of the claims for the 'electromagnetic brush' of Fig. 9.

'As lightning purifies the air so must electricity purify the blood': Not really—any large and uncontrolled application of electricity to the body is dangerous and often fatal. Some claim that exposure to very weak electromagnetic fields is also bad for health and the general public have been alarmed by these claims, particularly with reference to overhead power cables. I do not find the existing evidence for this viewpoint persuasive.

'Substitute for medicine': Unfortunately, externally applied electric and magnetic fields are not the panacea that some advertisements claim them to be. Electricity and magnetism have provided many benefits for mankind, but medical therapy has yet to be added to these, except at levels high enough to cause thermal effects or direct neuromuscular stimulation. I would personally welcome a cure for 'softening of the brain', and if you suffer from 'sudden death' (both in the small print of Fig. 9), a remedy would be most useful!

However, electricity and magnetism do have many medical applications. The body is an extraordinarily complex electrical machine that can be both interrogated, monitored and, under some circumstances affected using electromagnetism. Most of our major functions are electrically controlled and hence perhaps one claim in Fig. 9 is justified. Perhaps electricity is, in fact, life itself.

## Acknowledgements

I am indebted to Dr John Swanson for assistance in the preparation of part of this manuscript, the Magstim Company Ltd for the loan of equipment, and to the Central Sheffield University Hospitals Trust for its support and the use of its facilities.

ANTHONY T. BARKER

Born in 1950 in Yorkshire, he was educated at the University of Sheffield and after a period in the electronics industry, he joined the Royal Hallamshire Hospital in Sheffield, where he is now Head of Clinical Instrumentation and Specialist Patient Services. He researches on magnetic nerve stimulation, electrophysiology, and biological effects of electromagnetic fields. He is a Fellow of the Institute of Physics and Engineering in Medicine and of the Institution of Electrical Engineers, whose working party on the biological effects of low-level electromagnetic fields he chairs. He has published widely, is a past presenter of the Silvanus P. Thompson lecture series and was a co-presenter of the 1997 Faraday lecture.

# *Environmental campaigners and scientists—friends or foes?*

PETER MELCHETT

## Ludwig Mond

The invitation to write this chapter came to me 'as a grandson of Ludwig Mond', the centenary of whose benefaction to the Royal Institution was celebrated last year. So, as I may be here as—actually—the great great grandson of a famous chemist as much as in my role as Executive Director of Greenpeace UK, I thought I'd start with some personal observations about my own interest in environmental issues.

Apart from starting the chemical company which later became the basis for ICI, Ludwig's son and grandson, my great grandfather and grandfather, took a keen interest in the English countryside. They established large country estates, and particularly my grandfather became a keen blood sports enthusiast.

## The English countryside—decades of destruction

Thus I was brought up, in the Norfolk countryside, in a family atmosphere which celebrated science and its contribution to human development, and which also celebrated the upper class view of 'nature' in the English countryside.

I learned from ancient and apparently incredibly wise old gamekeepers about the mysteries of our native wildlife, the predatory destruction caused by vicious owls, the horrendous destructive capacity of the sinister, nocturnal badger, and the downright irrational ferocity of the fox. I learned that partridges mate for life, and that the weather during Ascot week was the only thing that had any significant impact on the number of grey partridges that survive to adulthood, apart, of course, from the effects of any predators that had survived the gamekeeper's gun, snare, poison, or trap.

All of this was, of course, rubbish. And it was science that revealed this to me; for example, from the work done by the Royal Society for the Protection of Birds and the Nature Conservancy Council, on the effects of organochlorine pesticides, such as DDT, I learned that the dead grey partridge chicks I'd been shown in a barley field by the old and wise gamekeeper had not died of drought, but of starvation, because the insects they depended on had been killed by pesticides.

Some years later, a 3-year study into factors affecting the survival rate of partridges, which was carried out on our family farm in north-west Norfolk, revealed that red leg partridges do not mate for life.

The owls I saw gamekeepers slaughtering as a child turned out to be migrating long-eared owls, posing no threat to any game bird in this country, or elsewhere. After we reintroduced badgers on the farm in Norfolk, in the very first year that they took up residence, a hen pheasant successfully brought off a clutch of over a dozen chicks from a nest a few yards away from the entrance to the badger set. And, again more significantly, through studying natural ecosystems, and relationships between different species, we've come to understand a little of the complexity of these relationships, and how inept and ignorant most attempts at human interference and 'management' of such systems have been in the past.

I think I learned three lessons from this period of my life, being brought up as a boy in Norfolk, and seeing, during the 1950s, 60s, and 70s, the massive changes that took place in the English countryside.

## Commercial interests and science

First, about how commercial interests affect science. In the 1970s I met one of the scientists who had been involved in the discussions about the use of DDT, which followed the publication of Rachel Carson's *Silent Spring*, and the discovery of wild birds scattered dead over farmland in this country. As the RSPB and others catalogued the catastrophic decline of predatory birds at the top of the food chain, such as sparrow hawks and peregrines, the evidence, at least in their eyes, became overwhelming that certain pesticides were causing these problems. I remember talking to the then Director of the RSPB, the late Ian Prest, about his experiences at that time. He came up against the industrial lobby that was there to protect the interests of the chemical companies making these pesticides. For them, no evidence was sufficient, no evidence was conclusive, no evidence justified action. They wanted the dead bird at the end of the pesticide sprayer nozzle before action was justified.

Ian Prest was a decent and honourable man, a scientist who believed in science and its integrity. The experience left a lasting impression on him because of the extraordinarily crude way in which this argument divided into one between scientists representing commercial interests, and the rest.

In the 12 years I've worked for Greenpeace, I have certainly come across this phenomenon many times, and it is something I will return to later.

## Reductionist science and ecological complexity

Second, I gradually understood how complex were the factors which had led to these massive changes. As you know, during this period, the United Kingdom lost about half of its ancient forests, well over 90% of many rare, semi-natural habitats, such as chalk grassland and thousands of miles of hedgerow. All of this was wrong, as is now almost universally agreed, but at the time, conservationists were fighting on many different fronts.

Scientists working for agrochemical companies and others with economic interests in these changes, which were all designed to intensify agriculture and forestry, leading to greater inputs of chemical fertilizers and pesticides, argued that the immediate consequences of growing crops or trees in this way were not as damaging as conservationists suggested.

But these scientists were looking at the effects, for example, of a particular herbicide or insecticide on an arable ecosystem—an ecosystem of relatively little interest to nature conservationists. Conservationists on the other hand, were looking at the loss of the original habitat, caused by conversion to arable farming in the first place, and at the effects of intensive agriculture not just on the arable field but on the surrounding countryside.

Looking back now, it is clear that even conservationists missed some of the longer-term, more insidious consequences of these changes. During this period, the decline in numbers of popular, and what had been considered common, bird species, such as skylarks and grey partridges was astonishing (see Fig. 1). Both species are now on a red list, targeted by

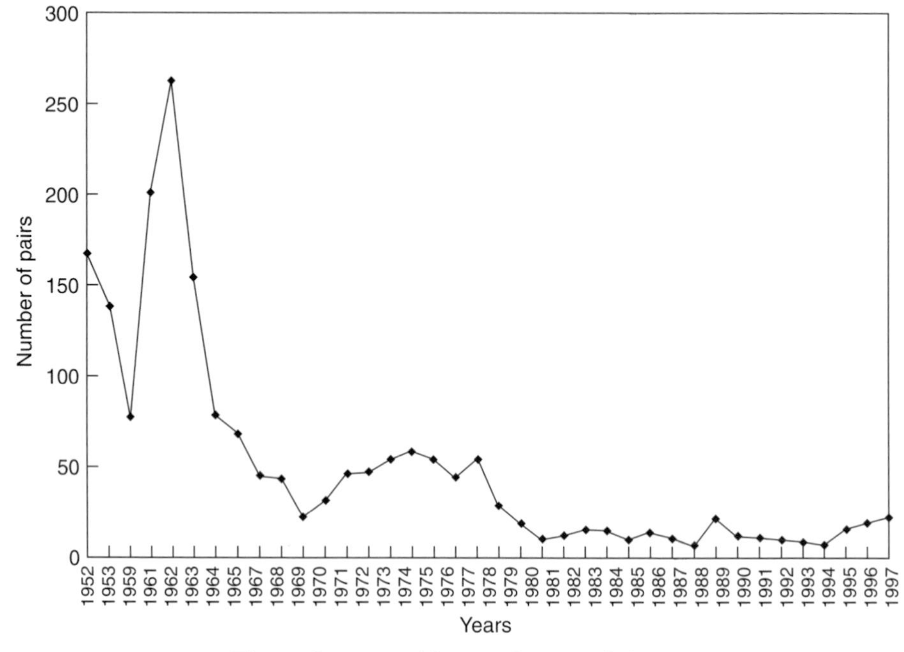


**Fig. 1** Grey partridges at Courtyard Farm.

the Government and others for urgent action to arrest their decline. I recently had the profoundly disturbing experience of discovering that our farm in Norfolk is one of the best places to see overwintering tree sparrows in East Anglia. This is not because the farm is a Site of Special Scientific Interest, or indeed special in any particular way. It's because we're aiming to manage part of the farm that is in set-aside in a way that benefits species such as skylarks and tree sparrows, and indeed hares and grey partridge. More disturbingly, it's because tree sparrows have declined so catastrophically in the farmed countryside generally.

With the controversy that surrounded these changes, how could this have been allowed to happen? I think the answer is that nobody really foresaw what the consequences would be of the whole range of the changes that were taking place in agriculture. It was obvious that if you put in a system which canalized a river, installed mechanized pump drainage for the surrounding wet meadows, put field drainage in that land, and then converted it to arable, that there would be dramatic changes in the flora and fauna that had previously existed. What was less well understood was that if you combined these changes in land use with the technical changes that were being made with plant breeding in crops like wheat and barley, along with many other changes, the negative effects on many species would be much more wide-ranging.

Looking at each particular change in isolation—taking a case by case approach that governments are so keen to do on so many environmental issues, did not alert people to the actual impact.

## Framing the debate to determine the outcome

The third lesson I think I learned was to experience the extent to which the answers scientists come up with are influenced by the questions they are given to answer, and by the culture in which they operate. In looking at the role that science had played in promoting, explaining, and ultimately playing some part in slowing down these destructive changes in our countryside, I came to understand that some issues were excluded from public debate, either because they didn't fall within the relevant scientific expertise of a particular committee or organization, or indeed government department, or because they were considered 'unrealistic', not because they were unscientific, or not factually correct in some way, but rather because they were simply not considered to be in the realm of what was 'possible' or practicable' by those in authority.

This was certainly something I frequently experienced in the four and a half years I spent as a Government Minister in the last Labour Government.

## Greenpeace—thinking the unthinkable and doing the impossible

So maybe it is no surprise that I've ended up working for an organization that tries to shift agendas, so the 'unthinkable' or 'impossible' change which is needed to protect the environment shifts on to the agenda rather than remaining off it. An organization which tries to look at change in a holistic way, and think about the trajectories that we're on, rather than looking at the detailed implications of small changes to part of the picture, or concentrating on trying to protect and defend one particular site or species.

Greenpeace tries to work by creating the conditions in which change can take place, and then by forcing such change to happen.

Scientists frequently ask people like me who work for campaigning groups what right we feel we have to try and alter the course of human affairs. The question is usually posed in the form 'who does Greenpeace represent?' or 'what legitimacy does Greenpeace have?' I also find that the questioner often assumes, or appears to assume, that it will be Greenpeace that takes decisions about what happens—the assumption seems to be that Greenpeace will decide whether nuclear weapons will continue to be tested, redundant oil installations sunk or recycled, or new nuclear power stations built. We won't make these decision, nor, or course, should we. Such decisions should be made by democratically elected governments.

The role of an organization, like Greenpeace is to defend nature. We do so on behalf of the individuals who directly support us. We exist only because we receive such public support—we don't accept funds from any other source than individual supporters, and without such support, we cease to exist. Of course, if our campaigns are successful, they're successful because they generate support from a far wider constituency, ultimately, normally, from an overwhelming majority.

But I find that something that's rarely understood about campaigning groups, is the time-scales over which we are prepared to operate, certainly compared with the time-scales of many of those with whom we interact. We all know that a week is a long time in politics, but the media are interested in a new story every day, and business, particularly in this country, has a time horizon limited by their next quarterly report to their shareholders, and the daily movement of their share price. By contrast, most of our successful campaigns have run for at least a decade from start to finish, and the campaign that Greenpeace started with, to end the testing of nuclear weapons, took 25 years to reach a successful conclusion, with the UN agreement to a Comprehensive Test Ban Treaty last year.

In that period, of course, we sometimes received significant public support for our campaign, and sometimes very little. On nuclear testing in the Pacific, we received a great deal of support in Australia, New Zealand, and Japan. We received rather less in France, particularly at the time that the French Government sank Greenpeace's flagship, the Rainbow Warrior, when it was on its way to protest against French nuclear weapons testing at Moruroa in the Pacific.

## Nuclear weapons—right or wrong?

Another question asked of Greenpeace from time to time is whether we have ever 'got it wrong' in the sense that we had started to campaign against something which we subsequently realized wasn't so bad after all. The answer is 'no', but underlying this question is, I think, a deeper misunderstanding about the nature of most of our campaigns. Again, it might be useful to look at this question in the context of our longest running campaign. Here, science played a part. In the early years of the campaign, government scientists insisted that the alleged dangers of the atmospheric testing of nuclear weapons were being overstated by anti nuclear groups, such as Greenpeace. When testing went underground, scientists working for the two democratic governments that had nuclear weapons test sites, the USA and France, insisted that the tests were 'safe'. So there was a scientific debate throughout this campaign about the safety or otherwise of testing nuclear weapons. From time to time, Greenpeace got involved in that debate, as did others. For example, the late Jacques Cousteau went to Moruroa to see if the coral atoll was splitting, and radiation was leaking from the test site. French Government scientists have consistently insisted that this was not happening, Greenpeace raised concerns that it was, and Jacques Cousteau, while not supporting the French Government's view, was not able to come to a clear view about radiation leaking from the atoll either.

But of course the question being asked of science in this case was not the question being raised by the campaign. Campaigners like Greenpeace who wanted to end nuclear weapons testing also wanted to abolish nuclear weapons. The dangers we were concerned about went well beyond the immediate impact of testing new weapons, to the implications of possible accidents, to say nothing of the deliberate use of such weapons. A debate about the risk of nuclear weapons testing rather missed the point, namely whether it was necessary to use this terrible technology at all.

## Greenpeace needs and supports democracy and free speech

Throughout the 25 years of this campaign, while we focused primarily on French nuclear weapons testing in the Pacific, we also campaigned against the nuclear weapons tests carried out by the UK and the USA at the Nevada test site in the USA, and against nuclear weapons testing being carried out by China and the then USSR. Nevertheless, the scope for a campaigning organization in, for example, the former USSR, was limited. Greenpeace did establish a national organization in the Soviet Union, and sailed on campaigning voyages to Novaya Zemlya on more than one occasion, being arrested by the KGB each time we got anywhere near the Soviet test site. We have also carried out at least two antinuclear weapons testing protests on mainland China. Nevertheless, pressure groups which have arisen in Western democracies depend for their existence on freedom of speech, and free democratic institutions of state. Greenpeace currently has over 30 national offices world-wide, but only two in non-democratic countries, Tunisia and China.

As an international campaigning group, Greenpeace has a rather harder job than people working primarily within a national context, such as a national politician, national media, and indeed many scientists, at least in their public role, if they have one. Greenpeace has to adopt the same policy, and the same arguments about issues, whether we're addressing an audience in Tokyo or Washington, in Helsinki or Rio de Janeiro, Moscow, or Sydney. What appears completely normal, natural, and acceptable to people in one country can, in another country, appear to be a controversial, unsupported assertion—indeed maybe even a deliberate attack on the national interest.

## Genetic engineering—the next disaster?

I think I've said enough about Greenpeace as an organization, and I want to return to some of the themes that I outlined at the start of this paper. I talked about the controversy over chemicals like DDT in the 1960s and 1970s in the UK, and the way in which the scientific debate was framed, and the many years that passed during which scientists arguing on behalf of industrial and commercial concerns insisted that no ban should be introduced without evidence that a dead body would provide—thinning eggshells were simply not enough, at least for many years.

Today, we face the same sorts of arguments over genetic engineering. Here is a new technology being introduced where the assumption being made by the commercial interests involved, and those representing the public interest, such as governments and regulatory agencies, is that things should proceed unless a dead body turns up. This raises a number of issues. First, and most significantly, as was the case with changes in agricultural practice and policy in the decades following the last war, the wider picture is not part of the agenda of debate. What will be the consequences of the introduction of this technology on agriculture? On the way people feel about food and food production? On people's relationship with nature? On people's perceptions of what is natural and safe? None of these questions are being considered by anybody that claims in any way to represent public interest, be it political, scientific, commercial, or whatever. The furthest we ever get with a discussion of these issues is the grossly oversimplistic assertion which appears in speeches by those who favour genetic engineering, that it will 'feed the world', or at least contribute to reducing hunger and starvation world-wide.

Although widely acknowledged, certainly in private, by scientists and others involved in this field, the wider and longer-term consequences are simply not part of any official regulatory agenda. Intergenerational effects are excluded, as they were when DDT was licensed. The hugely complex interactions which take place in the natural world are not, indeed cannot, be considered by regulatory agencies, just as they were not when DDT was licensed.

## Improper claims for 'proper' licensing systems

These are blatant shortcomings in the regulatory process. In the wake of BSE and Mad Cow disease, they should be obvious to everyone, if they weren't before. But scientists, commercial interests, and politicians continue to rely on the mantra that products have been through a 'proper' licensing system.

I don't think I'm alone in finding all of this slightly incredible. I remember the excitement with which the idea of nuclear power generating electricity that would be too cheap to meter was greeted.

Today, similar claims for genetic engineering—food too cheap to pay for—would be greeted with the derision that they would deserve. Indeed, such derision should be the fate of other, similarly unjustified claims for genetic engineering, such as the 'feed the world' claim.

The public's increased scepticism, or I would say realism based on real experience, about such claims is one of the more welcome developments that I have seen during my own lifetime. So, for example, when a nuclear scientist says that an accident like Chernobyl can only happen once in a thousand years, or whatever it was, and it happens during your own lifetime, during a period of less than 80 years, claims about the likelihood of similar catastrophic events endorsed by the same people, whether it's a Royal Commission, a similar group of scientists, or the political establishment, will, quite reasonably, be greeted with a degree of scepticism.

## The public perception of risk—rationality based on real experiences

Research has also shown that the public's perception of risk is based on different but entirely rational factors compared with the way in which risk is seen by many technical and scientific experts. For example, a factor which ordinary people consider to be of great importance is the degree to which a risk is accepted voluntarily, or imposed involuntarily. So the risks involved in sport parachuting, bungee jumping, or mountain climbing, while extraordinarily high compared with many environmental risks imposed on people, are considered entirely acceptable. Even the risks involved in doing some jobs are considered acceptable—being a formula one racing driver for example. However, the risk of dying of lung cancer through the consequences of passive smoking because a colleague at work was encouraged to take up smoking by the advertising on the side of the formula one racing car, is considered by many to be unacceptable.

Ordinary people also make more sophisticated judgements about risks and benefits than is generally accepted, in my experience, by scientific or political commentators. Recent research into public attitudes on genetic engineering, carried out by Lancaster University, indicated that people were prepared to see significant risks run if they could be sure what the benefits were (*Uncertain World—Genetically Modified Organisms, Food and Public Attitudes in Britain*, Associations of the Centre for Environmental Change, Lancaster Univ., March 1997). So, for example, people do not object to genetically engineered medicines, particularly medicine that might counteract life-threatening diseases. They quite reasonably make the judgement that if a disease is going to kill someone, it's reasonable for that individual to run very, very significant risks to try and cure the disease. But when asked whether they agree with genetically engineering tomatoes so that they don't rot so quickly, or soya beans that can be grown and sprayed with chemical produced by the same company that's also producing the genetically engineered bean, people say—What's the benefit? The answers are, of course, that the benefits, beyond the commercial interests of the company concerned, are hard to discern, and certainly are not clear. And the risk is being run by the person eating the food product at the end of the chain. The risks, however slight, are nevertheless, sensibly and rationally, unacceptable.

If the public had been asked similar questions about feeding the brains of other ruminant animals to cows, to provide a source of protein in their diet, similar questions would, I think, have been raised. People might reasonably have expected that there would be some possibility of a risk to those eating the final products of the cow, beef, or dairy product.

So the question is, where was the benefit. Certainly, the benefit didn't accrue to the ultimate consumer. Nevertheless, the process of feeding animal brains to cows passed through all the necessary regulatory frameworks, and scientists regularly told politicians that there was no evidence that this was causing any harm.

In this instance, once again, the question being asked of scientists was extremely narrowly framed—Is there evidence of harm to human beings or not? The regulatory framework, indeed the very strong political dynamic at the time, really precluded any consideration about whether the Ministry of Agriculture should intervene in the animal food chain for reasons other than the discovery of the dead body that I referred to before.

## The Brent Spar—a significant Greenpeace victory

In much the same way, the regulatory system which governs the disposal of redundant oil installations never considered whether the public in the north-east Atlantic region would prefer oil installations to be brought ashore and recycled or re-used.

Thus, Shell went through the regulatory procedures when they were considering what to do with their redundant oil storage buoy, the Brent Spar. These regulatory procedures didn't consider social factors, aesthetic values, nor indeed longer-term issues of economic or industrial policy, but looked very narrowly at the immediate impact of one redundant oil installation on the deep ocean. It has to be said that it did even that extremely poorly, failing, for example, to consult the Scottish Association of Marine Scientists, who included some of the few scientists who knew anything about the deep Atlantic where the proposed dump site was located.

Nevertheless, whole areas of public concern, including many environmental and social issues, were simply excluded from the process. As subsequent events proved, this left Shell vulnerable to a campaign by Greenpeace, which ultimately succeeded in getting the decision to dump the Brent Spar changed.

As often happens with our campaigns, the historical context in which we work was ignored by all of those looking at things in a much more short-term framework. Greenpeace had campaigned against the dumping of various forms of waste in the oceans, and in particular in the north-east Atlantic region, for at least a decade. Some of Greenpeace UK's first campaign work was aimed at stopping the dumping of barrels of low-level radioactive waste in the north-east Atlantic. After many years of campaigning, we were successful, and the practice is now outlawed by international agreement. We had similar campaigns, and ultimately similar successes, with the dumping of industrial wastes and then sewage sludge in the seas of the north-east Atlantic region. Again, both practices are now banned by international agreement. In that context, the first proposal to dump a redundant oil installation in the north-east Atlantic was bound to be controversial. Six months before any public campaigning took place, Greenpeace alerted Shell and the then government to our opposition, with the publication of a detailed technical and scientific report.

If the Shell company in Germany had been responsible for the Brent Spar they would have taken that report seriously. I think a different government in the UK would have taken it seriously. In this case, scientific, corporate, and political thinking were shaped by our culture.

Historically, the UK has seen the oceans as somewhere where we can safely and cheaply dispose of much of our waste. As we know, attitudes in other European countries, particularly the Netherlands, Germany, and in Scandinavia are very different.

## 'Sound' science—keeping Sir Humphrey happy

As an aside, one of the things that the UK Government and Shell insisted on throughout our campaign, and indeed since, was that the proposal to dump the Brent Spar was based on 'sound' science. I'm not sure if I'm the only person who finds this frequently used adjective—sound—when applied to anything, including science, extraordinarily inappropriate. I know that the expression 'sound science' has international pedigree, but I'm afraid I can't help thinking of '*Yes Minister*' and Sir Humphrey, every time I hear the word 'sound' used in this context.

You may remember that in the episode in question Sir Humphrey was advising the Minister to appoint a scientist to look into a particular problem, as a way of dealing with it—shelving the issue. The Minister protested that sooner or later the person he appointed would come up with a report. Sir Humphrey reassured the Minister, by explaining that the scientist in question was entirely 'sound'. On being asked to explain what this meant, Sir Humphrey indicated, as diplomatically as possible, that this meant that the scientist in question would come up with the answer the Minister wanted without even having to be told to do so.

I find it hard to believe that I'm the only person who has to suppress a chuckle every time I hear someone defending something because it is based on 'sound' science. I also wonder whether anyone has given any thought to defining 'unsound' science, which I assume must exist somewhere, although I must say I've never seen it identified, nor have I ever seen someone claiming to be an 'unsound' scientist, come forward to defend the unsoundness of what they do. At best, this terminology seems designed simply to indicate that those that agree with you support you, and that those that don't, don't.

## Honesty the only policy for Greenpeace

To return to the Brent Spar, I suppose it may be that Greenpeace is one of those organizations that's considered guilty of being 'unsound', and in the sense in which Sir Humphrey used it, we are, and proud to be so. We were certainly accused of many errors in the Brent Spar campaign, and we admitted to one mistake, when we got wrong where a sample that was taken on the Brent Spar had been extracted from—it came from a pipe leading to a storage tank, rather than the tank itself. This had nothing to do with the basis of our campaign, which in terms of the Brent Spar was launched 6 months before the sampling results were released, and which as I said earlier, formed part of a 10-year-old campaign against ocean dumping. Nor, in practice, did the sampling error have any impact on the outcome of the campaign, given that it was released only 4 days before Shell announced their final decision not to dump the Spar.

Nevertheless, I take it as a compliment to campaigning groups generally, and to Greenpeace in particular, that when we make a mistake it is headline news, whereas the numerous mistakes and indeed deliberate distortions, made by the Government do not rate a single press mention. For example, Greenpeace said that the Brent Spar would create a precedent, and that if it was dumped in the Atlantic, other oil installations, and indeed other waste, could follow.

The Government, and indeed Shell, firmly denied this. They said that Brent Spar was unique, that they had to look at each installation on a 'case by case' basis. They were wrong, we were right. Previously secret correspondence disclosed last year indicates that indeed the Government were planning to use the site identified for dumping the Brent Spar as a site that would be suitable more generally for future dumping of 'bulky wastes' in the ocean

## The consequences of the Brent Spar victory

More positively, we were right about the Brent Spar as a precedent. Since Shell took the decision not to dump the Brent Spar, twelve other redundant oil installations have reached the end of their lives in the North Sea, and all have been brought ashore for recycling or re-use (see Table 1). Many would have been, whatever happened to the Brent Spar, but not all.

In this sense, the Brent Spar campaign changed industrial policy. As the new Labour Government now states:

> We are making it clear that Britain accepts the presumption against sea disposal of redundant oil and gas installations.... There was a major issue in 1994 about Brent Spar. That would now be resolved in a different way.
> **The Rt Hon Michael Meacher MP** (speaking on BBC Radio 4 *Today* programme, 2 September 1997)

**Table 1.** North Sea oil installations decommissioned since 1995

| Nort Sea installation | Operator | Type of installation | Water depth (m) | Substructure tonnage | Topside tonnage | Production start date | Decommission date | Disposal mehod |
|---|---|---|---|---|---|---|---|---|
| GORDON BW | Hamilton (BHP) | satellite platform | 17 | 857 | 2163 | 1985 | 1996 | ONSHORE 90% so far re-used/recycled, aim of 100% |
| ESMOND CW | Hamilton (BHP) | wellhead platform | 30 | 1049 | 543 | 1985 | 1996 | ONSHORE 90% so far re-used/recycled, aim of 100% |
| ESMOND CP | Hamilton (BHP) | central production | 31 | 1912 | 5960 | 1985 | 1996 | ONSHORE 90% so far re-used/recycled, aim of 100% |
| EMERALD | Midland & Scottish Energy | FPF/FSU/subsea equipment | 150 | | | 1992 | 1996 | ONSHORE entirely removed |
| VIKING AD | Conoco | drilling platform | 24 | 714 | 570 | 1972 | 1996 | ONSHORE 99.7% reused/recycled |
| VIKING AP | Conoco | production platform | 24 | 625 | 2151 | 1972 | 1996 | ONSHORE 99.7% reused/recycled |
| VIKING AR | Conoco | riser platform | 24 | 450 | 550 | 1972 | 1996 | ONSHORE 99.7% reused/recycled |
| VIKING AC | Conoco | compression platform | 24 | 650 | 2500 | 1976 | 1996 | ONSHORE 99.7% reused/recycled |
| NE FRIGG FCS | Elf | concrete gravity base | 102 | 7480 | 4960 | 1983 | 1996 | ONSHORE >90% of FCS reused |
| FRIGG FP | Elf | flare platform | 105 | 2800 (total) | | 1975 | 1996 | ONSHORE |
| LEMAN BK | Shell | compression platform | 36 | 609 | 5000 | 1975 | 1996/97 | ONSHORE estimated recycling of >99% |
| ODIN | Esso | drilling, production, accommodation platform | 103 | 6150 | 7300 | 1984 | 1996/97 | ONSHORE total removal expected by October 1997 |

Source: collated from OPL and ODCP information

The previous case by case approach had no presumption against sea dumping. It ignored what has been government policy on waste for many years, namely that there's a hierarchy of desirable approaches, with waste elimination and then minimization at the top, followed by re-use and recycling, with various disposal means at the bottom. There should, therefore, be a presumption against disposal at sea, whatever the waste, and in favour of re-use and recycling.

But this policy question was not captured by any of the detailed scientific studies that Shell and their environmental and academic consultants carried out prior to reaching a decision to dump the Brent Spar. That issue was simply not on the agenda.

In this respect, I think probably the most useful thing that emerged from the Brent Spar campaign came in a report by a Committee established by the Natural Environment Research Council (NERC) after the Brent Spar campaign, at the request of the then Energy Minister at the Department of Trade and Industry. One of the Committee's conclusions was that 'Any decision to proceed, or not to proceed, with dumping oil structures or other wastes in the ocean involves social, economic, ethical, and aesthetic considerations which are outside the competence of the group, and judgements in which the technical assessment of the environmental impact is only one factor, and not necessarily the most important one' (report by NERC for the Department of Trade and Industry on Decommissioning Offshore Structures, April 1996).

## Climate change—science fails to carry the day

In this final section, I want to talk about the dominant environmental concern facing our planet at the present time, namely climate change. In December 1997 governments will be meeting in Kyoto in Japan to discuss a framework agreement to combat the threat of climate change. This environmental issue illustrates a rather different relationship with science than those I have discussed so far.

First, as with other changes to the world's atmosphere like the thinning of the ozone layer, climate change is a threat first identified by scientists, not environmentalists. In most early environmental campaigns, from the campaign against commercial whaling to campaigns against felling old-growth forests, environmentalists have generally been at the forefront of the argument, alongside concerned scientists, and occasionally ahead of them. In the case of ozone depletion and climate change, the problems were identified by scientists, argued over by scientists, and indeed in the case of climate change, politicians like

Mrs Thatcher were alerted to the issue before most of the environmental community became engaged in the discussion at all.

If the role of a campaigning group like Greenpeace was simply to raise the alarm—to raise issues—we would have no part to play in the climate change debate at all. As I've said, the potential problem was identified by scientists, a number of powerful politicians were concerned enough to initiate an intergovernmental process, spanning the world, to reach a scientific consensus on the nature and scale of the problem, and to discuss and agree the necessary action.

As you will know, gradually, a scientific consensus about the likely nature and impact of climate change has emerged, through the International Panel on Climate Change involving over 2000 scientists world-wide. At the Earth Summit in Rio 6 years ago, governments agreed to take action based on this scientific consensus.

Almost nothing happened. The UK and one or two other countries met the commitments that they made at Rio to stabilize $CO_2$ emissions at 1990 levels. As is widely accepted, the UK met this target because of the last Government's struggle and ultimate victory over the National Union of Mineworkers, and the subsequent closing of much of the British coal industry, the fossil fuel responsible for the highest pro rata emissions of $CO_2$, and its replacement with gas.

In practice, of course, campaigning groups do have a part to play. Scientific consensus and significant political support is not enough. Because other forces come into play, not least the enormously powerful lobbying efforts of the fossil fuel industry, represented by the Global Climate Coalition and many others.

## Politicians need an effective campaign

Over the last 5 years, I've lost count of the number of powerful, democratically elected politicians that have said to me or other colleagues in Greenpeace, that without effective environmental campaigning to alert the public to the dangers we face, and to convince the public to be worried and angry enough to support radical action, the politicians are powerless to act.

I think this is sometimes an uncomfortable truth for scientists to discover. Most believe that simply presenting overwhelming evidence of a likely impending catastrophe of planetary proportions would be enough to ensure remedial action. It is not.

So Greenpeace is active in identifying the impacts of climate change, from Antarctica to the Arctic. But, in line with much of what I've written in this chapter, we also see our role as reframing the debate about climate change. I hope that I've already made clear just how important framing the agenda for discussion can be in determining outcomes of so many environmental debates. This is no less true of climate change.

## Reframing the climate debate

Up until very recently, the debate has been framed entirely in terms of setting limits on $CO_2$ and other climate change gas emissions. But there is another, far starker and harder way of framing the discussion. This starts with looking at eco-limits, by which is meant the maximum temperature change, and the maximum rate of change, that ecosystems are likely to be able to withstand without suffering substantial damage (see Table 2). As a non-scientist, I don't intend to go into the science of this, but it is science that shows that the old concerns about oil running out are outdated. In truth, we cannot afford to burn the fossil fuels we have in reserves and keep the climate within safe limits.

If there are ecological limits, it's possible to calculate the amount of $CO_2$ and other climate change gases which can be emitted without breaching such limits. Of course such calculations are subject to wide margins of error, but nevertheless such calculations are possible and are reasonably robust.

## The world's carbon budget

Greenpeace has done these calculations, and we estimate that the world can afford to burn about 225 billion tonnes of fossil fuel reserves without breaching the United Nations Environment Programme's ecological limit of a 1° average temperature rise, and a 0.1°C rate of temperature change (see Fig. 2).

**Table 2.** Ecological limits

| Global ecological targets |
| --- |
| • Limit the long-term committed increase of temperature to less than 1°C above pre-industrial global average temperature.<br>• Bring the rate of change to below 0.1°C per decade as fast as possible i.e. within a few decades.<br>• Limit the long-term sea-level rise to less than 20 cm.<br>• Limit the rate of sea-level rise to below a maximum of 20 mm/decade. |

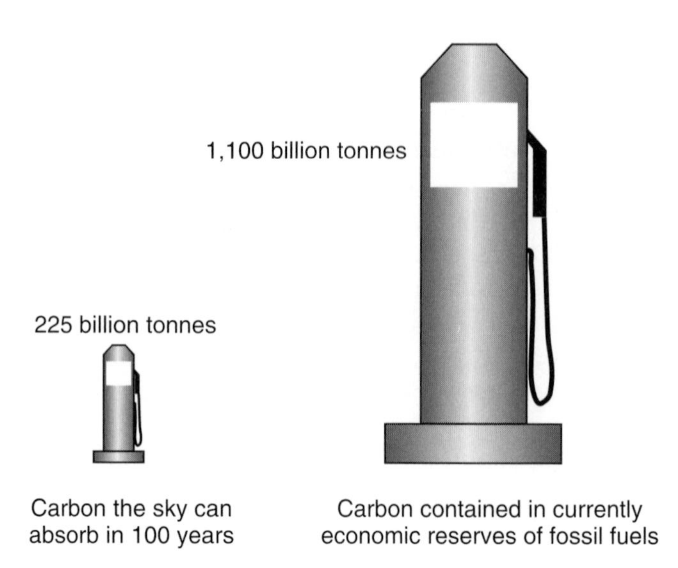


**Fig. 2** Carbon budget.

As a result of our work, Shell have done some other calculations. They come up with different figures, based as they are on a 2° change in average global temperatures, but nevertheless they accept the concept of ecological limits, and conclude that the world can afford to burn about 300 billion tonnes of fossil fuels.

This is what we would call the carbon budget—the carbon the world can afford to burn. Greenpeace's carbon budget represents a tiny proportion, about 5% of the known fossil fuel reserves and resources—that is both the currently exploitable reserves, and the fossil fuel resources which have been identified. Our carbon budget would not even allow us to burn the known reserves of oil and gas (see Figs 3 and 4). Not altogether surprisingly, Shell's carbon budget would allow us to burn existing reserves of conventional oil and gas, but only if all world-wide burning of other fossil fuels, in particular coal and firewood, stopped tomorrow, completely.

## Rapid change is needed—and possible

The scale of the changes that are going to be required world-wide by the carbon budget and the carbon logic which follows from it, may

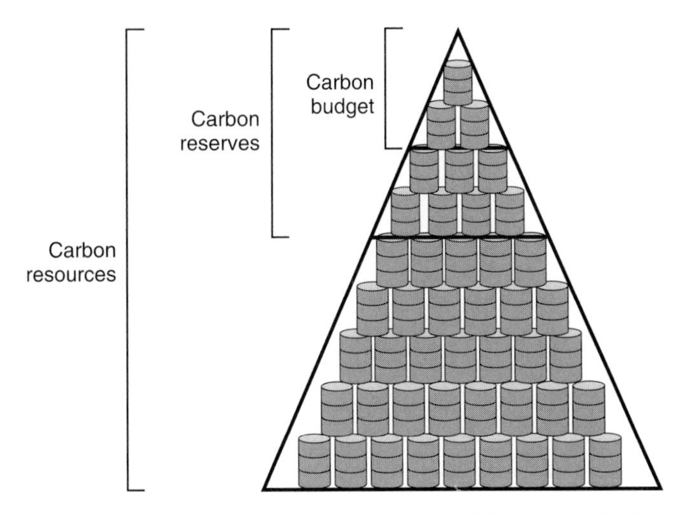


**Fig. 3** Fossil fuel reserves and resources and the carbon budget.

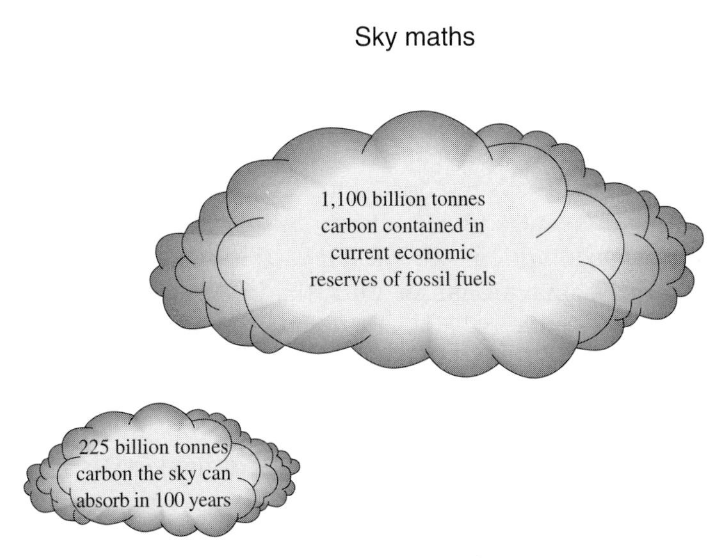


**Fig. 4** Running out of sky.

seem daunting. I don't think so. I have no doubt that such rapid change can take place, indeed has, within the very industry, the oil industry, which must now look to a rapid phase-out of all its operations. You will be familiar with much of this. For example, Thomas Eddison demonstrated the first electric light bulb in 1882; by 1885 a quarter of a million light bulbs were in use; by 1892, the number had reached 18 million. What is often forgotten in recounting such dramatic progress of a new technology is what had happened to the old technologies during the same period. In those 10 years, kerosene, gas and candles had disappeared on an extraordinary scale to be replaced by the new electric light.

Similar stories can be told in other areas where the oil industry was involved. For example, there were 8000 cars in the USA at the beginning of this century, in 1900. By 1916, there were 3.4 million, by 1945 26 million, and by 1950 40 million. In the USA, gasoline sales rose by 42% between 1945 and 1950.

In the beginning of the 1950s in Japan, oil provided 7% of total energy, with coal over 50%—oil provided less energy than firewood in Japan in the early 1950s. By the end of the 1960s, oil provided 70% of the total energy consumed in Japan.

In the process, of course, many coal merchants, and businesses dependent on coal, went out of business, and new oil companies were formed, and multinational oil companies found new markets.

It seems to me that these examples of the extraordinarily rapid expansion of new sources of energy, in this case oil, and new means of transportation, should be a source of hope for the future. Rapid changes can take place in technology, they have done over the last hundred years, and will—most people assume at a more rapid pace, during the next 100 years.

And changes we have seen in energy use have not all involved the growth of oil and growth in consumption. In 1975, the USA legislated for a doubling of average fuel efficiency in new cars, and as a result by 1985 reduced consumption by 2 million barrels of oil per day from what it would otherwise have been. In 1985, the USA was 25% more energy efficient and 32% more oil efficient than it had been in 1973. I'm not suggesting that it was anything like as efficient enough, nor that the profligate consumption of fossil fuels in the USA is not a terrible threat to the survival of our planet, but change on this scale gives some pointers to what might be possible in future. Indeed, the lessons from Japan, which became 31% more energy efficient and 51% more oil efficient in the years between 1973 and 1985, without anything other than positive effects on the Japanese economy, indicate that the sorts of changes that the carbon budget and carbon logic will demand of the world are likely to be positive rather than negative in their impacts.

## Greenpeace is founded on optimism

So I want to end on an optimistic note. To be a campaigner for change, it is necessary to be optimistic. Certainly, Greenpeace is an organization founded on optimism. We believe change for the better is not only possible, but that it's possible to make it happen. We believe that tiny organizations, of just a few individuals, can play a part in bringing about change for the better. Greenpeace is a very small organization, in terms of our annual turnover, the whole of Greenpeace's international organization, including all 32 national offices, is about the same size as the All-England Tennis Club. In the UK, in terms of the top 1000 companies, Greenpeace UK doesn't make the list. We are dwarfed by the *Bristol Evening Telegraph*, the 1000th company on the top 1000 list, with a turnover and staffing considerably greater than Greenpeace UK's.

Greenpeace believes that new, environmentally benign technologies will replace fossil fuels. We believe that this can and must happen within the next few decades, in order to prevent the possibility of catastrophic climate change. We believe that human beings have the good sense, the ingenuity, and the capacity to make such changes. We have been campaigning, world-wide, to promote one such new technology, solar photovoltaics, over the last 2 or 3 years. In that very short time, we have seen dramatic progress.

Both BP and Shell have committed to expand their share of the solar photovoltaics market. The Japanese, German, and American governments are all taking action to expand solar electricity generation. We have seen the first power station which was planned to be a conventional fossil fuel station replaced by renewable energy alternatives, involving energy conservation, some wind power, and a major solar photovoltaic installation by the US renewable energy company, ENRON. Greenpeace played a major part in encouraging this switch. There are many other examples of the development of solar power which I could give. The key thing is that this technology is available, and is competitive in many markets, for example all off-grid installations are probably cheaper if solar PV is used as opposed to connection to a grid. The technology is improving in efficiency all the time, and the price is falling rapidly. A study by BP recently showed that with significant investment in a major solar PV manufacturing plant, the price of solar PV would, through that single action, fall to be competitive with electricity delivered through the grid.

There is an old-fashioned, and I think always fairly inaccurate view of environmentalists as being opposed to technology, opposed to change, and opposed to business. None of that is true in general, and certainly not true of Greenpeace. We want to see massive change, in this case in energy production and consumption. Many of these changes will have wider benefits. For example, once solar photovoltaics are competitive with the price of grid-delivered electricity, people will be able to invest in their electricity source, put it on the roof of their house, and then forget about it. Their power bills will drop to a low level, and if they've invested in good solar panels towards the end of their working life, they can face a retirement without any concern about future power bills. I have a vision of those terrible pylons which desecrated so much of the English countryside during my childhood, and were a major source of concern and angry debate at the time, starting to be dismantled as renewable energy, and more localized and dispersed forms of energy generation, take a grip of the market. These are positive developments, which will have profound impacts on the way we live our lives. Similar changes are likely to take place in patterns of development and transport, indeed the intention is already there. These may seem daunting, even impossible, changes. They may seem to be entirely unjustified, given the inevitable scientific uncertainty which surrounds the science of climate change. But they will have benefits beyond reducing the emission of climate change gases—a reduction in the local pollution caused by the burning of fossil fuels, a reduction in the pollution caused by the mining or extraction of fossil fuels, and there will be others.

## Exercising judgement over the fate of our interconnected planet

So if I were to finish where I began, with a reference to my great, great grandfather, Ludwig Mond, I think it would be to emphasize a belief in the capacity of human beings to harness science and technology to bring benefits to humanity and the planet. What has changed since Ludwig was alive is that, as a result of scientific study and monitoring, we have learned so much more about the nature of the planet on which we live, its fragility and its interconnectedness. We have learned that we cannot, or at least should not, allow scientific and technological developments to take place simply because the technology is there and the discoveries have been made. We have to look at the impacts that our actions will have, not just on ourselves, or within our own community or country, but on the planet as a whole. We have to be more sophisticated in the way that we discuss these questions. Adapting what the NERC Committee said about the Brent Spar—we must be guided by 'social, economic, ethical and aesthetic considerations—not simply by what technology is available, and what seems inevitable.

We can, we must, exercise more judgement over our future, and the fate of our planet, than we have managed so far.

PETER MELCHETT

Peter Melchett was a government minister in the last Labour Government, at the Departments of the Environment and Industry and was a Minister of State in Northern Ireland. He has been a trustee of the World Wildlife Fund UK, President of the Ramblers Association, a council member of the Royal Society for the Protection of Birds, and Chair of Wildlife Link, the national liaison committee for over 30 wildlife organizations. He was appointed Executive Director of Greenpeace UK in January 1989.

# *Pollution: a local, national, and world problem*

LORD LEWIS OF NEWNHAM

My own interest in the environment and environmental problems arose in the late fifties. At this time environmental problems were not of general concern and yet as a young chemist I was often aware of the disposal problems and techniques that were employed in industry to get rid of their chemicals and general waste. These often led to major environmental problems that were not even recognized. A large hole in the ground referred to as the 'lagoon' was often the most ready way of disposing of waste chemicals in industry. There was a total lack of appreciation of the dangers of contamination of ground waters and the atmosphere. It is of course very easy to be critical of these procedures today, but this merely reflected the lack of appreciation of environmental pollution and the vast influences that pollution could have on all aspects of community living. The conditions in which communities associated with industrial development lived was often extreme from the point of view of health. Industrial smog and waste was considered an acceptable liability of living in these areas. The general impact of industry and industrial pollution, although recognized by many, was subservient to the economic return and the contribution to the cost-benefit analysis of environmental damage was totally ignored, as in many instances it also is today. It is perhaps salutary to appreciate that the recent earthquake at Kobe in Japan led to an increase in the gross domestic product as the environmental damage was zero rated.

If one views the environmental scene today there have been major advances in our treatment and consideration of the environment particularly as viewed from the chemical industry and industry in general. This contrasts markedly with the situation in eastern Europe. On a recent visit I was amazed at the degree of environmental pollution and the general acceptance of this by the community. It equated to being placed in a time capsule and moving back 30 years as far as the UK was concerned. It is perhaps also a fair indication of how pollution could have affected this country if the problems had not been addressed when they were. In my experience one of the greatest impacts on environmental thinking arose with the publication of the *Silent Spring* by Rachel Carson in 1962. The book was concerned with the effect of DDT on the environment and emphasized the potential long-term difficulties of pesticides within the community. This raised the problem of persistent and trans-boundary pollution problems in a very graphic manner and caught the imagination of the public in a very dramatic way. Since this time the discussion and appreciation of environmental pollution has moved to centre stage. The environmental difficulties were recognized as involving considerations at the local, nation, and global levels. Many of the questions posed in so many of these areas have now been translated into legislative procedures.

However, the over-riding problems still equate to those associated with global issues and pollution in developing countries and this is to a large extent reflected the difficulty in implementing legislation at the international level. There is often a facile acceptance on the part of many countries involved in international negotiations to accept obligations and limitations in pollution control which are not implemented, and appear as a coating of environmental respectability. The important feature of environmental thinking over the last decade is the recognition that the science and detection of environmental problems is the start rather than the end of the problem. The considerations of the economic and social implications on the control and removal of the pollution leads to the necessity of enforcement and legislation to be put in place. This clearly has implications at both the national and international level depending on the nature and source of the pollution involved.

Although, as we have discussed above, the main recognition of environmental problems was the 1970s, many of the basic problems were fully appreciated well before this. As was remarked by Ehrlich in the sixties: 'The casual chain of the deterioration is easily followed to its source. Too many cars, too many factories, too much detergent, too much pesticide... inadequate sewage treatment plants, too little water, too much carbon dioxide—all can be traced to too many people'.

This particular quotation encapsulates many of the main concerns in present day thinking in the environmental debate. It is perhaps salutary that these problems have only just come to the forefront in legislation in many parts of the world and once again emphasizes the great gap that still exists between the developing and developed world. The major problem of the population explosion is still one that avoids real solution and, as is implicit in the Ehrlich quotation is the main problem within society. Although there have been significant reductions in the population growth in the developed countries, for many of the developing countries a large family is considered a desirable safeguard for the future. It will not be easy to reverse this basic social view of life.

In more recent times the general situation has been further complicated by the recognition of the problems that have arisen with the breakdown of the Soviet Union. These include such features as the disposal of chemical and nuclear weapons within these countries. Clearly, the success of the negotiating table did not appreciate the extent to which the disposal of nuclear and chemical weapons was a major problem and how in many instances the disposal to sea dumping of nuclear weapons and the use of landfill sites for chemical weapons were providing major ecological problems. It has also been possible to document for the first time some of the major ecological disasters that have arisen in Russia because of the agricultural programmes of the 1940–60 period.

This chapter will be devoted to a consideration of some of the issues discussed above and will use as a major theme the problems associated with water. The initial discussion will be directed to examples involving global pollution, such as trans-boundary pollution and the general difficulties that can arise in the control of river basins. In particular we may consider an example of the difficulties that arise from upper river demands on the river flows and the effects that this can have on communities further down the river course. This is a problem that is as old as time and history is dominated by problems that are associated with this concern. The Nile and the Rhine are two examples of river courses that have been dominated with these sort of difficulties and even today are still providing major concerns in river basin control.

As an example of this we will first consider the position over the Aral Sea. This is now a classical ecological examples which emphasizes that issues often arise from a central policy that has not fully recognized the environmental implications at the local level. This particular instance illustrates tragically how variation in regional agricultural policy affected the total environment and resulted in knock-on effects that clearly had not been anticipated and had implications over vast distances which reflected the size of the water basin being influenced.

The Aral Sea is a major inland sea which has become radically affected by general agricultural developments in eastern parts of Russia. The Aral Sea, in 1960, was a major fishing source that covered 67 000 km$^2$ and supported a significant population involved in the fishing industry. The sea is served by two main river sources, the Amu Dar'ya and the Syr

Dar'ya which flow from Central Asia. During the 1940–60 period there was a major change in the farming policy for the lands through which the two rivers flowed. This involved a change to the growing of rice and cotton on lands that had been previously been primarily devoted to maize and cereal crops. Cotton was considered a very attractive alternative crop, even being called 'white gold', as this was a very lucrative economic commodity. However, both rice and cotton are very much more demanding on water than the cereal crops they replaced and in order to satisfy the enhanced demands for water, the irrigation systems that existed were replaced by systems that delivered larger quantities of water but for which recovery was less efficient. This resulted in the loss of considerable amounts of water from the rivers and as a result to less water into the Aral Sea. As the quality of the land on which the cotton was grown decreased, greater use was made of fertilizers and pesticides to maintain the yields. The situation was further complicated by the use of defoliants in the harvesting of the cotton. The cocktail of chemicals that flowed back into the rivers from the run-off from the land and hence eventually to the Aral Sea, made the water from the sea undrinkable, and led to major problems for the local peoples who utilized the sea as the primary water supply.

Present day statistics of anaemia in this area indicate that this is a major medical concern, particularly in the female population, and this coupled with corresponding high figures for infant mortality have both been associated with the presence of the high pesticide and fertilizer concentrations in the rivers and the sea.

The increase in the salinity of the water associated with the decrease in the volume of waters in the sea due to the decrease in the water flow in the rivers, has completely destroyed the fish and the fishing industry. The resultant decrease in the size of the sea has isolated communities that were on the sea to inland sites. In fact the decrease in the river flows reached such a limit that the sea was reduced in size such that it divided into two small seas which in 1993 had areas of 29 500 and 2500 $km^2$ and are projected in the year 2000 to have areas of 20 000 and 3000 $km^2$

The example illustrates the long-term effects of the change in use of rivers and extraction policy on downstream ecology. This is a basic concern that is clearly in no way restricted to the present example. The influence on any community of the water rites and their variation by upstream neighbours has been a feature of life from the very beginning.

The position has been made even more complicated in the case of the Aral Sea with the division of the Soviet Union into separate and individual states. Each state now has its own interpretation and demands on the rivers that flow through its territory. This decentralization of the states has led to the removal of control from Moscow, which in principle at least had the potential to acknowledge and rectify a situation of this nature. This has been replaced by a series of authorities each with its own particular concerns and views on the control over the water rites. This has compounded what was a difficult situation into what is becoming an impossible one.

One of the more important and major problems in water management applies to cities and for the undeveloped countries this is often a particular problem. Many of the projected developments and expansions that are occurring in cities are as a result of the population explosion coupled with the preference of people to move from rural to urban areas. This is leading to a great increase in the population of many of the bigger cities and major problems in serving the communities that are present with basic facilities. Water is one of these problems. For instance the position is acute in China where there is predicted to be a major problem in servicing the cities, with water, in the next decade. For coastal or near coastal developments the utilization of ground water supplies as a water source often leads to major pollution problems in saline drift from the sea. For instance in Jakarta, the sea has infiltrated into the land by a distance of up to 10 miles and this also leads to subsidence of the roads and buildings in the city.

The economic implications of supplying water to these communities have often not been thought through, and the prospect for long-term planning on the overall economy has not been considered. Thus, in Indonesia the purification procedure employed by the poorer section of the community is to boil the water. This costs about $50m per year. This amount of money could be better employed in providing a purification plant to deal with the problem at source as well reducing the pollution that occurs from the burning of fossil fuel used to heat the water. Often the richer sections of the community have facilities for the purification of water and the poorer members have to pay excessive amounts to people who will sell the purified water. Thus, in Jakarta it is estimated that water is sold by a vendor at about 40 times the cost to that in the homes with purification processes.

In many parts of the world, the agricultural community has by tradition been used to paying less for water supplies than the general domestic user. This often reflected the historically earlier demands by the agricultural community on water supplies and the contractual agreements that were made at these earlier times. In Arizona the cost of water for agricultural is approximately one-fifth of the cost in the cities. In many of the arid zones of the world much of the loss of water that occurs may be associated with poor irrigation processes in agriculture and again it is important to recognize that these difficulties are not only associated with the developing world. Recently in the Los Angeles area, a scheme was introduced that involved the replacement of part of the irrigation of the immediate agricultural areas with finance provided by the city. The driving force for this programme was the realization that water loss from the old irrigation scheme was reducing the water supply and hence raising the price of the water within the city. As anticipated, on completion of the scheme there was a significant reduction in the water costs for the city.

Many of the sewage problems associated with cities are also associated with water problems and in Mexico the sewage works are only operative for about 2% of the time, reflecting both the efficiency of the sewage plants and the lack of water. Mexico city itself has a major water problem, having over-pumped the Mexican Valley aquifer, and is now forced to pump its water supply a distance of 180 km and up to 1000 m from the Cutzmala River, resulting in much higher costs for the water. The city in addition appears to be in danger of exhausting this supply by the year 2000. In many parts of the developing world water problems have a major impact on the sewage problem and the efficient use of water and the conservation of the general supplies of water are of paramount concern to both the health and economics of a community. Although agriculture is a main consumer of water in many parts of the world, in some countries the biggest consumer of water supplies is the public sector. In the UK, for example, 50% of processed water is used within the public sector while industry accounts for 12% and the power generating industry 36%, leaving agriculture a mere 1% of the demand. Agriculture does in this case obtain a significant quantity of its water supply from unprocessed sources. Perhaps even more alarming is the consumption pattern within domestic premises. The average consumption of water per capita in a typical modern household is 150–200 litres per day. Of this 3–6 litres are used in drinking and cooking. For the remainder 3–10 litres are used in cleaning, 15–20 litres in washing and personal hygiene while by far the major demand is used for toilets, baths, washing machines and dish washers, at 83 litres, with a general loss of 50 litres via leakage and gardening. Although the general design of the water industry is directed towards drinking water only 3–6 litres of the 200 litres is used for drinking and cooking. This does pose the interesting question of whether this problem could be solved in another way and certainly should encourage methods of reducing the consumption patterns in the other areas of domestic usage.

In discussing the UK it is perhaps important to recognize that many of the problems associated with water and the purity of water can be attributed to agricultural usage. The presence and increase in concentration of pesticides and nutrients such as nitrate and phosphate in water has been associated with the increase in the usage of these in agricultural practice. However, many of the problems in this area are not solely the responsibility of agricultural practice but may also be associated with the use of these materials, particularly pesticides and herbicides, in other ways such as the control of weeds on golf courses or the clearing of vegetation from rail tracks. In the case of the latter, the growth of weeds on the edge of the tracks is often contained by the use of excessive amounts of herbicides, and it has been claimed that as much as a fivefold excess has been used, this excess having a high probability of ending in the water courses.

With nitrates and phosphates, which are used as fertilizers for agricultural land, problems occur with contamination of rivers through run-off from the land. This may lead to eutrophication (nutrient enrichment) of the rivers. This can be a difficult problem to deal with as it involves contamination from a diffuse source, in this case the run-off from a field adjacent to the river. This would contrast with a point source, which may be a drain running into the river and draining the field concerned. Eutrophication can have a major effect on the local fauna as it leads to a de-oxygenation of the water course. It often leads to the production of algal blooms which may be toxic to animals. This was the cause of the closure of some of the reservoirs in the Midlands in 1989, following the deaths of sheep and dogs that in drinking the water took in the algal which proved to be toxic. In fresh water the controlling factor in eutrophication is the phosphate concentration while for sea water it is nitrate controlled. It has been shown that in the UK about 35% of the phosphate in freshwaters comes from agricultural sources. However, the main source, i.e. 60%+, is from the effluent from sewage works. This arises from either the domestic use of detergents or human sources.

A further difficulty associated with the use of nitrate as a fertilizer is the resultant nitrate content of the river waters associated with run-off from the fields. This must not exceed the nitrate limits of the EC water directive of 50 mg/l. This is the concentration of nitrate in drinking water that is believed to be the limit above which there is danger of cancer of the bowel or of blue baby syndrome (methaemoglobinaemia). The statistical basis of 50 mg/l for both of these medical conditions is not, however, well founded, and has led to a lot of discussion of the permissible concentration in drinking water. This figure has a very significant financial implication. This arises from the policy of the creation of the so-called 'nitrate-free zones', which corresponds to those lands that cause the related water courses to exceed the nitrate limitation of 50 mg/l. In the case of the UK this involves about 600 000 ha of land and this can be compared with the figure of 200 000 ha, which is the estimated figure for contaminated land in the UK. Clearly this is a great deal of land and this policy is of major concern to the farming community as this potentially restricts their productivity. A reduction in the control by increasing the nitrate figure to 70 mg rather than 50 mg/l would make a very significant reduction in the area of the nitrate vulnerable zones.

The details of the role and dangers in the use of nitrates within the agricultural–water systems is a problem that still needs to be resolved. However, if the concern of the community is the intake of nitrate, water is by no means the only major potential source; considerable quantities of nitrate are consumed from vegetables in particular lettuce and spinach provide a rich source of nitrate equivalent to the consumption of many litres of water. An interesting effect that has also been observed in lettuce is the greater concentration of nitrate in winter, rather than summer, lettuce. This is believed to be associated with the take up of nitrate by the vegetables which is accelerated by sunlight.

Two other potential areas of contamination of water arising from agricultural practice are slurry and silage effluent. Slurry produced from farm animals is 100 times more toxic than domestic sewage while silage liquor is 200 times more toxic. The dangers of these effluents on the fish population in rivers is a constant cause of concern. Both of these contaminents lead to de-oxygenation of the river with concomitant losses of fish.

One further recent development is the change that has occurred in animal husbandry where there have been very large increases in the unit size employed. An example that is giving rise to major concern is the growth in both the number and size of pig farms. Many of these units may involve up to 600 pigs per unit. The slurry from these provides a contamination hazard that equates to those associated with medium-sized industrial complexes, and yet the environmental controls applicable to an industrial concern are not in place for the equivalent agricultural process.

An importance consideration in any discussion of the environment are the problems that arise in air quality from motor vehicles. As was recognized by Ehrlich in the 1960s, one of the major polluters within society is the motor car. The exhaust from the car leads to air contamination from a variety of chemicals the main components being carbon dioxide, carbon monoxide, oxides of nitrogen, and solid particulates. The solid particulates have a varied composition dependent on the fuel and the particle size. The oxides of nitrogen provide the source for the production of ozone from the oxygen in the air via a series of photochemical reactions dependent on the intensity of the sunlight. It has been noted in a survey carried out by the RAC that for the 60 000 cars whose exhausts were tested 12% caused 55% of the pollution. One poorly maintained car was equivalent to 40 well maintained vehicles in terms of pollution. In general the older the car the more pollution of the atmosphere occurs.

The problem is further compounded for diesel engines. The efficiency and life time of the diesel engine is so high compared with the petrol engine that often buses are running with diesel engines 20–25 years old with the corresponding technology of 20–25 years ago. The modern diesel is very much more benign in terms of pollution emissions, but with the long life of the diesel engine the emissions from many of the present vehicles on the road corresponds to the previous technology.

One solution for the reduction in emissions has been the fitting of catalytic converters to motor cars. This has significantly cut down the emissions of oxides of nitrogen, one of the major contributors to air pollution; however, the efficiency of the catalyst is critically dependent on the temperature at which it operates. The temperature build up is related to the heat of the exhaust gases and this requires the engine to be running for some time, corresponding to about a 2–3 mile journey to attain the required temperature. In many instances, particularly for town traffic, the journeys are of shorter duration and the effectiveness of the converter to reduce the emission is therefore much reduced.

The problem associated with particulate emissions from motor vehicles has been primarily associated with the diesel engine. The characteristic black smoke that issues, especially from old engines, has as a major component particulate matter. This is considered to be dangerous to health and certainly leads to contamination of buildings in towns. The amount of the emissions is generally specified in terms of the particle size; $PM_{12}$, corresponds to particles in the range of 12 $\mu$m. In the case of diesel emissions this appears to be the size range that causes a problem, although particles of greater size are emitted these are readily deposited. The 12 $\mu$m particles are suspended in the atmosphere for hours before being deposited. These particles therefore have opportunity to be inhaled, with the potential medical problems that may arise.

It was considered until recently that the particulate problem was one that was much more acute with the diesel rather than the petrol engine. However, it has now been recognized that there is an even greater problem associated with petrol engine emissions. For petrol engines the main particle size has now been established as being in the 2.5 $\mu$m range. Particles of this size are even more penetrating into the lung and as such are considered to be more dangerous, particularly as it has been shown that these have a much higher persistence in the atmosphere, the residence time being of the order of 10–30 days. This poses a whole new set of difficulties in the long-term problems associated with the control of particulates in vehicle emissions.

Many solutions to problems in the control of environment difficulties give rise to alternative problems. The control of traffic flow and reduction in speeds of cars by the use of 'traffic calming' procedures has become very popular in many urban communities. The reduced speed of vehicles by this means certainly tends to inhibit many motorists from using the roads, but this is often countered by the increase in the pollution that occurs when a car reduces its speed from 30 m.p.h. to 15 m.p.h. The particulates, nitrogen oxide and carbon monoxide levels in the emissions being doubled. It then becomes a matter of deciding which provides the greater problem, the density of traffic flow or the emission profile.

Finally, may I turn to perhaps one of the most difficult aspects in any consideration of environmental problems, and that is risk assessment. A major feature of any environmental analysis involves the assessment of the risk for a particular situation. Risk tolerance is variable and difficult to analyse. We may illustrate the difficulties by the consideration of one or two examples. There appears to be a much higher tolerance by the public to traffic accidents than, say, to contamination of foods. Although the dangers from pesticides are given a high profile in the press, the detailed data of the dangers and effects of pesticides are not well documented and will undoubtedly vary from compound to compound; many of the plants that we eat contain their own pesticides. This lack of knowledge on the behaviour of pesticides has led the EC to include in its water directive extremely low concentrations for all pesticides. The initial limits really reflect the limits of chemical detection of these compounds.

The difficulties in dealing with risk analysis in many instances reflect the degree of personal control that the public feels it has in assessing the problem. The example of the traffic accident appears to be within the control of the drivers whereas the contamination of foods is generally considered to be outside the customer's control. This is then viewed as involving a much bigger risk. The problem in many instances is further complicated by the multivariable nature of exposure to risk and any cocktail effect that may be associated with the presence of many variables in the polluting source. In tobacco smoke there are of the order of 4000 chemicals, for coffee of the order of 500 and in wood smoke many thousands of different chemical compounds. The effects of these mixtures of chemicals can lead to situations that may vary from acute to chronic, depending upon the response and sensitivity of the person concerned. The actual composition of the mixtures of chemical compounds may also vary with the type of coffee, tobacco, or wood that is used. This makes the study of both toxicology and epidemiology of long-term exposures to, say, a mixture of chemicals of this nature difficult to assess in terms of environmental risk. The difficulties in the acceptance by the public of the relationship between cancer and long-term exposure to cigarette smoking is well known, but it took many years to establish this relationship to the satisfaction of the courts. This is a relatively easy problem to document for individual smokers and makes any attempt to try to determine the effects of, say, nitrate in water on cancer of some organ in the human body a momentous task.

Many of the difficulties that one encounters in environmental pollution are associated with the difficulty of getting sufficient data to assess the problem in scientific terms. A general approach that is very often employed is the so-called 'Vorsorgeprinzip' or precautionary principle. Essentially, this implies that considerations of environmental problems should err on the side of caution. Although this is a very defensible and to many a desirable approach, it can and has been used as a means of censure on many developments in industry. This is because of the difficulties in accumulating enough information to establish the absolute knowledge necessary to satisfy all the potential environmental implications in a given situation. It is perhaps important to emphasize that the scientific approach, the one that is often applied to the assessment of the potential risk in these situations, can never prove a theory but only disprove one. This makes discussion of problems in this area open to doubt and sensitive to any interpretation involving the 'Vorsorgeprinzip'. I believe that perhaps a more realistic approach is summarized in a statement attributed to Kant: 'It is often necessary to make a decision on the basis of knowledge sufficient for action, but insufficient to satisfy the intellect.'

The solution to environmental problems and in particular the assessment of the risk involved is a very complicated procedure. In many instances the solutions suggested lead to other, often unforeseen difficulties. However, the important factor is that the potential of environmental damage by society is now appreciated and under constant surveillance. Although it would be dangerous to prejudge the outcome of the environmental debate, there is a general attempt being made by nations to control the problems of pollution and to try to ensure that the next generation will not suffer by the mistakes of the present one.

## THE LORD LEWIS

He was born in 1928 in Barrow-in-Furness. In 1970 he was elected to the first Chair of Inorganic Chemistry in the University of Cambridge. In 1975 he was appointed the first Warden of Robinson College—and was closely involved in its Foundation. He is the immediate past-Chairman of the Royal Commission on Environmental Pollution and is at present Chairman of Sub-Committee C of the House of Lords' European Communities Committee which scrutinizes all EC environmental directives.

# THE ROYAL INSTITUTION

The Royal Institution of Great Britain was founded in 1799 by Benjamin Thompson, Count Rumford. It has occupied the same premises for nearly 200 years and, in that time, a truly astounding series of scientific discoveries has been made within its walls. Rumford himself was an early and effective exponent of energy conservation. Thomas Young established the wave theory of light; Humphry Davy isolated the first alkali and alkaline earth metals, and invented the miners' lamp; Tyndall explained the flow of glaciers and was the first to measure the absorption and radiation of heat by gases and vapours; Dewar liquefied hydrogen and gave the world the vacuum flask; all who wished to learn the new science of X-ray crystallography that W.H. Bragg and his son had discovered came to the Royal Institution, while W.L. Bragg, a generation later, promoted the application of the same science to the unravelling of the structure of proteins. In the recent past the research concentrated on photochemistry under the leadership of Professor Sir George (now Lord) Porter, while the current focus of the research work is the exploration of the properties of complex materials.

Towering over all else is the work of Michael Faraday, the London bookbinder who became one of the world's greatest scientists. Faraday's discovery of electromagnetic induction laid the foundation of today's electrical industries. His magnetic laboratory, where many of his most important discoveries were made, was restored in 1972 to the form it was known to have had in 1854. A museum, adjacent to the laboratory, houses a unique collection of original apparatus arranged to illustrate the more important aspects of Faraday's immense contribution to the advancement of science in his fifty years at the Royal Institution.

## Why the Royal Institution Is Unique

It provides the only forum in London where non-specialists may meet the leading scientists of our time and hear their latest discoveries explained in everyday language.

It is the only Society that is actively engaged in research, and provides lectures covering all aspects of science and technology, with membership open to all.

It houses the only independent research laboratory in London's West End (and one of the few in Britain)—the Davy Faraday Research Laboratory.

## What the Royal Institution Does for Young Scientists

The Royal Institution has an extensive programme of scientific activities designed to inform and inspire young people. This programme includes lectures for primary and secondary school children, sixth form conferences, Computational Science Seminars for sixth-formers and Mathematics Masterclasses for 12–13 year-old children.

## What the Royal Institution Offers to its Members

Programmes, each term, of activities including summaries of the Discourses; synopses of the Christmas Lectures and annual Record.
Evening Discourses and an associated exhibition to which guests may be invited.
An annual volume of the *Proceedings of the Royal Institution of Great Britain* containing accounts of Discourses.
Christmas Lectures to which children may be introduced.
Meetings such as the RI Discussion Evenings; Seminars of the Royal Institution Centre for the History of Science and Technology, and other specialist research discussions.
Use of the Libraries and borrowing of the books. The Library is open from 9 a.m. to 9 p.m. on weekdays.
Use of the Conversation Room for social purposes.
Access to the Faraday Laboratory and Museum for themselves and guests. Invitations to debates on matters of current concern, evening parties and lectures marking special scientific occasions.
Royal Institution publications at privileged rates.
Group visits to various scientific, historical, and other institutions of interest.

### *Evening Discourses*

The Evening Discourses have been given regularly since 1826. They cover all aspects of science and technology (with regular ventures into the arts) in a form suitable for the interested layman, and many scientists use them to keep in touch with fields other than their own. An exhibition, on a subject relating to the Discourse, is arranged each evening, and light refreshments are available after the lecture.

### *Christmas Lectures*

Faraday introduced a series of six Christmas Lectures for children in 1826. These are still given annually, but today they reach a much wider audience through television. Titles have included: 'The Languages of Animals' by David Attenborough, 'The Natural History of a Sunbeam' by Sir George Porter, 'The Planets' by Carl Sagan and 'Exploring Music' by Charles Taylor.

### *The Library*

The Royal Institution library reflects the functions and the activities of the RI. The subject coverage is science, its history, its role in society including education, and its interaction with religion, literature, and the arts. The emphasis is on the popular science books, the history of science, and the research monographs of interest to the research group in the Davy Faraday Research Laboratories.

It is probably the only library of its kind specializing in the public understanding of science, that is science for the non-specialist. It also has a junior section.

### *Schools' Lectures*

Extending the policy of bringing science to children, the Royal Institution provides lectures throughout the year for school children of various ages, ranging from primary to sixth-form groups. These lectures, attended by thousands, play a vital part in stimulating an interest in science by means of demonstrations, many of which could not be performed in schools.

### *Seminars, Masterclasses, and Primary Schools' Lectures*

In addition to educational activities within the Royal Institution, there is an expanding external programme of activities which are organized at venues throughout the UK. These include a range of seminars and master classes in the areas of mathematics, technology and, most recently, computational science. Lectures aimed at the 8–10 year-old age group are also an increasing component of our external activities.

### *Teachers' Workshops*

Lectures to younger children are commonly accompanied by workshops for teachers which aim to explain, illustrate, and amplify the scientific principles demonstrated by the lecture.

## Membership of the Royal Institution

### *Member*

The Royal Institution welcomes all who are interested in science, no special scientific qualification being required. By becoming a Member of the Royal Institution an individual not only derives a great deal of personal benefit and enjoyment but also the satisfaction of helping to support the unique contribution made to our society by the Royal Institution.

### *Family Associate Subscriber*

A Member may nominate one member of his or her family residing at the same address, and not being under the age of 16 (there is no upper age limit), to be a Family Associate Subscriber. Family Associate Subscribers can attend the Evening Discourses and other lectures, and use the Libraries.

### *Associate Subscriber*

Any person between the ages of 16 and 27 may become an Associate Subscriber. Associate Subscribers can attend the Evening Discourses and other lectures, and use the Libraries.

### *Junior Associate*

Any person between the ages of 11 and 15 may become a Junior Associate. Junior Associates can attend the Christmas Lectures and other functions, and use the Libraries. There are also visits organized during Easter and Summer vacations.

### *Corporate Subscriber*

Companies, firms and other bodies are invited to support the work of the Royal Institution by becoming Corporate Subscribers; such organizations make a very valuable contribution to the income of the Institution and so endorse its value to the community. Two representatives may attend the Evening Discourses and other lectures, and may use the Libraries.

## *College Corporate Subscriber*

Senior educational establishments may become College Corporate Subscribers; this entitles two representatives to attend the Evening Discourses and other lectures, and to use the Libraries.

## *School Subscriber*

Schools and Colleges of Education may become School Subscribers; this entitles two members of staff to attend the Evening Discourses and other lectures, and to use the Libraries.

**Membership forms** can be obtained from: The Membership Secretary, The Royal Institution, 21 Albemarle Street, London W1X 4BS. Telephone: 0171 409 2992. Fax: 0171 629 3569.

# *Discourses*

**Faraday in the Pits, Faraday at Sea** *26 January 1996*
*Frank James, MSc, PhD, DIC, FRAS*

**Tick, Tick, Tick Pulsating Star: How We Wonder What You Are** *2 February 1996*
*S. Jocelyn Bell Burnell, PhD*

**Pondering on Pisa** *9 February 1996*
*John B. Burland, DSc, FEng, FICE, MIStructE*

**Pollution: a Local, National and World Problem** *16 February 1996*
*The Lord Lewis, Kt, FRS*

**The Colossus of Bletchley Park** *23 February 1996*
*Anthony E. Sale, FBCS*

**To Divide or Not to Divide?** *1 March 1996*
*Paul Nurse, FRS*

**Lasers—Making Light Work** *8 March 1996*
*Colin Webb, FRS*

**There or Thereabouts—the Story of Measurement, Past, Present and Future** *15 March 1996*
*Andrew Wallard, PhD*

**On the Air, the Constituents of the Atmosphere, their Separation and Uses (The BOC Lecture)** *3 May 1996*
*Michael E. Garrett*

**Sheathing the Two-Edged Sword: One Hundred Years of Radioactivity** *10 May 1996*
*Peter Christmas, MA, DPhil*

**Human Use and Abuse of Alcohol, a Historical Perspective** *17 May 1996*
*Bert L. Vallee, MD*

**The Crucible of Creation: the Burgess Shale and the Rise of Animals (The Woodhull Lecture)** — *24 May 1996*
*Simon Conway Morris, FRS*

**Meteorites: Messengers from the Past** — *18 October 1996*
*Monica Grady, PhD*

**Asthma and Allergy: Disorders of Civilisation** — *25 October 1996*
*Stephen T. Holgate*

**Tower Bridge: Enduring Monument or Victorian Folly** — *1 November 1996*
*Roy Aylott, CEng, FICE, FIHT*

**Insects and Chemical Signals: A Volatile Situation** — *8 November 1996*
*John Pickett*

**Evidence and Speculation in Astronomy** — *15 November 1996*
*Sir Martin Rees, FRS*

**Metals Combined—or, Chaos Vanquished** — *22 November 1996*
*Robert W. Cahn, FRS*

**Disappearing Species: Problems *and* Opportunities** — *29 November 1996*
*Norman Myers, PhD*

**Television Beyond the Millennium** — *6 December 1996*
*Will Wyatt*

**Magellen Looks at Venus** — *24 January 1997*
*D. P. McKenzie, FRS*

**The Search for Extraterrestrial Life—And the Future of Life on Earth** — *31 January 1997*
*Sir Arnold Wolfendale, FRS*

**Molecular Information Processing: Will It Happen?** — *7 February 1997*
*Peter Day, FRS*

**One Hundred Years of the Electron** — *14 February 1997*
*Sir Brian Pippard, FRS*

**Self-Assembly: Nature's Way To Do It** — *21 February 1997*
*Kuniaki Nagayana, PhD*

**J. J. Thomson, Discoverer of the Electron: The Man in His Setting** — *28 February 1997*
*David Thomson*

**An Arts/Science Interface: Mediaeval Manuscripts, Pigments and Spectroscopy** — *7 March 1997*
*Robin J. H. Clark, FRS*

**The Coleridge Experiment** *14 March 1997*
*Richard Holmes, OBE, FRSSL*

**Carbon Nanotubes, the Tiniest Man-made Tubes** *9 May 1997*
*Sumio Iijima, PhD*

**The Biology of Nitric Oxide** *16 May 1997*
*Salvador Moncada, FRCP, FRS*

**The Epidemic of Mad Cow Disease (BSE) in the United Kingdom** *23 May 1997*
*Roy M. Anderson, FRS*

**To See a World in a Grain of Sand** *30 May 1997*
*Sir John Meurig Thomas, FRS*

**Electricity, Magnetism and the Body** *17 October 1997*
*Anthony T. Barker, PhD, FIEE, FIPEM*

**Risky Business** *24 October 1997*
*Ben Selinger, FTS, FRACI*

**Vegetation and Climate: Virtually Worlds in Themselves** *31 October 1997*
*P.A. Spicer, PhD*

**MRI: A Window into the Human Body** *7 November 1997*
*Laurie Hall, FRS(Can)*

**Environmental Campaigners and Scientists—Friends or Foes** *14 November 1997*
*Lord Melchett*

**Fuel Cells for the Future** *21 November 1997*
*Andrew Hamnett, DPhil*

**Gardening as Ecology** *28 November 1997*
*Stefan Buczacki, DPhil*

**Killers in the Brain: New Discoveries in Neurodegenerative Diseases** *5 December 1997*
*Nancy Rothwell, PhD, DSc*

**The Human Singing Voice** *30 January 1998*
*David M. Howard, PhD, FIOA*

**El Niño and its Significance** *6 February 1998*
*Sir Crispin Tickell, GCMG, KCVO*

**Mendeleev's Palette** *13 February 1998*
*Anthony K. Cheetham, FRS*

**Destructive Rust, Creative Rust:** *20 February 1998*
**From the Pantheon to the Millennium Dome**
*Jack Harris, MBE, FEng, FRS*

**The Science of Murphy's Law** *27 February 1998*
*Robert Matthews, MA, CPhys, MInstP, FRAS*

**FLIMsy Constructions** *6 March 1998*
*David Phillips, BSc, PhD, FRSC*

**Every Drop to Drink** *13 March 1998*
*K. J. Ives, CBE, DSc, FEng*

**God, Time, and Cosmology** *8 May 1998*
*F. Russell Stannard, PhD, FInstP, Cphys*

**The Royal Institution as a Mayfair Landholder** *15 May 1998*
*H. J. V. Tyrrell, MA, Dsc, FRSC*

**Solar Fuel from Chemistry** *22 May 1998*
*Harry B. Gray*

**A Philosopher's Tree: Michael Fara*day* To*day*** *29 May 1998*
**at the Royal Institution**
*Peter Day, FRS*